Swarm
Intelligence

The Morgan Kaufmann Series in Evolutionary Computation
Series Editor: David B. Fogel

Swarm Intelligence
James Kennedy and Russell C. Eberhart, with Yuhui Shi

Illustrating Evolutionary Computation with Mathematica
Christian Jacob

Evolutionary Design by Computers
Edited by Peter J. Bentley

Genetic Programming III: Darwinian Invention and Problem Solving
John R. Koza, Forrest H. Bennett III, David Andre, and Martin A. Keane

Genetic Programming: An Introduction
Wolfgang Banzhaf, Peter Nordin, Robert E. Keller, and Frank D. Francone

FOGA Foundations of Genetic Algorithms Volumes 1–5

Proceedings

GECCO—Proceedings of the Genetic and Evolutionary Computation Conference, the Joint Meeting of the International Conference on Genetic Algorithms (ICGA) and the Annual Genetic Programming Conference (GP)
GECCO 2000
GECCO 1999

GP—International Conference on Genetic Programming
GP 4, 1999
GP 3, 1998
GP 2, 1997

ICGA—International Conference on Genetic Algorithms
ICGA 7, 1997
ICGA 6, 1995
ICGA 5, 1993
ICGA 4, 1991
ICGA 3, 1989

Forthcoming

FOGA 6
Edited by Worthy N. Martin and William M. Spears

Creative Evolutionary Systems
Edited by Peter J. Bentley and David W. Corne

Evolutionary Computation in Bioinformatics
Edited by Gary Fogel and David W. Corne

Swarm
Intelligence

James Kennedy

Russell C. Eberhart
*Purdue School of Engineering and Technology,
Indiana University Purdue University Indianapolis*

with Yuhui Shi
Electronic Data Systems, Inc.

MORGAN KAUFMANN PUBLISHERS

AN IMPRINT OF ACADEMIC PRESS
A Harcourt Science and Technology Company
SAN FRANCISCO   SAN DIEGO   NEW YORK   BOSTON
LONDON   SYDNEY   TOKYO

Senior Editor   Denise E. M. Penrose
Publishing Services Manager   Scott Norton
Assistant Publishing Services Manager   Edward Wade
Editorial Coordinator   Emilia Thiuri
Cover Design   Chen Design Associates, SF
Cover Photography   Max Spector/Chen Design Associates, SF
Text Design   Rebecca Evans & Associates
Technical Illustration and Composition   Technologies 'N Typography
Copyeditor   Ken DellaPenta
Proofreader   Jennifer McClain
Indexer   Bill Meyers
Printer   Courier Corporation

Designations used by companies to distinguish their products are often claimed as trademarks or registered trademarks. In all instances where Morgan Kaufmann Publishers is aware of a claim, the product names appear in initial capital or all capital letters. Readers, however, should contact the appropriate companies for more complete information regarding trademarks and registration.

ACADEMIC PRESS
A Harcourt Science and Technology Company
525 B Street, Suite 1900, San Diego, CA 92101-4495, USA
http://www.academicpress.com

Academic Press
Harcourt Place, 32 Jamestown Road, London, NW1 7BY, United Kingdom
http://www.academicpress.com

Morgan Kaufmann Publishers
340 Pine Street, Sixth Floor, San Francisco, CA 94104-3205, USA
http://www.mkp.com

Library of Congress Cataloging-in-Publication Data

Kennedy, James
   Swarm intelligence : collective, adaptive/James Kennedy, Russell C. Eberhart, with Yuhui Shi
    p. cm.
   Includes bibliographical references and index.
   ISBN 1-55860-595-9
    1. Swarm intelligence. 2. Systems engineering. 3. Distributed artificial intelligence. I. Eberhart, Russell C. II. Shi, Yuhui. III. Title

Q337.3 .K45 2001
006.3—dc21   00-069641

This book is printed on acid-free paper.

Significant strangers and friends who have probably forgotten me have contributed to this book in ways unknown to them. I owe inexpressible gratitude to Cathy, Bonnie, and Jamey for continuing to love me even through the many hours I sat at the computer ignoring them.

–Jim Kennedy

This book is dedicated to Francie, Mark, and Sean, and to Professor Michael S. P. Lucas. The support of the Purdue School of Engineering and Technology at Indiana University Purdue University Indianapolis, especially that of Dean H. Öner Yurtseven, is gratefully acknowledged.

–Russ Eberhart

Contents

Preface

At this moment, a half-dozen astronauts are assembling a new space station hundreds of miles above the surface of the earth. Thousands of sailors live and work under the sea in submarines. Incas jog through the Andes. Nomads roam the Arabian sands. *Homo sapiens*—literally, "intelligent man"—has adapted to nearly every environment on the face of the earth, below it, and as far above it as we can propel ourselves. We must be doing something right.

In this book we argue that what we do right is related to our sociality. We will investigate that elusive quality known as intelligence, which is considered first of all a trait of humans and second as something that might be created in a computer, and our conclusion will be that whatever this "intelligence" is, it arises from interactions among individuals. We humans are the most social of animals: we live together in families, tribes, cities, nations, behaving and thinking according to the rules and norms of our communities, adopting the customs of our fellows, including the facts they believe and the explanations they use to tie those facts together. Even when we are alone, we think about other people, and even when we think about inanimate things, we think using language— the medium of interpersonal communication.

Almost as soon as the electronic computer was invented (or, we could point out, more than a century earlier, when Babbage's mechanical analytical engine was first conceived), philosophers and scientists began to ask questions about the similarities between computer programs and minds. Computers can process symbolic information, can derive conclusions from premises, can store information and recall it when it is appropriate, and so on—all things that minds do. If minds can be intelligent, those thinkers reasoned, there was no reason that computers could not be. And thus was born the great experiment of artificial intelligence.

To the early AI researchers, the mark of intelligence was the ability to solve large problems quickly. A problem might have a huge number of possible solutions, most of which are not very good, some of which are passable, and a very few of which are the best. Given the huge number of possible ways to solve a problem, how would an intelligent computer program find the best choice, or at least a very good one? AI researchers thought up a number of clever methods for sorting through the possibilities, and shortcuts, called *heuristics,* to speed up the process. Since logical principles are universal, a logical method could be developed for one problem and used for another. For instance, it is not hard to see that strings of logical premises and conclusions are very similar to tours through cities. You can put facts together to draw conclusions in the same way that you can plan routes among a number of locations. Thus, programs that search a geographical map can be easily adapted to explore deductive threads in other domains. By the mid-1950s, programs already existed that could prove mathematical theorems and solve problems that were hard even for a human. The promise of these programs was staggering: if computers could be programmed to solve hard problems on their own, then it should only be a short time until they were able to converse with us and perform all the functions that we the living found tiresome or uninteresting.

But it was quickly found that, while the computer could perform superhuman feats of calculation and memory, it was very poor—a complete failure—at the simple things. No AI program could recognize a face, for instance, or carry on a simple conversation. These "brilliant" machines weren't very good at solving problems having to do with real people and real business and things with moving parts. It seemed that no matter how many variables were added to the decision process, there was always something else. Systems didn't work the same when they were hot, or cold, or stressed, or dirty, or cranky, or in the light, or in the dark, or when two things went wrong at the same time. There was always something else.

The early AI researchers had made an important assumption, so fundamental that it was never stated explicitly nor consciously acknowledged. They assumed that cognition is something inside an individual's head. An AI program was modeled on the vision of a single disconnected person, processing information inside his or her brain, turning the problem this way and that, rationally and coolly. Indeed, this is the way we experience our own thinking, as if we hear private voices and see private visions. But this experience can lead us to overlook what should be our most noticeable quality as a species: our tendency to associate with one another, to socialize. If you want to model human intelligence, we argue here, then you should do it by modeling individuals in a social context, interacting with one another.

In this regard it will be made clear that we do not mean the kinds of interaction typically seen in multiagent systems, where autonomous subroutines perform specialized functions. Agent subroutines may pass information back and forth, but subroutines are not changed as a result of the interaction, as people are. In real social interaction, information is exchanged, but also something else, perhaps more important: individuals exchange rules, tips, and beliefs about how to process the information. Thus a social interaction typically results in a change in the thinking processes—not just the contents—of the participants.

It is obvious that sexually reproducing animals must interact occasionally, at least, in order to make babies. It is equally obvious that most species interact far more often than that biological bottom line. Fish school, birds flock, bugs swarm—not just so they can mate, but for reasons extending above and beyond that. For instance, schools of fish have an advantage in escaping predators, as each individual fish can be a kind of lookout for the whole group. It is like having a thousand eyes. Herding animals also have an advantage in finding food: if one animal finds something to eat, the others will watch and follow. Social behavior helps individual species members adapt to their environment, especially by providing individuals with more information than their own senses can gather. You sniff the air and detect the scent of a predator; I, seeing you tense in anticipation, tense also, and grow suspicious. There are numerous other advantages as well that give social animals a survival advantage, to make social behavior the norm throughout the animal kingdom.

What is the relationship between adaptation and intelligence? Some writers have argued that in fact there is no difference, that intelligence *is* the ability to adapt (for instance, Fogel, 1995). We are not in a hurry to take on the fearsome task of battling this particular dragon at the moment and will leave the topic for now, but not without asserting that there is a relationship between adaptability and intelligence, and noting that social behavior greatly increases the ability of organisms to adapt.

We argue here against the view, widely held in cognitive science, of the individual as an isolated information-processing entity. We wish to write computer programs that simulate societies of individuals, each working on a problem and at the same time perceiving the problem-solving endeavors of its neighbors, and being influenced by those neighbors' successes. What would such programs look like?

In this book we explore ideas about intelligence arising in social contexts. Sometimes we talk about people and other living—carbon-based—organisms, and at other times we talk about silicon-based entities, existing in computer programs. To us, a mind is a mind, whether embodied in protoplasm or semiconductors, and intelligence is intelligence. The important thing is that minds arise from interaction with other minds. That is not to say that we will dismiss the question casually. The interesting relationship between human minds and simulated minds will keep us on our toes through much of the book; there is more to it than meets the eye.

In the title of this book, and throughout it, we use the word *swarm* to describe a certain family of social processes. In its common usage, "swarm" refers to a disorganized cluster of moving things, usually insects, moving irregularly, chaotically, somehow staying together even while all of them move in apparently random directions. This is a good visual image of what we talk about, though we won't try to convince you that gnats possess some little-known intelligence that we have discovered. As you will see, an insect swarm is a three-dimensional version of something that can take place in a space of many dimensions—a space of ideas, beliefs, attitudes, behaviors, and the other things that minds are concerned with, and in spaces of high-dimensional mathematical systems like those computer scientists and engineers may be interested in.

We implement our swarms in computer programs. Sometimes the emphasis is on understanding intelligence and aspects of culture. Other times, we use our swarms for optimization, showing how to solve hard engineering problems. The social-science and computer-science questions are so interrelated here that it seems they require the same answers. On the one hand, the psychologist wants to know, how do minds work and why do people act the way they do? On the other, the engineer wants to know, what kinds of programs can I write that will help me solve extremely difficult real-world problems? It seems to us that if you knew the answer to the first question, you would know the answer to the second one. The half-century's drive to make computers intelligent has been largely an endeavor in simulated thinking, trying to understand how people arrive at their answers, so that powerful electronic computational devices can be programmed to do the hard work. But it seems researchers have not understood minds well enough to program one. In this volume we propose a view of mind, and we propose a way to implement that view in computer programs—programs that are able to solve very hard mathematical problems.

In *The Computer and the Brain*, John von Neumann (1958) wrote, "I suspect that a deeper mathematical study of the nervous system . . . will affect our understanding of the aspects of mathematics itself that are involved. In fact, it may alter the way in which we look on mathematics and logics proper." This is just one of the prescient von Neumann's predictions that has turned out to be correct; the study of neural systems has opened up new perspectives for understanding complex systems of all sorts. In this volume we emphasize that neural systems of the intelligent kind are embedded in sociocultural systems of separate but connected nervous systems. Deeper computational studies of biological and cultural phenomena are affecting our understanding of many aspects of computing itself and are altering the way in which we perceive computing proper. We hope that this book is one step along the way toward that understanding and perception.

A Thumbnail Sketch of Particle Swarm Optimization

The field of evolutionary computation is often considered to comprise four major paradigms: genetic algorithms, evolutionary programming, evolution strategies, and genetic programming (Eberhart, Simpson, and Dobbins, 1996). (Genetic programming is sometimes categorized as a subfield of genetic algorithms.) As is the case with these evolutionary computation paradigms, particle swarm optimization utilizes a "population" of candidate solutions to evolve an optimal or near-optimal solution to a problem. The degree of optimality is measured by a fitness function defined by the user.

Particle swarm optimization, which has roots in artificial life and social psychology as well as engineering and computer science, differs from evolutionary computation methods in that the population members, called *particles,* are flown through the problem hyperspace. When the population is initialized, in addition to the variables being given random values, they are stochastically assigned velocities. Each iteration, each particle's velocity is stochastically accelerated toward its previous best position (where it had its highest fitness value) and toward a neighborhood best position (the position of highest fitness by any particle in its neighborhood).

The particle swarms we will be describing are closely related to *cellular automata* (CA), which are used for self-generating computer graphics movies, simulating biological systems and physical phenomena, designing massively parallel computers, and most importantly for basic research into the characteristics of complex dynamic systems. According to mathematician Rudy Rucker, CAs have three main attributes: (1) individual cell updates are done in parallel, (2) each new cell value depends only on the old values of the cell and its neighbors, and (3) all cells are updated using the same rules (Rucker, 1999). Individuals in a particle swarm population can be conceptualized as cells in a CA, whose states change in many dimensions simultaneously.

Particle swarm optimization is powerful, easy to understand, easy to implement, and computationally efficient. The central algorithm comprises just two lines of computer code and is often at least an order of magnitude faster than other evolutionary algorithms on benchmark functions. It is extremely resistant to being trapped in local optima.

As an engineering methodology, particle swarm optimization has been applied to fields as diverse as electric/hybrid vehicle battery pack state of charge, human performance assessment, and human tremor diagnosis. Particle swarm optimization also provides evidence for theoretical perspectives on mind, consciousness, and intelligence. These theoretical views, in addition to the implications and applications for engineering and computer science, are discussed in this book.

What This Book Is, and Is Not, About

Let's start with what it's not about. This book is not a cookbook or a how-to book. In this volume we will tell you about some exciting research that you may not have heard about—since it covers recent findings in both psychology and computer science, we expect most readers will find something here that is new to them. If you are interested in trying out some of these ideas, you will either find enough information to get started or we will show you where to go for the information.

This book is not a list of facts. Unfortunately, too much science, and especially science education, today has become a simple listing of research findings presented as absolute truths. All the research described in this volume is ongoing, not only ours but others' as well, and all conclusions are subject to interpretation. We tend to focus on issues; accomplishments and failures in science point the way to larger theoretical truths, which are what we really want. We will occasionally make statements that are controversial, hoping not to hurt anyone's feelings but to incite our readers to think about the topics, even if it means disagreeing with us.

This book is about *emergent behavior (self-organization)*, about simple processes leading to complex results. It's about the whole being more than the sum of its parts. In the words of one eminent mathematician, Stephen Wolfram: "It is possible to make things of great complexity out of things that are very simple. There is no conservation of simplicity."

We are not the first to publish a book with the words "swarm intelligence" in the title, but we do have a significantly distinct viewpoint from some others who use the term. For example, in *Swarm Intelligence: From Natural to Artificial Systems,* by Bonabeau, Dorigo, and Theraulaz (1999), which focuses on the modeling of social insect (primarily ant) behavior, page 7 states:

> It is, however, fair to say that very few applications of swarm intelligence have been developed. One of the main reasons for this relative lack of success resides in the fact that swarm-intelligent systems are hard to "program," because the paths to problem solving are not predefined but emergent in these systems and result from interactions among individuals and between individuals and their environment as much as from the behaviors of the individuals themselves. Therefore, using a swarm-intelligent system to solve a problem requires a thorough knowledge not only of what individual behaviors must be implemented but also of what interactions are needed to produce such or such global behavior.

It is our observation that quite a few applications of swarm intelligence (at least our brand of it) have been developed, that swarm intelligent systems are quite easy to program, and that a knowledge of individual behaviors and interactions is not needed. Rather, these behaviors and interactions emerge from very simple rules.

Bonabeau et al. define swarm intelligence as "the emergent collective intelligence of groups of simple agents." We agree with the spirit of this definition, but prefer not to tie swarm intelligence to the concept of "agents." Members of a swarm seem to us to fall short of the usual qualifications for something to be called an "agent," notably autonomy and specialization. Swarm members tend to be homogeneous and follow their programs explicitly. It may be politically incorrect for us to fail to align ourselves with the popular paradigm, given the current hype surrounding anything to do with agents. We just don't think it is the best fit.

So why, after all, did we call our paradigm a "particle *swarm?*" Well, to tell the truth, our very first programs were intended to model the coordinated movements of bird flocks and schools of fish. As the programs evolved from modeling social behavior to doing optimization, at some point the two-dimensional plots we used to watch the algorithms perform ceased to look much like bird flocks or fish schools and started looking more like swarms of mosquitoes. The name came as simply as that.

Mark Millonas (1994), at Santa Fe Institute, who develops his kind of swarm models for applications in artificial life, has articulated five basic principles of *swarm intelligence:*

- The *proximity* principle: The population should be able to carry out simple space and time computations.
- The *quality* principle: The population should be able to respond to quality factors in the environment.
- The principle of *diverse response:* The population should not commit its activity along excessively narrow channels.
- The principle of *stability:* The population should not change its mode of behavior every time the environment changes.
- The principle of *adaptability:* The population must be able to change behavior mode when it's worth the computational price.

(Note that stability and adaptability are the opposite sides of the same coin.) All five of Millonas' principles seem to describe particle swarms; we'll keep the name.

As for the term *particle,* population members are massless and volumeless mathematical abstractions and would be called "points" if they stayed still; velocities and accelerations are more appropriately applied to particles, even if each is defined to have arbitrarily small mass and volume. Reeves (1983) discusses *particle systems* consisting of clouds of primitive particles as models of diffuse objects such as clouds, fire, and smoke within a computer graphics framework. Thus, the label we chose to represent the concept is *particle swarm.*

Assertions

The discussions in this book center around two fundamental assertions and the corollaries that follow from them. The assertions emerge from the interdisciplinary nature of this research; they may seem like strange bedfellows, but they work together to provide insights for both social and computer scientists.

I. *Mind is social.* We reject the cognitivistic perspective of mind as an internal, private thing or process and argue instead that both function and phenomenon derive from the interactions of individuals in a social world. Though it is mainstream social science, the statement needs to be made explicit in this age where the cognitivistic view dominates popular as well as scientific thought.

A. *Human intelligence results from social interaction.* Evaluating, comparing, and imitating one another, learning from experience and emulating the successful behaviors of others, people are able to adapt to complex environments through the discovery of relatively optimal patterns of attitudes, beliefs, and behaviors. Our species' predilection for a certain kind of social interaction has resulted in the development of the inherent intelligence of humans.

B. *Culture and cognition are inseparable consequences of human sociality.* Culture emerges as individuals become more similar through mutual social learning. The sweep of culture moves individuals toward more adaptive patterns of thought and behavior. The emergent and immergent phenomena occur simultaneously and inseparably.

II. *Particle swarms are a useful computational intelligence (soft computing) methodology.* There are a number of definitions of "computational intelligence" and "soft computing." Computational intelligence and soft computing both include hybrids of evolutionary computation, fuzzy logic, neural networks, and artificial life. Central to the concept of computational intelligence is system adaptation that enables or facilitates intelligent behavior in complex and changing environments. Included in soft computing is the softening "parameterization" of operations such as AND, OR, and NOT.

A. *Swarm intelligence provides a useful paradigm for implementing adaptive systems.* In this sense, it is an extension of evolutionary computation. Included application areas are simulation, control, and diagnostic systems in engineering and computer science.

B. *Particle swarm optimization is an extension of, and potentially important new incarnation of, cellular automata.* We speak of course of topologically structured systems in which the members' topological positions do not vary. Each cell, or location, performs only very simple calculations.

Organization of the Book

This book is intended for researchers; senior undergraduate and graduate students with a social science, cognitive science, engineering, or computer science background; and those with a keen interest in this quickly evolving "interdiscipline." It is also written for what is referred to in the business as the "intelligent layperson." You shouldn't need a Ph.D. to read this book; a driving curiosity and interest in the current state of science should be enough. The sections on application of the swarm algorithm principles will be especially helpful to those researchers and engineers who are concerned with getting something that *works*. It is helpful to understand the basic concepts of classical (two-valued) logic and elementary statistics. Familiarity with personal computers is also helpful, but not required. We will occasionally wade into some mathematical equations, but only an elementary knowledge of mathematics should be necessary for understanding the concepts discussed here.

Part I lays the groundwork for our journey into the world of particle swarms and swarm intelligence that occurs later in the book. We visit big topics such as life, intelligence, optimization, adaptation, simulation, and modeling.

Chapter 1, Models and Concepts of Life and Intelligence, first looks at what kinds of phenomena can be included under these terms. What is life? This is an important question of our historical era, as there are many ambiguous cases. Can life be created by humans? What is the role of adaptation in life and thought? And why do so many natural adaptive systems seem to rely on randomness?

Is cultural evolution Darwinian? Some think so; the question of evolution in culture is central to this volume. The Game of Life and cellular automata in general are computational examples of emergence, which seems to be fundamental to life and intelligence, and some artificial life paradigms are introduced. The chapter begins to inquire about the nature of intelligence and reviews some of the ways that researchers have tried to model human thought. We conclude that intelligence just means "the qualities of a good mind," which of course might not be defined the same by everybody.

Chapter 2, Symbols, Connections, and Optimization by Trial and Error, is intended to provide a background that will make the later chapters meaningful. What is optimization and what does it have to do with minds? We describe aspects of complex fitness landscapes and some methods that are used to find optimal regions on them. Minds can be thought of as points in high-dimensional space: what would be needed to optimize them? Symbols as discrete packages of meaning are contrasted to the connectionist approach where meaning is distributed across a network. Some issues are discussed having to do with numeric representations of cognitive variables and mathematical problems.

Chapter 3, On Our Nonexistence as Entities: The Social Organism, considers the various zoom angles that can be used to look at living and thinking things. Though we tend to think of ourselves as autonomous beings, we can be considered as macroentities hosting multitudes of cellular or even subcellular guests, or as microentities inhabiting a planet that is alive. The chapter addresses some issues about social behavior. Why do animals live in groups? How do the social insects manage to build arches, organize cemeteries, stack woodchips? How do bird flocks and fish schools stay together? And what in the world could any of this have to do with human intelligence? (Hint: It has a lot to do with it.)

Some interesting questions have had to be answered before robots could do anything on their own. Rodney Brooks' subsumption architecture builds apparently goal-directed behavior out of modules. And what's the difference between a simulated robot and an agent? Finally, Chapter 3 looks at computer programs that can converse with people. How do they do it? Usually by exploiting the shallowness or mindlessness of most conversation.

Chapter 4, Evolutionary Computation Theory and Paradigms, describes in some detail the four major computational paradigms that use evolutionary theory for problem solving. The fitness of potential problem solutions is calculated, and the survival of the fittest allows better solutions to reproduce. These powerful methods are known as the "second-best way" to solve any problem.

Chapter 5, Humans—Actual, Imagined, and Implied, starts off musing on language as a bottom-up phenomenon. The chapter goes on to review the downfall of behavioristic psychology and the rise of cognitivism, with social psychology simmering in the background. Clearly there is a relationship between culture and mind, and a number of researchers have tried to write computer programs based on that relationship. As we review various paradigms, it becomes apparent that a lot of people think that culture must be similar to Darwinistic evolution. Are they the same? How are they different?

Chapter 6, Thinking Is Social, eases us into our own research on social models of optimization. The adaptive culture model is based on Axelrod's culture model—in fact, it is exactly like it except for one little thing: individuals imitate their neighbors, not on the basis of similarity, but on the basis of their performance. If your neighbor has a better solution to the problem than you do, you try to be more like them. It is a very simple algorithm with big implications.

Part II focuses on our particle swarm paradigm and the collective and individual intelligence that arises within the swarm. We first introduce the conceptually simplest version of particle swarms, binary particle swarms, and then discuss the "workhorse" of particle swarms, the real-valued version. Variations on the basic algorithm and the performance of the particle swarm on benchmark functions precede a review of a few applications.

Chapter 7, The Particle Swarm, begins by suggesting that the same simple processes that underlie cultural adaptation can be incorporated into a computational paradigm. Multivariate decision making is reflected in a binary particle swarm. The performance of binary particle swarms is then evaluated on a number of benchmarks.

The chapter then describes the real-valued particle swarm optimization paradigm. Individuals are depicted as points in a shared high-dimensional space. The influence of each individual's successes and those of neighbors is similar to the binary version, but change is now portrayed as movement rather than probability. The chapter concludes with a description of the use of particle swarm optimization to find the weights in a simple neural network.

Chapter 8, Variations and Comparisons, is a somewhat more technical look at what various researchers have done with the basic particle swarm algorithm. We first look at the effects of the algorithm's main parameters and at a couple of techniques for improving performance. Are particle swarms actually just another kind of evolutionary algorithm? There are reasons to think so, and reasons not to. Considering the similarities and differences between evolution and culture can help us understand the algorithm and possible things to try with it.

Chapter 9, Applications, reviews a few of the applications of particle swarm optimization. The use of particle swarm optimization to evolve artificial neural networks is presented first. Evolutionary computation techniques have most commonly been used to evolve neural network weights, but have sometimes been used to evolve neural network structure or the neural network learning algorithm. The strengths and weaknesses of these approaches are reviewed. The use of particle swarm optimization to replace the learning algorithm and evolve both the weights and structure of a neural network is described. An added benefit of this approach is that it makes scaling or normalization of input data unnecessary. The classification of the Iris Data Set is used to illustrate the approach. Although a feedforward neural network is used as the example, the methodology is valid for practically any type of network.

Chapter 10, Implications and Speculations, reviews the implications of particle swarms for theorizing about psychology and computation. If social interaction provides the algorithm for optimizing minds, then what must that be like for the individual? Various social- and computer-science perspectives are brought to bear on the subject.

Chapter 11, And in Conclusion . . . , looks back at some of the motifs that were woven through the narrative.

Appendix A, Statistics for Swarmers, is where we review some methods for scientific experimental design and data analysis. The discussion is a high-level overview to help researchers design their investigations; you should be conversant with these tools if you're going to evaluate what you are doing with particle swarm optimization—or any other stochastic optimization, for that matter. Included are sections on descriptive and inferential statistics, confidence intervals, student's t-test, one-way analysis of variance, factorial and multivariate ANOVA, regression analysis, and the chi-square test of independence. The material in this appendix provides you with sufficient information to perform some of the simple statistical analyses.

Appendix B, Genetic Algorithm Implementation, explains how to use the genetic algorithm software distributed at the book's web site. The program, which includes the famous Fisher Iris Data Set, is set up to optimize weights in a neural network. You can experiment with various parameters described in Chapter 4 to see how they affect the ability of the algorithm to optimize the weights in the neural network, to accurately classify flowers according to several measurements taken on them. The source code is also available at the book's web site and can be edited to optimize any kind of function you might like to try.

Software

The software associated with this book can be found on the Internet at *www.engr.iupui.edu/~eberhart/web/PSObook.html.* The decision to use the Internet as the medium to distribute the software was made for two main reasons. First, by *not* including it with the book as, say, a CD-ROM, the cost of the book can be lower. And we hope more folks will read the book as a result of the lower price. Second, we can update the software (and add new stuff) whenever we want—so we can actually do something about it when readers let us know about the (inevitable?) software critters known as bugs. Some of the software is designed to be run online from within your web browser; some of it is downloadable and executable in a Windows environment on your PC.

Definitions

A few terms that are used at multiple places in the book are defined in this section. These terms either do not have universally accepted definitions or their definitions are not widely known outside of the research community. Throughout the book, glossary terms are italicized and will be defined in the back of the book. Unless otherwise stated, the following definitions are to be used throughout the book:

Evolutionary computation comprises machine learning optimization and classification paradigms roughly based on mechanisms of evolution such as biological genetics and natural selection (Eberhart, Simpson, and Dobbins, 1996). The evolutionary computation field includes genetic algorithms, evolutionary programming, genetic programming, and evolution strategies, in addition to the new kid on the block: particle swarm optimization.

Mind is a term we use in the ordinary sense, which is of course not very well defined. Generally, mind is "that which thinks." David Chalmers helps us out by noting that the colloquial use of the concept of mind really contains two aspects, which he calls "phenomenological" and "psychological." The phenomenological aspect of mind has to do with the conscious experience of thinking, what it is like to think, while the psychological aspect (as Chalmers uses the term, perhaps many psychologists would disagree) has to do with the function of thinking, the information processing that results in observable behavior. The connection between conscious experience and cognitive function is neither simple nor obvious. Because consciousness is not observable, falsifiable, or provable, and we are talking in this book about computer programs that simulate human behavior, we mostly ignore the phenomenology of mind, except where it is relevant in explaining function. Sometimes the experience of being human makes it harder to perceive functional cognition objectively, and we feel responsible to note where first-person subjectivity steers the folk-psychologist away from a scientific view.

A *swarm* is a population of interacting elements that is able to optimize some global objective through collaborative search of a space. Interactions that are relatively local (topologically) are often emphasized. There is a general stochastic (or chaotic) tendency in a swarm for individuals to move toward a center of mass in the population on critical dimensions, resulting in convergence on an optimum.

An *artificial neural network* (ANN) is an analysis paradigm that is roughly modeled after the massively parallel structure of the brain. It simulates a highly interconnected, parallel computational structure with many relatively simple individual *processing elements* (PEs) (Eberhart, Simpson, and Dobbins, 1996). In this book the terms *artificial neural network* and *neural network* are used interchangeably.

Acknowledgments

We would like to acknowledge the help of our editor, Denise Penrose, and that of Edward Wade and Emilia Thiuri, at Morgan Kaufmann Publishers. Special thanks goes to our reviewers, who stuck with us through a major reorganization of the book and provided insightful and useful comments. Finally, we thank our families for their patience for yet another project that took Dad away for significant periods of time.

part
one

Foundations

chapter
one

Models and Concepts of Life and Intelligence

This chapter begins to set the stage for the computational intelligence paradigm we call "particle swarm," which will be the focus of the second half of the book. As human cognition is really the gold standard for intelligence, we will, as artificial intelligence researchers have done before us, base our model on people's thinking. Unlike many previous AI researchers, though, we do not subscribe to the view of mind as equivalent to brain, as a private internal process, as some set of mechanistic dynamics, and we deemphasize the autonomy of the individual thinker. The currently prevailing cognitivist view, while it is extreme in its assumptions, has taken on the mantle of orthodoxy in both popular and scientific thinking. Thus we expect that many readers will appreciate our setting a context for this new perspective. This introductory discussion will emphasize the adaptive and dynamic nature of life in general, and of human intelligence in particular, and will introduce some computational approaches that support these views.

We consider thinking to be an aspect of our social nature, and we are in very good company in assuming this. Further, we tend to emphasize the similarities between human social behavior and that of other species. The main difference to us is that people, that is, minds, "move" in a high-dimensional abstract space. People navigate through a world of meaning, of many distinctions, gradations of differences, and degrees of similarity. This chapter then will investigate some views of the adaptability of living things and computational models and the adaptability of human thought, again with some discussion of computational instantiations. ■

The Mechanics of Life and Thought

From the beginning of written history there has been speculation about exactly what distinguished living from nonliving things. The distinction seemed obvious, but hard to put a finger on. Aristotle believed:

> What has soul in it differs from what has not, in that the former displays life . . . Living, that is, may mean thinking or perception or local movement and rest, or movement in the sense of nutrition, decay, and growth . . . This power of self-nutrition . . . is the originative power, the possession of which leads us to speak of things as living.

This list of attributes seemed to summarize the qualities of living things, in the days before genetic engineering and "artificial life" computer programs were possible; Aristotle's black-and-white philosophy defined orthodox thought for a thousand years and influenced it for another thousand.

It does not seem that the idea was seriously entertained that living bodies were continuous with inorganic things until the 17th century, when William Harvey discovered that blood circulates through the body; suddenly the heart was a pump, like any other pump, and the blood moved like any other fluid. The impact was immediate and profound. The year after the publication of Harvey's *On the Motion of the Heart and Blood in Animals,* Descartes noted: "Examining the functions which might . . . exist in this body, I found precisely all those that might exist in us without our having the power of thought, and consequently without our soul—that is to say, this part of us, distinct from the body, of which it has been said that its nature is to think." So in the same stroke with which he noted—or invented—the famous dichotomy between mind and body, Descartes established as well the connection between living bodies and other physical matter that is perhaps the real revolution of the past few centuries. Our living bodies are just like everything else in the world. Where earlier philosophers had thought of the entire human organism, mind and body, as a living unity distinct from inanimate matter, Descartes invited the domain of cold matter up into the body, and squeezed the soul back into some little-understood abstract dimension of the universe that was somehow—but nobody knew how—connected with a body, though fundamentally different from it. It was not that Descartes invented the notion that mental stuff was different from physical stuff—everybody already thought that. It was that he suggested that living bodies were the same as all the other stuff in the world. Minds stayed where they were: different.

Though he knew it must be true, even Charles Darwin found it hard to accept that living matter was continuous with inanimate matter: "The most humble organism is something much higher than the inorganic dust under our feet; and no one with an unbiased mind can study any living creature, however humble, without being struck with enthusiasm at its marvelous structure and properties." Indeed it seems that a hallmark of life is its incredible complexity. Even the smallest, most primitive microbe contains processes and structures that can only be described as amazing. That these phenomena were designed by chance generation and selection is so different from the way we ordinarily conceive design and creation that people have difficulty even imagining that life could have developed in this way, even when they know it must be true.

In considering a subtle aspect of the world such as the difference between living and nonliving objects, it seems desirable, though it may turn out to be impossible, to know whether our distinctions are based on the qualities of things or our attributions about them. A major obstacle is that we are accustomed to thinking of ourselves as above and beyond nature somehow; while human accomplishments should not be trivialized, we must acknowledge (if this discussion is going to continue) that some of our feelings of grandeur are delusional—and we can't always tell which ones. The taxonomic distinction between biological and other physical systems has been one of the cornerstones of our sense of being special in the world. We felt we were divine, and our flesh was the living proof of it. But just as Copernicus bumped our little planet out of the center of the universe, and Darwin demoted our species from divinity to beast, we live to witness modern science chipping away these days at even this last lingering self-aggrandizement, the idea that life itself contains some element that sets it above inanimate things. Today, ethical arguments arise in the contemplation of the aliveness of unborn fetuses, of comatose medical patients, of donor organs, of tissues growing in test tubes, of stem cells. Are these things alive? Where is the boundary between life and inanimate physical objects, really? And how about those scientists who argue that the earth itself is a living superorganism? Or that an insect colony is a superorganism—doesn't that make the so-called "death" of one ant something less than the loss of a life, something more like cutting hair or losing a tooth? On another front, the creation of adaptive robots and lifelike beings in computer programs, with goal-seeking behaviors, capable of self-reproduction, learning and reasoning, and even evolution in their digital environments, blurs the division as well between living and nonliving systems. Creatures in *artificial life* programs may be able to do all the things that living things do. Who is to say they are not themselves alive?

And why does it matter? Why should we have a category of things we call "alive" and another category for which the word is inappropriate? It seems to us that the distinction has to do with a moral concern about killing things—not prohibition exactly, but concern. It is morally acceptable to end some kinds of dynamic processes, and it is not acceptable to end others. Termination of a living process calls for some special kinds of emotion, depending mainly on the bond between the living thing and ourselves. For whatever reasons, we humans develop an empathic relationship with particular objects in the world, especially ones that interact with us, and the concept of "living" then is hopelessly bound up in these empathic relations. The tendency to distinguish something vibrant and special in living things is part of our sociality; it is an extension of our tendency to fraternize with other members of our species, which is a theme you will encounter a lot in this book.

There may be good, rational reasons to draw a line between living and other things. Perhaps a definition of life should include only those objects that possess a particular chemical makeup or that have evolved from a single original lineage. The organisms we count as living are based on hydrocarbon molecules, albeit of wide varieties, and self-reproducing DNA and RNA are found throughout the biological kingdoms. Is this what is meant by life? We doubt it. The concept of extraterrestrial life, for instance, while hard to imagine, is nonetheless easy to accept. What would we say if there were silicon-based beings on another planet? What if we discovered Plutonians who had evolved from silica sand into a life-like form with highly organized patterning—which we would recognize immediately as microchips—and intricate, flexible, adaptive patterns of cognition and behavior? If we ran into them in the *Star Wars* bar, would we consider them alive? The answer is plainly yes. Does it make any sense to allow that alien computers can be alive, while terrestrial ones cannot? The answer has to be no.

Possibly the reluctance to consider computer programs, robots, or molecularly manufactured beings as living stems from the fact that these things are man-made—and how can anything living be man-made? This argument is religious, and we can't dispute it; those who believe life can come only from deity can go ahead and win the argument.

Biologists with the Minimal Genome Project have been investigating the bottom line, trying to discern what is the very least information that needs to be contained in the genes for an organism (or is it?) to perpetuate its dynamics, to be called "alive" (Hutchinson et al., 1999). By methodically altering the genes of some very simple one-celled organisms and by comparing the genomes of some very simple bacteria, these scientists, led by J. Craig Venter, have been able to identify which ones are necessary, and which are not, for the organisms' survival. They have determined that 265 to 350 of *Mycoplasma genitalium's* 480 genes are essential for life under laboratory conditions (interestingly, the biologists don't know what 111 of those genes do—only that they are necessary). With the minimal subset, the organism may not be able to survive outside the warm comfort of its petri dish, but in carefully controlled conditions it should be able to stay alive. As we are talking about a sequence of only a few hundred genes, and molecular engineering is an everyday activity, the question arises, what will you call it when—not if—some laboratory mechanics put together a package of molecules that can reproduce and metabolize nutrients in a carefully controlled synthetic environment? Do you call it life?

In all ways known, biological organisms are like other machines (e.g., Wooldridge, 1968). The operation of muscle, of digestive enzyme, of neuron, of DNA—as Descartes observed, all these things are explainable in technological terms. We know now that there is no special problem with replacing body parts with man-made machinery, once the details of their processes are known. If our bodies may be a kind of ordinary hardware, what about our minds? Knowing that brain injury can result in mental anomalies, and knowing that electrical and chemical activities in the brain correlate with certain kinds of mental activities, we can correctly conclude that brains provide the machinery of minds. We will not attempt to tackle the problem of the relationship of minds and brains in this introductory section—we'll work up to that. For now we only note that brains are physical objects, extremely complex but physical nonetheless, bound by the ordinary laws of physics; the machinery of thought is like other machinery.

As for the question of whether man-made artifacts can really be alive, we have encountered a situation that will appear at several points in this book, so we may as well address it now. In science as in other domains of discourse, there may be disagreements about how things work, what they are made of, whether causes are related to effects, about the true qualities of things—questions about the world. These kinds of questions can be addressed through observation, experiment, and sometimes deduction. Sometimes, though, the question is simply whether something belongs to a semantic category. Given a word that is the label for a category, and given some thing, idea, or process in the world, we may argue about whether that label should properly be attached to the object. There is of course no way to prove the answer to such a question, as categories are linguistic conventions that derive their usefulness exactly from the fact that people agree on their usage. Is *X* an *A?* If we agree it is, then it is: a tomato is a vegetable, not a fruit. If we disagree, then we may as well go on to the next thing, because there is no way to prove that one of us is correct. One of the fundamental characteristics of symbols, including words, is that they are arbitrary; their meaning derives from common usage. In the present case, we will wiggle away from the dispute by agreeing with the scientific convention of calling lifelike beings in computer programs "artificial life" or "Alife." Some later questions will not offer such a diplomatic resolution.

Computers are notorious for their inability to figure out what the user wants them to do. If you are writing a document and you press ALT instead of CTRL, the chances are good that your word processor will do some frustrating thing that is entirely different from what you intended. Hitting the wrong key, you might end up deleting a file, editing text, changing channels, navigating to some web site you didn't want to go to—the computer doesn't even try to understand what you mean. This kind of crisp interpretation of the world is typical of "inanimate" objects. Things happen in black and white, zero and hundred percents. Now and then a software product will contain a "wizard" or other feature that is supposed to anticipate the user's needs, but generally these things are simply based on the programmer's assumptions about what users might want to do—the program doesn't "know" what you want to do, it just does what it was programmed to do and contains a lot of if-then statements to deal with many possibilities. Some of them are quite clever, but we never feel guilty shutting down the computer. People who shut down other people are called "inhuman"—another interesting word. You are not considered inhuman for turning off your computer at the end of the day. It is not alive.

Contrast the crispness of machines with animate things. Last week there was a dog on television who had lost both his rear legs. An animate system, for instance, the two-legged dog, adapts to novelty and to ambiguity in the environment. This dog had learned to walk on its forepaws perfectly well. The television program showed him chasing a ball, following his master, eating, doing all the things we expect a dog to do, balanced on his forepaws with his tail up in the air. For all he knew, dogs were supposed to have two legs.

Stochastic Adaptation: Is Anything Ever Really Random?

There was a word in the previous paragraph that will become a core concept for everything we will be talking about. The word is *adapt*. It has been argued that an important dimension of difference between animate and inanimate things is the ability to adapt. Later we will consider an argument that intelligence is a propensity or ability to adapt, as well. So it will be good to dwell on the term for a moment, to consider what it means.

The word "adaptation" comes from a Latin root meaning "to fit to." Thus right from the beginning the word implies that there are two things: something that adapts and something it adapts to. Among living things, we say that organisms or species adapt to their environments. Of course, interacting species adapting to their environments end up changing the environments and adapting to the changed niches, which include other adapting organisms, and so on—adaptation can be endlessly dynamic. It is important to keep in mind that an entity is adaptive in relation to some criterion.

Adaptation in nature is almost always a *stochastic* process, meaning that it contains randomness; the word usually refers to a phenomenon that is probabilistic in nature. Most of the paradigms discussed here will have a random component. By one definition, randomness exists when repeated occurrences of the same phenomenon can result in different outcomes. Almost always, things that appear random turn out to be *deterministic* (they follow certainly from causes), except that we don't know the chains of causality involved. For instance, people may lament after an accident, "If only I had taken Third Street instead of Main," or "If only I had never introduced her to him." These "if only's" belie our inability to perceive the chains of causes that reach from an early state of a system to a later state.

"Random numbers" in a computer are of course not random at all, though we have little trouble referring to them as such. Given the same starting point, the random number generator will always produce exactly the same series of numbers. These *quasirandom* processes appear random to us because the sequence is unpredictable, unless you happen to know the formula that produces it. Most of the time, when we say "random" we really mean "unpredictable," or even just "unexpected;" in other words, we are really describing the state of our understanding rather than a characteristic of the phenomena themselves. Randomness may be another one of those things that don't exist in the world, but only in our minds, something we attribute to the world that is not a quality of it. The attribution of randomness is based on the observer's inability to understand what caused a pattern of events and is not necessarily a quality of the pattern itself. For instance, we flip a coin in order to introduce randomness into a decision-making process. If the direction and magnitude of the force of the thumb against the coin were known, as well as the mass of the coin and the distribution of density through its volume, relevant atmospheric characteristics, and so on, the trajectory of the coin could be perfectly predicted. But because these factors are hard to measure and control, the outcome of a flip is unpredictable, what we call "random"—close enough.

Sociobiologist E. O. Wilson (1978) proposed an interesting elaboration on the coin-flipping example. He agrees with us that if all the knowledge of physical science were focused on the coin flip, the outcome could be perfectly predictable. So, he suggests, let us flip something more interesting. What would happen if we flipped something more complicated— perhaps a bee? A bee has memory, can learn, reacts to things, and in fact would probably try to escape the trajectory imposed on it by the flipper's thumb. But, as Wilson points out, if we had knowledge of the nervous system of the bee, the behaviors of bees, and something of the history of this bee in particular, as well as the other things that we understood in the previous paragraph about forces acting on a flipped object, we might be able to predict the trajectory of the bee's flight quite well, at least better than chance. Wilson makes the profound point that the bee, kicking and flapping and twitching as it soars off the thumb, has "free will," at least from its own point of view. From the human observer's perspective, though, it is just an agitated missile, quite predictable.

Of course it is easy to extend Wilson's provocative example one little step further and imagine flipping a human being. Perhaps a mighty creature like King Kong or a more subtle Leviathan—perhaps social forces or forces of nature—could control the actual trajectory of a human through space. Would that human being have free will? He or she would probably think so.

Let's insert a social observation here. In a later chapter we will be discussing coordination games. These are situations involving two or more participants who affect one another's, as well as their own, outcomes. Game-theory researchers long ago pointed out that it is impossible to devise a strategy for successfully interacting with someone who is making random choices. Imagine someone who spoke whether the other person was talking or not, and whether the subject was related to the previous one; such unpredictable behavior would be unconscionably rude and unnerving. Being predictable is something we do for other people; it is a service we provide to enable other people to deal with us. Random be-havior is rude. Even contrary behavior, where the person does the op-posite of what we want, is at least predictable, and though we might not appreciate them, we can accept rebels who behave contrary to our expec-tations. But behavior performed without any reference to our expecta-tions, that is, random behavior, is dangerous; the "loose cannon" is a danger. Some writers, such as cultural psychologist Michael Tomasello (1999), neuroscientist Leslie Brothers (1997), and social psychologist Tom Ostrom (1984), have argued for the primacy of social cognition: cat-egorization and other cognitive processes are simply extensions of social information-processing techniques to apply to nonsocial objects. Our or-dinary interpretation of randomness and its opposite, the identification of simple kinds of order, might very well be a form of social thinking.

For most of the 20th century it was thought that "true" randomness existed at the subatomic level. Results from double-slit experiments and numerous thought experiments had convinced quantum physicists that subatomic entities such as photons should be conceptualized both as particles and as waves. In their wave form, such objects were thought to occupy a state that was truly stochastic, a probability distribution, and their position and momentum were not fixed until they were observed. In one of the classic scientific debates of 20th-century physics, Niels Bohr argued that a particle's state was truly, unknowably random, while Ein-stein argued vigorously that this must be impossible: "God does not play dice." Until very recently, Bohr was considered the winner of the dispute, and quantum events were considered to be perhaps the only example of true stochasticity in the universe. But in 1998, physicists Dürr, Nonn, and Rempe (1998) disproved Bohr's theorizing, which had been based on Heisenberg's uncertainty principle. The real source of quantum "ran-domness" is now believed to be the interactions or "entanglements" of particles, whose behavior is in fact deterministic.

The basis of observed randomness is our incomplete knowledge of the world. A seemingly random set of events may have a perfectly good ex-planation; that is, it may be perfectly compressible. Press, Teukolsky, Vetterling, and Flannery's bible of scientific programming, *Numerical Recipes in C* (1993), shows a reasonably good random number generator that can be written with one line of code. If we don't know what the un-derlying process is that is generating the observed sequence, we call it random. If we define "randomness" as ignorance, we can continue to use the term, in humility.

In a future section we will consider cellular automata, computer pro-grams whose rules are fully deterministic, but whose outputs appear to be random, or sometimes orderly in an unpredictable way. In discussions of randomness in this book, we will assume that we are talking about something similar to the unpredictability that arises from complex systems such as cellular automata. This kind of randomness—and let's go ahead and call it that—is an integral aspect of the environment; life and intelligence must be able to respond to unexpected challenges.

We have noted that adaptation usually seems to include some random component, but we have not asked why. If we consider an adaptive system as one that is adjusting itself in response to feedback, then the question is in finding the appropriate adjustment to make; this can be very difficult in a complex system, as we will see shortly, because of both external demands and the need to maintain internal consistency. In adaptive computer programs, randomness usually serves one of two functions. First, it is often simply an expression of uncertainty. Maybe we don't know where to start searching for a number, or where to go next, but we have to go somewhere—a random direction is as good as any. A good, unbiased quasirandom number can be especially useful in areas where people have known predispositions. Like the drunk who looks for his keys under the streetlight, instead of in the bushes where he dropped them, "because there's more light here," we often make decisions that reflect our own cognitive tendencies more than the necessities of the task at hand. A random choice can safeguard against such tendencies. The second important function of random numbers is, interestingly, to introduce creativity or innovation. Just as artists and innovators are often the eccentrics of a society, sometimes we need to introduce some randomness just to try something new, in hopes of improving our position. And lots of times it works.

The "Two Great Stochastic Systems"

Stochastic adaptation is seen in all known living systems. Probabilistic choices allow creative, innovative exploration of new possibilities, especially in a changing environment. There are other advantages as well—for instance, perfectly predictable prey would be a little too convenient for predators. In general, organisms can adapt by making adjustments within what Gregory Bateson (1979) called the "two great stochastic systems." These systems are evolution and mind.

Later we will go into evolutionary and genetic processes in more detail. For the present, we note that evolution operates through variation and *selection*. A variety of problem solutions (chromosomes or patterns of features) are proposed and tested; those that do well in the test tend to survive and reproduce, while those that perform poorly tend to be eliminated. This is what is meant by "survival of the fittest." Looking at it from the population level, for instance, looking at the proportion of individuals with a phenotypic trait, we see a kind of stochastic change, as the probabilities of various traits increase and decrease in a population over time, and the mean values of quantitative features, for instance, height or weight, shift along their continua.

The second "great stochastic system" is the mind. Contemporary cognitive scientists see the mind as a stochastic system of neurons, adjusting their firings in response to stimuli that include other neurons. The brain is an excellent exemplar of the concept of a complex adaptive system— but this is not what Bateson meant when he described mind as a great stochastic system, and it is not what we mean. The distinction between brain and mind is extremely important and not subtle at all, though most people today fail to observe it. We are talking about minds here, not brains.

Some theorists argue that mental processes are very similar to evolutionary ones, where hypotheses or ideas are proposed, tested, and either accepted or rejected by a population. The most prevalent opinion along these lines is the "memetic" view, proposed by Dawkins (1976) in *The Selfish Gene,* which suggests that ideas and other cultural symbols and patterns, called *memes,* act like genes; they evolve through selection, with mutation and recombination just like biological genes, increasing their frequency in the population if they are adaptive, dying out if they're not.

Dawkins points out that genes propagate through replication, making copies of themselves with slight differences from one "generation" to the next. Now the evolution of species, which crawls along over the eons, has created a kind of environment for the propagation of a new faster kind of replicator. In particular, the human brain provides a host environment for memetic evolution. Memes are not restricted to brains, though; they can be transmitted from brains to books and to computers, and from books to computers, and from computer to computer. The evolution of self-reproducing memes is much faster than biological evolution, and as we will see, some memes even try to dominate or control the direction of human genetic evolution.

Dawkins asserts that memes replicate through imitation, and in fact he coined the word from the ancient Greek *mimeme,* meaning to imitate. Ideas spread through imitation; for instance, one person expresses an idea in the presence of another, who adopts the idea for his or her own. This person then expresses the idea, probably in their own words, adding the possibility of mutation, and the meme replicates through the population. As if manifesting some kind of self-referential amplification, the theory of memes itself has spread through the scientific community with surprising efficiency; now there are scientific journals dedicated to the topic, and a large number of recent books on scientific and computational subjects have endorsed the view in some form. The view that mind and evolution operate by similar or identical processes is very widely accepted by those who should know.

The similarities between the two stochastic systems are significant, but the analogy becomes ambiguous when we try to get specific about the evolution of raw abstractions. The ancient question is, what, really, is an "idea?" You could take the view that ideas exist somewhere, in a Platonic World of Forms, independent of minds. Our insights and understanding, then, are glimpses of the ideal world, perceptions of shadows of pure ideas. "New" ideas existed previously, they had just not been known yet to any human mind. Next we would have to ask, how does evolution operate on such ideas? No one has proposed that ideas evolve in their disembodied state, that truth itself changes over time, that the World of Forms itself is in a dynamical process of evolving. The same mathematical facts, for instance, that were true thousands of years ago are equally true today—and if they are not true now, they never were. It has only been proposed that ideas evolve in the minds of humans; in other words, *our knowledge* evolves. The distinction is crucial.

The other view is that ideas have no existence independent of minds. According to this view, ideas are only found in the states of individual minds. At its most extreme, this position holds that an idea is nothing more than a pattern of neuronal connections and activations. Memetic evolution in this view then consists in imitating neuronal patterns, testing them, and accepting them if they pass the test. The problem here is that the same idea is almost certainly embodied in different neuronal patterns in different people—individuals' brains are simply wired up differently.

Luckily for us, we don't have to resolve millennia-old questions about the preexistence of ideas. For the present discussion it is only important to note that for memetic evolution to act upon ideas, they have to become manifest somehow in minds. They have to take on a form that allows mental representation of some type (a controversial topic in itself), and if they are to propagate through the population, their form must permit communication. The requirement of communication probably means that ideas need to be encoded in a symbol system such as language.

Our argument is that cultural evolution should be defined, not as operations on ideas, but as operations on minds. The evolution of ideas involves changes in the states of minds that hold ideas, not changes in the ideas themselves; it is a search—by minds—through the universe of ideas, to find the fitter ones. This will become more important when we discuss mental activity, intelligence, and culture.

We will emphasize now, and later as well, that cognition is a different process from genetic evolution. The great stochastic systems are different from one another: one uses selection, removing less fit members from the population, and the other adapts by changing the states of individuals who persist over time. These are two different kinds of adaptation. This is not to deny that ideas evolve in the minds of humans. The ideas expressed by people certainly change over time, in an adaptive way. We are suggesting a change in emphasis, that a scientific view of the evolution of ideas should look at changes of states of individuals, rather than at the ideas themselves.

There is something difficult about thinking of minds in populations. We are used to thinking of ourselves as autonomous thinkers, sometimes accepting beliefs and processes that others around us hold, but most of the time figuring out things on our own. Our experience is that we are in control of our cognitive systems, perhaps with some exceptions, such as when we are surprised or overcome with emotions; we experience our thoughts as controlled and even logical. If we are to consider the evolution of ideas through a population, though, we will need to transcend this illusion of autonomy and observe the individual in the context of a society, whether it is family, tribe, nation, culture, or pancontinental species. We are not writing this book to justify or romanticize commonly held beliefs about mankind's important place in the universe; we intend to look unblinking at the evidence. In order to do that, we need to remove self-interest and sentimental self-aggrandizement from the discussion.

In *Mind and Nature*, Bateson detailed what he considered to be the criteria of mind, qualities that were necessary and sufficient for something to be called a mind:

1. *A mind is an aggregate of interacting parts or components.*

2. *The interaction between parts of mind is triggered by difference.* For instance, perception depends on changes in stimuli.

3. *Mental process requires collateral energy.* The two systems involved each contribute energy to an interaction—as Bateson says, "You can take a horse to water, but you cannot make him drink. The drinking is his business" (p. 102).

4. *Mental process requires circular (or more complex) chains of determination.* The idea of reciprocal causation, or feedback, is a very important one and is fundamental to mental processes.

5. *In mental process, the effects of difference are to be regarded as transforms (i.e., coded versions) of the difference which preceded them.* Effects are not the same as their causes; the map is not the same as the territory.

6. *The description and classification of these processes of transformation discloses a hierarchy of logical types immanent in the phenomena.*

We will not attempt here to explain or elaborate Bateson's insightful and subtle analysis. We do want to note though that he hints—in fact it is a theme of his book—that biological evolution meets these criteria, that nature is a kind of mind. This seems a fair and just turnaround of the currently prevalent opinion that mind is a kind of evolution. We will see that evolutionary processes can be encoded in computer programs used to solve seemingly intractable problems, where the problem is defined as the analogue of an ecological niche, and recombined and mutated variations are tested and competitively selected. In other words, we can capitalize on the intelligence, the mental power, of evolution to solve many kinds of problems. Once we understand the social nature of mental processes, we can capitalize on those as well.

It appears to some thinkers that mind is a phenomenon that occurs when human beings coexist in societies. The day-to-day rules of living together are not especially complicated—some relatively straightforward scraps of wisdom will get you through life well enough. But the accumulated effect of these rules is a cultural system of imponderable depth and breadth. If we could get a sufficiently rich system of interactions going in a computer, we just might be able to elicit something like human intelligence. The next few sections will examine some ways that scientists and other curious people have tried to understand how to write computer programs with the qualities of the great stochastic systems.

The Game of Life: Emergence in Complex Systems

We have described inanimate things as being intolerant of variation in the environment and have used everyday commercial computer programs as an example of this. But it doesn't need to be that way: researchers have developed methods of computing that *are* tolerant of errors and novel inputs, computer programs that show the kinds of adaptation that we associate with living things.

The Game of Life

An arbitrary starting point for the story of the modern paradigm known as artificial life is an idea published in *Scientific American* in October 1970, in Martin Gardner's "Mathematical Games" column: mathematician John Conway's "Game of Life" (Gardner, 1970). The Game of Life is a grid of binary elements, maybe checkerboard squares or pixels on a screen, arranged on a two-dimensional plane, with a simple set of rules to define the state of each element or "cell." Each cell is conceptualized as belonging to a neighborhood comprising its eight immediate neighbors (above, below, to the sides, and diagonal). The rules say that

- If an occupied cell has fewer than two neighbors in the "on" state (which we can call "alive"), then that cell will die of loneliness—it will be in the "off" or "dead" state in the next turn.
- If it has more than three neighbors in the on state, then it will die of overcrowding.
- If the cell is unoccupied and it has exactly three alive neighbors, it will be born in the next iteration.

The Game of Life can be programmed into a computer, where the rules can be run iteratively at a high speed (unlike in Conway's time, the 1960s, when the game had to be seen on a checkerboard or Go board, or other matrix, in slow motion). The effect on the screen is mesmerizing, always changing and never repeating, and the program demonstrates many of the features that have come to be considered aspects of artificial life.

A common way to program it is to define cells or pixels on the screen as lit when they are in the alive state, and dark when they are not, or vice versa. The program is often run on a *torus grid,* meaning that the cells at the edges are considered to be neighbors to the cells at the opposite end. A torus is shaped like a doughnut; if you rolled up a sheet of paper so that the left and right edges met, it would be a tube or cylinder, and if you rolled that so that the open ends met, it would be in a torus shape. A torus grid is usually displayed as a rectangle on the screen, but where rules in the program require a cell to assess the states of its neighbors, those at the edges look at the cells at the far end.

When the Game of Life runs, cohesive patterns form out of apparent randomness on the grid and race around the screen. Two patterns colliding may explode into myriad new patterns, or they may disappear, or they may just keep going right through one another. Some simple patterns just sit there and blink, while some have been likened to cannons, shooting out new patterns to career around the grid. A kind of Game of Life lore has developed over the years, with names for these different patterns, such as "blinkers," "b-heptaminos," "brains" and "bookends," "gliders" and "glider guns," and "r-pentaminos." Some of the names are scientifically meaningful, and some are just fun. Figure 1.1 shows the glider pattern, which moves diagonally across the screen forever, unless it hits something.

The Game of Life is fascinating to watch. Writer M. Mitchell Waldrop (1992) has called it a "cartoon biology" (p. 202). When it is programmed to start with a random pattern, it is different every time, and sometimes it runs for a long time before all the cells die out. Artificial life researchers have pointed out many ways that Conway's simple game is similar to biological life, and it has even been suggested that this algorithm, or one like it, might be implemented as a kind of universal computer. Mostly, though, there is a sense conveyed by the program that is chillingly familiar, a sense of real life in process.

Emergence

Perhaps the most obvious and most interesting characteristic of the Game of Life is a property called *emergence*. There is much discussion in the scientific community about a complete definition of emergence; at least in this simple instance we can see that the complex, self-organizing patterns on the screen were not written into the code of the program that produced them. The program only says whether you are dead or alive depending on how many of your neighbors are alive; it never defines blinkers and gliders and r-pentaminos. They *emerge* somehow from a lower-level specification of the system.

Emergence is considered to be a defining characteristic of a complex dynamical system. Consider for instance the economy of a nation. (Note that economies of nations interact, and that the "economy of a nation" is simply an arbitrary subset of the world's economy.) As a physical system, an economy consists of a great many independent pairwise interactions between people who offer services or goods and people who want them. The 18th-century Scottish economist Adam Smith proposed that

Time

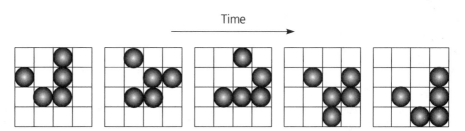

Figure 1.1 The ubiquitous "glider" pattern in the Game of Life.

there seems to be something like an "invisible hand" guiding the whole process, and in fact it is the absence of an invisible hand that makes the system interesting to us. Somehow, though no person controls an economy, it evolves toward and maintains an equilibrium; it is a relatively stable stochastic process, not really predictable but dependable. Prices and wages are consistent from place to place and over time. This does not mean they are the same everywhere, or at all times, but the differences are more or less consistent. The important point is that there is no central control. The stability or consistency of an economy at the large scale *emerges* from the qualities of the very many person-to-person interactions that make it up. (We will emphasize later the fact that the local individual interactions are affected by the global state of the system as well, a top-down phenomenon we call *immergence*.)

An ecology (note that all ecologies are interconnected) is another example of a complex system with emergent qualities. (Note that economies *and* ecologies are components of the same global system.) Predators and prey keep one another in check; trees and underbrush, hosts and parasites, heat and gases and fluids and soils interact incalculably to produce a system that persists very well over time in a kind of dynamic equilibrium. These complex systems are resilient; that is, if they are perturbed, they return to balance. This is a feature that we earlier described as "animate," a feature of life: *graceful degradation* or even complete recovery in the face of disruption. Any individual in an economy may experience failure. Any organism in an ecological environment might be uprooted or killed. Whole species may become extinct. Volcanoes, hurricanes, stock market crashes may result in devastation, but the system repairs itself and returns to a stable state—without central control.

Some have argued that emergence is simply a word, like "random," to cover up our ignorance about the true relationships between causes and effects. In this view, "emergent" really means that an effect was not predicted, implying that the person describing the system didn't understand it well enough to guess what was going to happen. On the other hand, emergence can be viewed as a kind of process in which the system returns robustly to an attractor that is an irresistible mathematical necessity of the dynamics that define it. These are not mutually exclusive definitions of emergence; an omniscient person would expect the emergent effect, but its relation to the lower-level events that produce it is often too complex for most of us to comprehend. In his book *Emergence: From Chaos to Order,* John Holland (1998) simply refuses to try to define the term, leaving us to understand it in both its senses, whichever is more appropriate in any instance. Whichever way you choose to consider emergence, the fact remains that very complex systems are able to maintain something like equilibrium, stability, or regularity without any invisible hand or central control.

An important aspect of the Game of Life is the fact that the rules as given by Conway cause the system to run for a long time without repeating. Other rules can be made up: for instance, a cell could stay alive only if all or none of its neighbors were alive. There are many freeware Game of Life programs that readers can acquire for experimenting with various rules. The finding will be that most rule sets are uninteresting. For most sets of rules the system simply stops after a few time steps, with all cells either on or off, or the screen fills with utter chaos. Only a few known sets of rules result in a system that continues without repetition. In order to understand how this occurs, we will have to move the discussion up a notch to discuss the superset of which the Game of Life is a member: cellular automata.

Cellular Automata and the Edge of Chaos

The cellular automaton (CA) is a very simple virtual machine that results in complex, even lifelike, behavior. In the most common one-dimensional, binary versions, a "cell" is a site on a string of ones and zeroes (a *bitstring*): a cell can exist in either of two states, represented as zero or one. The state of a cell at the next time step is determined by its state at the current time and by the states of cells in its neighborhood. A neighborhood usually comprises some number of cells lying on each side of the cell in question. The simplest neighborhood includes a cell and its immediately adjacent neighbors on either side. Thus a neighborhood might look like 011, which is the cell in the middle with the state value 1 and its two adjacent neighbors, one in state 0 and one in state 1.

Table 1.1 A set of rules for implementing a one-dimensional cellular automaton.

Neighbor 1	Cell now	Neighbor 2		Cell next
1	1	1	→	0
1	1	0	→	1
1	0	1	→	0
1	0	0	→	1
0	1	1	→	1
0	1	0	→	0
0	0	1	→	1
0	0	0	→	0

A neighborhood is embedded in a bitstring of some length greater than the neighborhood size. For instance, the neighborhood 011 occurs around the fifth position of this bitstring of length = 20:

01*001*100011011110111

The behavior of a CA is defined in a set of rules based on the pattern of states in the neighborhood. With a three-cell neighborhood, there are 2^3 = 8 possible neighborhood configurations. A CA rule set specifies the state of the cell at the next iteration for each of these eight configurations. For instance, a rule set might be given as shown in Table 1.1.

These rules are applied to every cell in the bitstring, the next state is calculated, and then the new bitstring is printed in the next row, or more likely, ones are indicated by lit pixels and zeroes by dark pixels. As the system iterates, the patterns of cells as they are affected by their rules over time are printed on the screen, each iteration producing the row of pixels directly under the preceding iteration's. Figure 1.2 shows some examples.

In an important early contribution to the study that became artificial life theory, Stephen Wolfram (1984/1994) noted that CA behaviors can be assigned to four classes:

1. Evolution leads to a homogeneous state, with all cells in the same state.

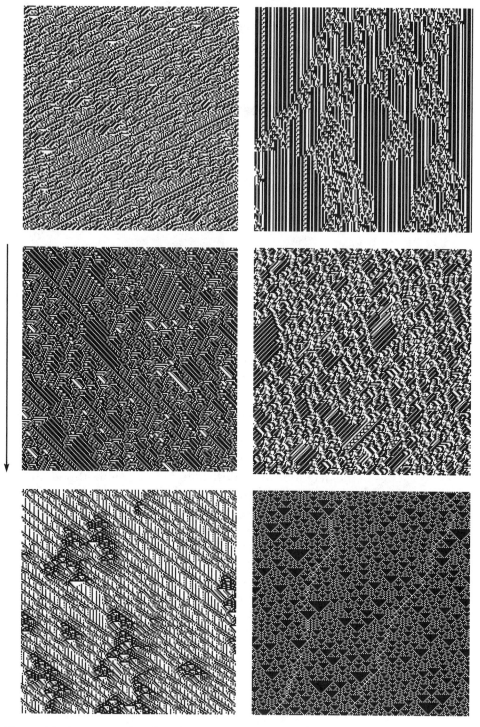

Figure 1.2 Some examples of one-dimensional binary cellular automata. Each row of pixels is determined by the states of the pixels in the row above it, as operated on by the particular set of rules.

2. Evolution leads to a set of simple stable or periodic structures.

3. Evolution leads to a chaotic pattern.

4. Evolution leads to complex localized structures, sometimes long-lived.

The first type of CA can be characterized as a *point attractor;* the dynamics of the system simply stop with all cells in the one state or all in the zero state. The second type results in a repeating, cyclic pattern, called a *periodic attractor,* and the third type results in a complex nonrepeating pattern with no apparent patterns to be seen: a *strange attractor.* Keep in mind that, though these strange system behaviors appear random, they are fully deterministic. The rules are clear and inviolable; there are no perturbations from outside the system or surprises introduced. The state of every cell can be predicted precisely from its neighborhood's state in the previous iteration, all the way back from the starting configuration.

Cellular automata of the fourth type are the interesting ones to us. Certain sets of rules result in patterns that run on and on, maybe forever, creating patterns that are characteristic and identifiable but that never repeat. These patterns *look* like something produced by nature, like the tangle of branches in a winter wood, the king's crowns of raindrops on a puddle, the wiggling ripples of wind blowing over a pond's surface, the fuzz on a thistle. Looking at the computer screen as the sequence of CA iterations unfolds, you feel a sense of recognizing nature at work.

John von Neumann (1951), who first proposed the idea of cellular automata a half-century ago, described an organism composed of cells, each one of which was a "black box," a unit with some kinds of unknown processes inside it. The term "cellular automata" was actually coined by Arthur Burks (1970). The state of each cell, for instance, a neuron, depends on the states of the cells it is connected to. The rules of a CA describe patterns of causality, such as exist among parts of many kinds of complex systems. The state of any element in the system—ecology, brain, climate, economy, galaxy—depends causally on the states of other elements, and not just on the states of individual elements, but on the *pattern* or combination of states. Life and mind are woven on just this kind of causal patterning, and it is a theme we will be dwelling on much in these pages.

Stephen Wolfram had shown that Type 4 cellular automata (the Game of Life is one) could function as universal computers. Langton demonstrated how CA gliders or moving particles can be used to perform computations, while stationary, period-two blinkers can store information. In one of the pioneering essays on artificial life, Christopher Langton (1991) argued that the unending complexity of the Type 4 CA, almost comprehensible but never predictable, may be directed and manipulated in such a way that it can perform any kind of computation imaginable. In his resonating phrase, this is "computation on the edge of chaos."

Why the "edge of chaos?" Langton compared "solid" and "fluid" states of matter, that is, static and dynamic molecular states, and suggested that these were two "fundamental universality classes" of dynamical behavior. A system that exhibits Type 1 or 2 behavior, that is, a static or periodic system, is predictable and orderly. Langton compared these kinds of systems to the behaviors of molecules in solids and noted that such simple and predictable systems are unable to either store or process information. Type 3 systems, similar to the gaseous states of matter, are essentially random. Whatever information they may contain is overpowered by the noise of the system. Langton argued that the interesting Type 4 dynamical systems, which are capable of both storing and processing information, resemble matter in between states, at the instant of a *phase transition.* For instance, molecules in frozen water are locked into their positions, unable to move, while molecules in a liquid move chaotically, unpredictably. Decreasing the temperature of a liquid can cause the molecules to change from the unpredictable to the predictable state, but right at the transition each molecule must make an either-or decision, to try to lock into place or to try to break free; at the same moment, that molecule's neighbors are making the same decision, and they may decide the same or different. The result is the formation of extended, complex Jack-Frost fingers or islands of molecules in the solid state, propagating through the liquid matrix, dissolving in some places even while new structures form in others. The behavior of the system at the edge of chaos is not predictable, and it's not random: it is *complex.*

Langton observed the appearance, at the edge of chaos, of simple linear patterns in the CA that persist over time, sometimes shifting position regularly so as to appear to move across the screen (see Figure 1.3); some of Conway's patterns are examples of these phenomena in two dimensions. Langton compared these to solitary waves and described them as "particle-like." It is as if a quality were transmitted across the field through changes at successive locations of a medium, though the emergent entity is composed only of changes of the states of those locations. These "particles" sometimes continue through time until they collide with other particles, with the result being dependent on the nature of the

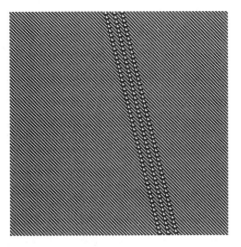

Figure 1.3 Particles in a cellular automaton. As the system steps through time, three patterns (in this case) are perpetuated.

particles, the angle of their impact, the state of the neighborhood in which they collided, and other factors.

Langton's early interest in cellular automata has evolved into his current involvement in the development of Swarm, a computer program that allows researchers to simulate the interactions of actors in a complex system. Langton and his colleagues at Santa Fe Institute and the Swarm Corporation use the word "swarm" to mean any loosely structured collection of agents that interact with one another. In common usage, the term *agent* refers to an object in a computer program that has an identity and performs some actions, usually with some presumption that the activities of the agent are somewhat autonomous or independent of the activities of other agents. An ant colony is a kind of swarm where the agents are ants, highway traffic can be conceptualized as a swarm whose agents are cars and drivers, and so on; all these kinds of swarms can be simulated using this CA-derived software.

Swarm researchers are especially interested in the interactions between individual- and group-level phenomena; for instance, a bird flock has properties over and above the properties of the birds themselves, though there is of course a direct relation between the two levels of phenomena. The Swarm simulation program allows a user to program hierarchical swarms; for instance, an economy might be made up of a swarm of agents who are people, while each person might be made up of a swarm that is their beliefs or even the neurons that provide the infrastructure for their beliefs—swarms within swarms. So far, uses of Swarm have been focused on providing simulations of higher-level phenomena that emerge from lower-level specification of rules, much as gliders and other CA particles emerge from the low-level rule sets specified in a cellular automaton.

One-dimensional cellular automata are the most commonly studied type. Where the two-dimensional Game of Life was pictured as a torus, the one-dimensional CA is simply a ring, with the first and last cells of a row joined. One row of binary cells iterates down the screen. In the Game of Life cells are defined on a plane rather than a line, and the state of each cell depends on the cells above and below as well as to the sides of it. That means that patterns run not only down the screen, but up it, and across it, and at weird unpredictable angles like a rat in a dairy.

Later we will be spending some time on the topic of dimensionality. Just as a teaser, think about what a three-dimensional Game of Life would look like. Gliders and other patterns can move not only in a rectangle on a plane, but above and below it; they can move up away from the surface of the environment as well as across it. Now the CA can be a bird flock, or a star cluster, or weather, extending in space.

It was easy to imagine a third dimension—now try a fourth. Mathematically there is no reason not to devise a rule set for a four-dimensional CA. What would it look like? Well, we can't say. Imagine a five-, six-, or seven-dimensional, try a one-hundred-dimensional vision of gliders, blinkers, patterns marching through the hyperspace (a *hyperspace* is a space of more than three dimensions). Perhaps we can eventually approach the emergent dynamics of the great stochastic systems. The cellular automaton, a simple machine brimming with interactions, gives us a way of thinking of randomness and stochastic behavior in the systems we will be discussing; everything follows the rules, but nothing is predictable.

Artificial Life in Computer Programs

We have mentioned that life may be a quality we attribute to some things, even more than it is a quality of the things themselves. This issue is perhaps most salient in the consideration of computer programs that emulate life processes. An artificial life program does not have to resemble any "real" living thing; it might be considered living by some set of criteria but seem, well, completely strange by earthly biostandards, by what we know from nature as it is on earth. Things might be able to learn, to reproduce, to evolve in an adaptive way, without resembling earthling life much at all. Conway's Game of Life opened the door for intense study of systems that exhibit lifelike behaviors.

And what *are* lifelike behaviors? It is sometimes said that a defining feature of life is its resistance to entropy. *Entropy* is disorder, which always increases in a system. Well-organized patterns fall into decay, disintegration, deterioration, as rust never sleeps. Yet life alone seems continually to renew itself. Wolfram's Type 4 CAs seem to have the ability to perpetually renew themselves, and it is for this reason that they have become the symbol, if not the seed, for artificial life research. They seem resistant to entropy.

It is possible to apply evolutionary algorithms to CA rules to produce behaviors of particular types, "breeding" interesting CAs. For instance, random mutations can be introduced in order to explore variations on CA rules by flipping a bit in the rule table occasionally—not too often, or you will destroy the behavior, but a low rate of mutation might allow the evolution of interesting dynamics. In this case we are considering the rule set as a kind of chromosome, a string of genes encoding the behavior of the cellular automata. This is a good analogy to the distinction between *genotype* and *phenotype* in nature: you can't see the genotype; it is an abstract coding scheme or program for the organism. What you see in nature is the phenotypic expression of the genotype. Just so in cellular automata, a researcher might encode the rule set in a string of ones and zeroes and mutate that string to evolve surprising behaviors in the visible cells.

One early program that could be called artificial life was produced as a sort of demonstration of the effects of genetic mutation on phenotypic forms. In *The Blind Watchmaker,* Richard Dawkins (1987) created graphical creatures he called *biomorphs* (see Figure 1.4), whose form was encoded in a chromosome of nine genes, each of which could take on a value from zero to nine. These genes encoded the rules for the development of the biomorph, for instance, the angle or length of a branch, and they could be "mutated" by one step. That is, a gene in state 5 can, through mutation, be changed to a 4 or a 6. The user could generate a population of biomorphs with random genes, then select one that looked interesting. Dawkins reported that, though he expected his organisms to look like trees, he discovered that insect forms were also possible. Thus, he might select a biomorph that looked particularly insect-like. Once selected, this biomorph reproduces to create a generation of children, each containing a random mutation that causes it to differ slightly from its parent. The user can then select an interesting-looking member of the offspring population to produce another generation of

Figure 1.4 Examples of biomorphs as described by Richard Dawkins.

mutations resembling *it,* and so forth. The result is incremental and un-predictable change in a desired direction: evolution.

In nature of course it is the environment that decides what individuals will produce the next generation. Individual organisms with some new genetic pattern, that is, a mutation, may or may not reproduce, depending on the effect of the mutation in combination with the rest of the genetic heritage in the environment. In Dawkins' program the researcher "plays God" with the population, deciding who lives and who dies, guiding evolution from step to step according to whatever whim or goal he or she has.

It is important to think about the size and structure of the *search space* for a population of biomorphs. With nine genes, each of which can exist in one of 10 states, we can see that there are 9^{10}, that is, 3,486,784,401, possible genetic combinations. These can be thought of as points in a nine-dimensional space. Though of course we can't visualize such a space, we can visualize objects with attributes selected from locations on each of those nine dimensions. If mutation consists of making a change on a single dimension, and if the changes are (as in Dawkins' program) simple moves to positions that are a single unit away on the dimension, then we can say that the child is *adjacent* to the parent. That means that conceptually, mathematically, they are near one another in the search space. This conceptualization turns out to be important when we want to search for a good point or region in the space, for instance, in searching for the solution to a problem. Whether the objects are continuous, binary, or—as in the current case—discrete, the concept of distance pertains in searching the space; in particular, the concept applies when we want to decide what size steps to take in our search. If we can't look at every point in a space, as in this case where there are billions of possibilities, then we must come up with a plan for moving around. This plan will usually have to specify how big steps should be in order to balance the need to *explore* or look at new regions, versus the need to *exploit* knowledge that has already been gained, by focusing on good areas. Dawkins' biomorphs search by taking small steps through the space of possible mutations, so that the creatures evolve gradually, changing one aspect at a time. In nature as well, adaptive changes introduced by mutation are generally small and infrequent.

Other artificial life researchers have taken a slightly less godlike approach to the evolution of computer-generated organisms. If the researcher provides information about the physics and other aspects of the environment, then life-forms can adapt to environmental demands and to one another. The artificial creatures developed by Karl Sims (e.g., 1994) are a beautiful example of this. Sims creates artificial three-dimensional worlds with physical properties such as gravity, friction, collision protection, and elasticity of objects so that they bounce realistically off one another upon colliding. "Seed" creatures are introduced into this world and allowed to evolve bodies and nervous systems. In order to guide the evolution, the creatures are given a task to measure their fitness; for instance, in one version the goal is to control a cube in the center of the world. The creatures are built out of rectangular forms whose size, shape, number, and arrangement is determined by evolution. Their "minds" are built out of artificial neurons composed of mathematical operations that are also selected by evolution.

The randomly initialized creatures in Sims' programs have been shown to evolve surprising strategies for controlling the cube, as well as innovative solutions to problems of locomotion on land and in water, with all forms of walking, hopping, crawling, rolling, and many kinds of fin waving, tail wagging, and body wriggling in a fluid environment. Some movies of these creatures are available on the web—they are well worth looking at.

It is possible then, in computer programs, to evolve lifelike beings with realistic solutions to the kinds of problems that real life-forms must solve. All that is needed is some way to define and measure fitness, some method for generating new variations, and some rules for capitalizing on adaptive improvements.

Christopher Langton (1988) has defined artificial life as "the study of man-made systems that exhibit behaviors characteristic of natural living systems." In the same paper, he states, "Life is a property of form, not matter . . ." If we accept Langton's premise, then we would have to admit that the similarity between an artificial-life program and life itself may be somewhat stronger than the usual kind of analogy. We will avoid declaring that computer programs live, but will note here and in other places throughout this book that it is unsettlingly difficult sometimes to draw the line between a phenomenon and a simulation of that phenomenon.

Intelligence: Good Minds in People and Machines

We have discussed the futility of arguing about whether an instance of something fits into one category or another. The next topic is perhaps the most famous example of this dilemma. *Intelligence* is a word usually used to describe the mental abilities of humans, though it can be applied to other organisms and even to inanimate things, especially computers and computer programs. There is very little agreement among psychologists and little agreement among computer scientists about what this word means—and almost no agreement between the two groups. Because the Holy Grail of computing is the elicitation of intelligence from electronic machines, we should spend some time considering the meaning of the word and the history of the concept in modern times. It isn't always such a warm and fuzzy story. Again we find that the Great Stochastic Systems, evolution and mind, are tightly linked.

Intelligence in People: The Boring Criterion

In this *Bell Curve* world (Herrnstein and Murray, 1994) there is a great amount of heated controversy and disagreement about the relationship between intelligence and heredity. Much of the discussion focuses on the issues of statistical differences among various populations on IQ test performance. It is interesting in this light to note that the modern concept of intelligence arose in association with heredity, and that attempts to understand intelligence and heredity have historically been intertwined. Following the world-shaking evolutionary pronouncements of Darwin, the *eugenicists* of the 19th and the first half of the 20th centuries had the idea that humans should practice artificial selection. According to Darwinist theory, natural selection is the process by which species are changed; adaptation is propelled by selection pressures applied by the environment. *Artificial selection* then would be a process whereby human authorities would decide which of their peers would be allowed to propagate and which would not. As Victoria Woodhull, who in 1872 was the first woman ever nominated for president of the United States, said, "If superior people are to be desired, they must be bred; and if imbeciles, criminals, paupers, and [the] otherwise unfit are undesirable citizens, they must not be bred." Thus was born the modern concept of intelligence: a criterion for selecting who should be allowed to reproduce.

The concept of intelligence has always been intended to distinguish "better" people from "worse." Intelligence can be simply defined by a set of measures that support the experts' opinion of what comprises a "good" mind. In 20th-century American and European society, the definition has included some widely agreed-upon qualities such as memory, problem-solving ability, and verbal and mathematical abilities. The fact is, though, that intelligent people, including intelligence experts, have different abilities from one another; as far as they attempt to define intelligence they end up describing their own favorite qualities—thus inevitable disagreement.

We noted above that there is very little consensus on a definition of intelligence. In psychology, the most quoted (off the record) definition of intelligence is that given in 1923 by E. G. Boring: intelligence is whatever it is that an intelligence test measures. The study of intelligence in psychology has been dominated by a focus on testing, ways to measure the trait, and has suffered from a lack of success at defining what it is.

Intelligence has always been considered as a trait of the individual; for instance, an underlying assumption is that measurements of a person's intelligence will produce approximately the same results at different times (an aspect of measurement known as "reliability"). This book is about collective adaptation, and we will tend to view intelligence from the viewpoint of the population. As noted in the first paragraphs of the Preface, humans have proven themselves as a species. While differences between individuals exist, the achievements of the outstanding individuals are absorbed into their societies, to the benefit of all members. We can't all be Sir Isaac Newton, but every high school offers courses in physics and calculus, and all of us benefit from Newton's discoveries. The achievements of outstanding individuals make us all more intelligent— we may not all rise to the standard of an Einstein, but by absorbing their ideas we raise our level of functioning considerably. Michael Tomasello (1999) calls this the *ratchet effect:* cumulative cultural evolution resulting from innovation and imitation produces an accumulation of problem solutions building on one another. Like a ratchet that prevents slippage backwards, cultures maintain standards and strive to meet or exceed them.

Intelligence in Machines: The Turing Criterion

So far we have been talking about intelligence as a psychological phenomenon, as it appears in humans. Yet for the past half-century or more there has been a parallel discussion in the computer-science community. It became immediately apparent that electronic computers could duplicate many of the processes associated with minds: they could process symbols and statistical associations, could reason and remember and react to stimuli, all things that minds do. If they could do things that our biological brains do, then perhaps we could eventually build computers that would be more powerful than brains and program them to solve problems that brains can't solve. There was even the suggestion that computers could become more intelligent than people.

In a nutshell: that hasn't happened. Modern computers have not turned out to be very good at thinking, at solving real problems for us. Perhaps part of the reason may be found in the way that computer scientists have defined intelligence, which is very different from how psychologists have defined it, and how they have implemented their beliefs about intelligence.

It is fair to say that any discussion of computer intelligence eventually comes around to the *Turing test* (Turing, 1950). The Turing test sounds simple enough. A subject is placed in a room with a computer keyboard and monitor, while in another room there is a computer and a person. The subject types questions into the keyboard and receives a reply from the other side. A simplistic summary of the test is this: if the subject can't tell if the computer's responses were generated by the human or the machine, then the computer is considered intelligent.

At first this sounds like a strange test of intelligence. It certainly is different from the IQ tests we give ourselves! The funny thing is, the puniest computer could do perfectly well on a human intelligence test. A computer has perfect recall, and a simple brute-force algorithm, for instance, one that tests every possible combination, can find the solution to any pattern-matching or arranging problem such as are found in IQ tests, if it's given enough time. A computer can be programmed with a huge vocabulary and some rules of grammar for the verbal scales, and of course the math test would be a snap. By the standard that we set for ourselves, computers are already very intelligent. So what we look for in computer intelligence (e.g., the Turing test) is something else entirely. We can only really be fooled into thinking a computer is a person if it makes mistakes, or takes a long time to do a math problem, or claims to have gaps in its knowledge. This is a confusing situation indeed.

For our discussion, the important thing is where the Turing test looks for intelligence. Turing did not suggest looking "inside the head" of the computer for a sign of intelligence; he suggested looking directly at the social interface between the judge and the subject. A program that was capable of forming a convincing bond was intelligent. It was not that the interaction was a "sign of" intelligence—it was in fact intelligence. Today's cognitivistic orthodoxy looks for intelligence inside the person, or in the information processing of the machine. Turing knew to look at the social nexus directly.

Evolutionary computation researcher David Fogel (1995) says that a good definition of intelligence should apply to humans and machines equally well, and believes that the concept should apply to evolution as well as to behaviors perceptible on the human time scale. According to him, intelligence is epitomized by the scientific method, in which predictions about the world are tested by comparison to measurements of real-world phenomena and a generalized model is proposed, refined, and tested again. Successful models might be adjusted, extended, or combined with other models. The result is a new generation of ideas that carry what David's father Larry Fogel has called "a heredity of reasonableness." The process is iterated, with the model successively approximating the environment.

The Fogels conclude that intelligence, whether in animate or inanimate contexts, can be defined as the "ability of a system to adapt its behavior to meet its goals in a range of environments" (Fogel, 1995, p. 24). Their research enterprise focuses on eliciting intelligence from computing machines by programming processes that are analogous to evolution, learning incrementally by trial and error, adapting their own mutability in reaction to aspects of the environment as it is encountered.

David Fogel (1995) theorizes, following Wirt Atmar (1976), that *sociogenetic learning* is an intelligent process in which the basic unit of mutability is the idea, with culture being the reservoir of learned behaviors and beliefs. "Good" adaptive ideas are maintained by the society, much as good genes increase in a population, while poor ideas are forgotten. In insect societies (as we will see) this only requires the evaporation of pheromone trails; in humans it requires time for actual forgetting.

For some researchers, a hallmark of the scientific method and intelligence in general is its reliance on *inductive* logical methods. Whereas *deductive* reasoning is concerned with beliefs that follow from one another by necessity, induction is concerned with beliefs that are supported by the accumulation of evidence. It comprises the class of reasoning techniques that generalize from known facts to unknown ones, for instance, through use of analogy or probabilistic estimates of future events based on past ones. Interestingly, inductive reasoning, which people use every day, is not logically valid. We apply our prior knowledge of the subject at hand, assumptions about the uncertainty of the facts that have been discovered, observations of systematic relationships among events, theoretical assumptions, and so on, to understand the world—but this is not valid logic, by a long shot.

The popular view of scientists as coldly rational, clear-thinking logicians is challenged by the observation that creativity, imagination, serendipity, and induction are crucial aspects of the scientific method. Valid experimental methods are practiced, as far as possible, to test hypotheses—but where do those hypotheses come from? If they followed necessarily from previous findings, then experimentation would not be necessary, but it is. A scientist resembles an artist more than a calculator.

chapter
two

Symbols, Connections, and Optimization by Trial and Error

This chapter introduces some of the technical concepts that will provide the foundation for our synthesis of social and computer scientific theorizing. Thinking people and evolving species solve problems by trying solutions, making adjustments in response to feedback, and trying again. While we may consider the application of logical rules to be a "higher" or more sophisticated kind of problem-solving approach, in fact the "lower" methods work extremely well. In this chapter we develop a vocabulary and set of concepts for discussing cognitive processes and the trial-and-error optimization of hard problems. What are some approaches that can be taken? What have computer scientists implemented in artificially intelligent programs, and how did it work? Human language—now *there's* a nice, big problem! How in the world can people (even people who aren't outstandingly intelligent in other ways) possibly navigate through a semantic jungle of thousands of words and constantly innovative grammatical structures? We discuss some of the major types of problems, and major types of solutions, at a level that should not be intimidating to the reader who is encountering these concepts for the first time. ■

Symbols in Trees and Networks

Though we are asserting that the source of human intelligence is to be found in the connections between individuals, we would not argue that no processing goes on within them. Clearly there is some ordering of information, some arranging of facts and opinions in such a way as to produce a feeling of certainty or knowing. We could argue (and later will develop the argument) that the feeling of certainty itself is a social artifact, based on the sense that the individual could, if asked, explain or act on some facts in a socially acceptable—that is, consistent—way. But first we will consider some of the ways that theorists have attempted to model the processes that comprise human thinking and have attempted to simulate, and even to replicate, human thinking. As we develop the swarm model of intelligence, we will need to represent cognitive processes in some way. Where representations are complex—and how could they not be?—it will be necessary to find optimal states of mind. We will find these states through social processes.

The earliest approaches to *artificial intelligence* (AI) assumed that human intelligence is a matter of processing symbols. Symbol processing means that a problem is embedded in a universe of symbols, which are like algebraic variables; that is, a symbol is a discrete unit of knowledge (a woefully underdefined term!) that can be manipulated according to some rules of logic. Further, the same rules of logic were expected to apply to all subject-matter domains. For instance, we could set up a syllogistic algorithm:

All A are C
If B is A
Then B is C

and plug words in where the letters go. The program then would be given the definitions of some symbols (say, A = "man," B = "Socrates," and C = "mortal") as facts in its "knowledge base"; then if the first two statements were sent to the "inference engine," it could create the new fact, "Socrates is mortal." Of course it would only know that Socrates was a B if we told it so, unless it figured it out from another thread of logic—and the same would be true for that reasoning; that is, either a human gave the program the facts or it figured them out from a previous chain of logically linked symbols. This can only go back to the start of the program, and it necessarily eventually depends on some facts that have been given it. This is an aspect of a question that has been called the *grounding prob-lem.* There is a problem indeed if facts depend on other facts that are in-evitably defined themselves by the facts that depend on them. There is necessarily an innate assumption, in the symbol-processing paradigm, that all the relevant facts are known and that their interrelationships are known.

There is another assumption, too, and one that was not recognized for a long time but that meant that the crisp symbol-processing para-digm could never really succeed. The assumption was revealed by Lotfi Zadeh (see Yager et al., 1987), a Berkeley computer-science professor, who forged a philosophy that revolutionized the way many computer scientists think about thinking. Zadeh pointed out that it is usually the case that "Some *A* are sort of *C.*" Truth, that is, is *fuzzy*—it has degree. Fuzzy logic is built on the presupposition that the truth of propositions can be rated on a scale from zero to one, so that a statement that is en-tirely false has a truth value of 0.0, a statement that is perfectly true has a value of 1.0, and most statements go somewhere in the middle. Zadeh methodically worked out the rules for reasoning with approximate facts. Fuzzy logic operates in parallel, unlike the earlier symbol-processing models, which evaluated facts in a linked series, one after the other; fuzzy facts are evaluated all at once. Some people think that "fuzzy logic" means something like fuzzy thinking, a kind of sloppy way of vaguely guessing at things instead of really figuring them out, but in fact the re-sults of fuzzy operations are precise. It is also sometimes alleged that fuzziness is another word for probability, but it is an entirely different form of uncertainty. For instance, take a second to draw a circle on a sheet of paper. Now look at it. Is it really a circle? If you measured, would you find that every point on the edge was exactly the same distance from the center? No—even if you used a compass you would find that varia-tions in the thickness of the line result in variations in the radius. So your circle is not really a circle, it is *sort of* a circle (see Figure 2.1). If you did re-ally well, it may be 99 percent a circle, and if you drew a square or a figure eight or something uncirclelike, it might be closer to 1 percent a circle—maybe it has roundish sides or encloses an area. It is not meaningful to say that it is "probably" a circle (unless you mean that you probably *meant to* draw a circle, which is an entirely different question). Fuzzy logic has turned out to be a very powerful way to control complicated processes in machines and to solve hard problems of many kinds.

Traditional (symbol-processing) AI commonly made inferences by ar-ranging symbols in the form of a *tree.* A question is considered, and de-pending on the answer to it another question is asked. For instance, you

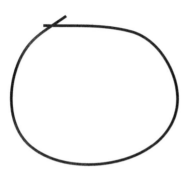

Figure 2.1 Is this a circle? Yes and no. In some ways it is a circle and in other ways it is not. It is a fuzzy circle.

might start by asking, "Is the door locked?" If the answer is yes, then the next thing is to find the key; if the answer is no, then the next step is to turn the doorknob and push the door open. Each step leads to another decision, and the answer to that leads to yet another, and so on. A graph of the domain of such a decision-making process looks like a tree, with ever more numerous branches as you progress through the logic. Fuzzy logic, on the other hand, assumes that, rather than making crisp decisions at each point, a reasonable person might "sort of" choose one alternative over another, but without absolutely rejecting another one, or a statement may be partly true and partly false at the same time. Instead of a tree, a fuzzy logic diagram might look more like ripples flowing over a surface or a storm blowing through the countryside, stronger in some areas and weaker in others. Rules are implemented in parallel, all at once, and the conclusion is obtained by combining the outputs of all the if-then rules and defuzzifying the combination to produce a crisp output.

One limitation of tree representations is the requirement that there be no feedback. *A might imply or cause B, but in a tree we must be sure that B does not imply or cause A in return, either directly or by acting through intermediate links. In reality though premises and conclusions imply one another all the time, and causes affect other causes that affect them back. We saw that Bateson considered this kind of reciprocal causation as a defining or necessary quality of mind. Not only are feedback loops common in nature, but some theorists think that feedback is really the basis for many natural processes.*

We should point out here, because we will be dealing with them extensively, the interchangeability of matrices and graphs. Suppose we have a table of numbers in rows and columns. We can say that the numbers indicate the strength of the effect of the row variable upon the column variable. Variables—which we might think of as symbols, or things, or events, or ideas, or people, for that matter—can be assumed to have positive or negative effects on one another.

When the system is considered as a tree, it really only makes sense to allow positive connections; zero and negative connection strengths would be logically indiscriminable. That is because each node represents a discrete decision to follow one route or another. An option unchosen is in the same state whether there is a vacuous link or a negative one.

In contrast to a tree, in a network matrix it is not hard to depict things in a feedback relationship, having effects upon one another. Further, effects in a network can be positive or negative; one thing can prevent another as easily as it can cause it. When two things have a positive causal effect on one another, we would say they have a positive feedback or autocatalytic effect. If they decrease one another, their relationship is one of negative feedback. In many of the examples throughout this book, autocatalysis is the basis for the dynamics that make a system interesting.

Interrelated variables in a matrix can be graphed as well (see Figure 2.2). Each variable is a node in the graph, and we can draw arrows between pairs of them where a table has nonzero entries, indicating causal (one causes the other) or implicative (one implies the other) effects between them. Now it should be clear that a graph and a table can contain exactly the same information. The connections among the nodes are sometimes called *constraints,* where the state of the node (for instance, off or on) is intended to satisfy the constraints upon it. Of course there are other terminologies, but for this discussion we will call them nodes and constraints, or sometimes just "connections."

Tree representations use only half of the matrix, either the half above or below the diagonal, depending on how you decide to implement it. In a tree structure, no node later in the chain of logic can affect any node earlier, so you don't need both halves of the matrix. A graph or matrix where nodes *can* affect one another would be called a *recurrent network.* Trees can be graphed as networks, but networks that contain feedback cannot be diagrammed as trees.

One common kind of network is depicted with symmetrical connections between nodes. Instead of arrows between pairs of nodes pointing from one to the other, the graph needs only lines; the implication is that things with positive connections "go together," and things with negative links do not. Figure 2.3 shows an example of such a network. In a matrix, the upper and lower diagonal halves of this kind of network are mirror

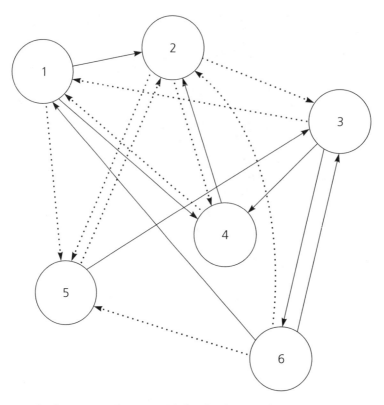

Figure 2.2 Graph of a pattern of causes with feedback. Dotted lines indicate negative causal force: the sending node prevents or inhibits the receiving node.

images of one another. This kind of graph is very common as a cognitive model in contemporary psychological theorizing, and it can be useful for problem solving in computer programs, too.

We will make the simplifying assumption here that nodes are binary, they are either on or off, yes or no, one or zero. In a constraint-satisfaction paradigm, the goal is to find a pattern of node states that best satisfies the constraints. Each node is connected to some other nodes by positive or negative links. The effect of an active node through a positive link is positive and through a negative link is negative, and the effect of an inactive node through a positive link is zero or negative, depending on how it is specified, and through a negative link is zero or positive. In other words, the effect is the product of the node state times the constraint value. A node with mostly positive inputs fits its constraints best when it is in the active state, and one with zero or negative inputs should be in the inactive state.

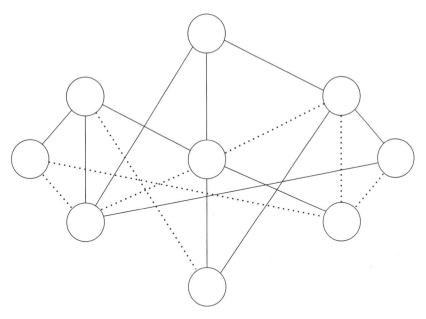

Figure 2.3 An example of a parallel constraint satisfaction network. Solid lines represent positive or excitatory connections, while dotted lines represent negative or inhibitory connections between pairs of nodes.

Paul Smolensky (1986) has noted that the optimal state of such a network, that is, the state where the most constraints are satisfied, can be found by maximizing the sum of the products of pairs of node states (0 and 1, or −1 and +1) times the value of the strength of the connection between them:

$$H = \sum_i \sum_j a_i w_{ij} a_j$$

where a_i is the activation of node i, w_{ij} is the strength of the connection between nodes i and j, and a_j is the activation of node j. The products of pairs of nodes times the weights of the connections between them should be as big as possible. The "harmony function" (Smolensky's cognitive theory was called "harmony theory") can be calculated over the entire network, or the optimality of each node can be calculated separately and the sum taken.

It has long been noted that people strive to maintain *cognitive consistency.* We want our thoughts to fit together with one another as well as with facts in and about the world. Leon Festinger (1957) theorized that people have a drive to reduce *cognitive dissonance,* defined as the clash among cognitive elements, whether they are beliefs, behaviors, or attitudes. Later we will stress that there is an important interpersonal component to this drive, but for the moment it is only necessary to point out that—for whatever reasons—people do strive to maintain consistency. Harmony maximization is a good way to represent this principle in a computer program, where inconsistencies between elements are depicted as negative connections.

We know that we want to maximize the harmony function over the network—but how do you do that? Any time we change the state of a node, we are about as likely to make it inconsistent with some other nodes as we are to improve its fit with some; if the network is not trivial, there are probably a number of conflicting constraints in it. John Hopfield had shown, in 1982, a method that usually arrived at the best state (a network's state is the pattern of its node activations). Hopfield actually had two methods, one for binary nodes and another for networks whose nodes can take on continuous values; here we will stay with the binary type (Hopfield, 1982, 1984). Hopfield's algorithm starts by randomly initializing nodes as zeroes or ones. Then a node is selected at random, and its inputs are summed; that is, the states of nodes connected to the selected one are multiplied by the strength of their connections, which can be positive or negative. If the sum of the inputs is greater than a threshold, then the node is assigned the active state; for instance, it is given a state value of one. If the sum is below the threshold, then the node is set to zero (or minus one), the inactive state. Then another node is selected at random and the process is repeated until all nodes are in the state they should be in.

Most parallel constraint satisfaction models use some version of Hopfield optimization, but we will be arguing that it is not a good model of the way people reduce conflicts among cognitive elements. The fact is, people talk with one another about their thoughts, and they learn from one another how to arrange their beliefs, attitudes, and behaviors to maximize their consistency, to reduce cognitive dissonance. We don't think that people organize their thoughts by considering them one at a time in the isolated privacy of their own skulls, and we don't think this is the best way for computers to do it, either.

The models we have discussed, whether crisp symbol processing, fuzzy logic, or constraint-satisfaction networks, all assume that somebody has given them knowledge. Given a set of connections—directed or symmetrical—between nodes in a graph, we have been trying to find node patterns that best fit. But where do the connections come from? Finding the pattern of connections among elements is called *learning*.

When two things change together, for instance, when they are both loud or both intense or both present whenever the other is present, or when they are only loud or intense or present when the other thing is in the opposite state, then we say the two things are *correlated*. Things may be positively correlated, like heat and sunshine or money and power, or pairs of things can be negatively correlated, for instance, wealth and starvation or aspirin and headaches. Notice that positive and negative correlations are the same, they are still correlations; you can always define a thing negatively—we could measure "poverty" instead of "wealth"—to change the sign of the correlation. Correlation is a form of statistical association between two variables that gives some evidence for predicting one from the other. Learning can be described in terms of finding correlations among things. The strengths of the connections in the graphs we have been looking at are a kind of correlation coefficient, showing whether the two nodes change together, how strong the relationship is, and whether the changes are positive or negative.

Every college student learns—or *should* learn—that "correlation does not imply causation." The fact that two things rise and fall together does not mean that one causes the other. A correlation matrix, just like the symmetrical table in a parallel constraint satisfaction network, is made up of mirror-image halves, as A's correlation with B is equal to B's correlation with A. In the network, a symmetrical connection weight between binary nodes indicates the likelihood of one being active when the other is.

The meaning of a connection weight is a little more complicated, or should we say interesting, when the connections are asymmetrical. In this case the weight indicates the likelihood or strength of activation of the node at the head of the arrow as a function of the state of the node at the arrow's origin. So if an arrow points from A to B, with a positive connection between them, that is saying that A's activation makes it more likely that B will be active or increases B's activation level: A causes B.

Let's consider an example that is simple but should make the situation clearer. We will have some binary causes, called $x1$, $x2$, and $x3$, and a binary effect, y (see Figure 2.4). In this simple causal network, we will say that y is in the zero state when the sum of its inputs is less than or equal to 0.5 and is in the active state when the sum is above that. We see that the connection from $x1$ to y has a weight of 1.0; thus, if $x1$ is active, then the input to y is $1 \times 1.0 = 1.0$, and y will be activated. The connection from $x2$ is only 0.5, though, so when it is active it sends an input of $1 \times 0.5 = 0.5$ to y—not enough to activate it, all by itself. The same holds for $x3$: it is not enough by itself to cause y, that is, to put y into its active state.

Note, though, that if $x2$ and $x3$ are both on, their summed effect is $(1 \times 0.5) + (1 \times 0.5) = 1.0$, and y will be turned on. Now the system of weighted connections, it is seen, can numerically emulate the rules of logic. The network shown in the top of Figure 2.4 can be stated in Boolean terms: IF ($x1$) OR ($x2$ AND $x3$) THEN y. Not only are a matrix and a graph interchangeable, but logical propositions are interchangeable with these as well.

More complex logical interactions might require the addition of *hidden nodes* to the graph. For instance, if we had more inputs, and the rule was IF ($x1$ AND $x2$) OR ($x3$ AND $x4$) THEN y, then we might need to link the first pair to a hidden node that sums its inputs, do the same for the second two, and create a logical OR relation between the hidden nodes and y, where each hidden node would have a 1.0 connection (or any strength above the threshold) to the output node y. To implement a NOT statement, we could use negative weights. All combinations are possible through judicious use of connection weights and hidden layers of nodes. A network where the information passes through from one or more inputs to one or more outputs is called a *feedforward network*.

In the previous example we used a crisp threshold for deciding if the output node y should be on or off: this a less-than-satisfying situation. If *some* of the causes of an effect, or premises of a conclusion, are present, then that should count for something, even if the activation does not reach the threshold. Say the inputs to y had added up to exactly 0.5, where we needed a "greater than 0.5" to flip its state. In reality things do not usually wait until the exactly perfect conditions are met and then go into a discretely opposite state. More commonly, the effect just becomes more uncertain up to a point near the decision boundary, then more certain again as the causes sum more clearly past the cutoff. If y were the kind of thing that was really binary, that is, it exists either in one state or the other and not in between, then we might interpret the sum of the inputs to indicate the *probability* of it being true or false. The more common instance, though, is where y is not binary but can exist to some degree, in which case the inputs to y indicate "how true" y is. If I wanted to know if "Joe is a big guy," the inputs to y might be Joe's height and perhaps Joe's weight or girth. If Joe is six feet tall and weighs, say, 200 pounds, then he might be a big guy, but if he is six foot, six inches tall and weighs 300 pounds, he is even more of a big guy. So the truth of the statement has degree, and we are back to fuzzy logic. In fact, many

IF (*x1*) OR (*x2* AND *x3*) then *y*

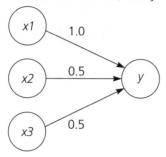

IF (*x1* AND *x2*) OR (*x3* AND *x4*) then *y*

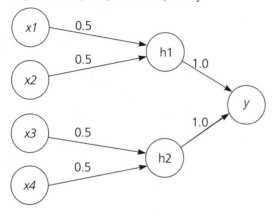

IF (*x1* OR *x2*) AND NOT (*x3* AND *x4*) then *y*

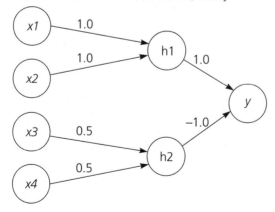

Figure 2.4 Logical propositions can be graphed as feedforward networks using binary nodes.

networks—and almost all feedforward networks—do have continuous-valued nodes, and the states of those nodes, usually in the range between zero and one, can be interpreted to indicate either probability or fuzzy truth values.

We should also mention that there are formulas for squashing the inputs into the tolerable range, usually zero to one, to make this business more manageable. The most common formula is referred to as the *sigmoid* function, since its output is S-shaped. Its equation is

$$s(x) = \frac{1}{1 + \exp(-x)}$$

and its graph is seen in Figure 2.5. When $x = 0$, $s(x) = 0.5$. Negative inputs result in a value that approaches zero as the x becomes more negative, and positive values approach a limit of one. The sigmoid function is very useful and will come up again later in this book.

Networks like the ones we have been describing are often called *neural networks* because they roughly resemble the way cells in the brain are hooked up, and because they are often able to categorize things and make decisions something like a brain does.

We still have not answered the question of learning: finding values for the connection weights that makes for the accurate estimation of outputs from inputs. This is another *optimization* problem. Optimization means that we are looking for input values that minimize or maximize the result of some function—we'll be going into the gory details of optimization later in this chapter. Normally in training a feedforward network we want to minimize *error* at the output; that is, we want the sums of products to make the best estimate of what y should be. Again, there is a traditional way to find optimal patterns of weights, and there is our way.

The traditional way is to use some version of gradient descent, a kind of method for adjusting numbers in a direction that seems to result in improvement. *Backpropagation of error* is the most standard way to adjust neural net weights. "Backprop," as it is called, finds optimal combinations of weights by adjusting its search according to a momentum term, so that shorter steps are taken as the best values are approached.

Neural networks were introduced to the psychological community by the PDP (Parallel Distributed Processing) Group headquartered at the University of California at San Diego, to suggest a theory of cognition that is called *connectionism* (Rumelhart, McClelland, and the PDP Group, 1986; McClelland, Rumelhart, and the PDP Group, 1986). Connectionist

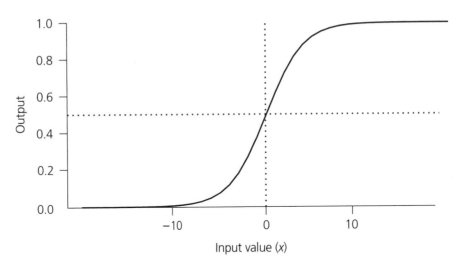

Figure 2.5 The sigmoid function takes numeric inputs and squashes them into the range between zero and one.

theory differs from the symbol-processing view in that it assumes that cognitive elements have a "microstructure": they are composed of patterns distributed over a network of connection weights. Both gradient descent and Hopfield optimization are part of the connectionist perspective. Connectionist approaches, though, just like their traditional AI predecessors, tend to miss something extremely crucial about human cognition: we don't do it alone. Learning often, if not usually, occurs in the form of adaptation to information received from other people, including both information about facts and how to make sense of them. Optimization of node activation patterns as well as patterns of connection weights can be an extremely complicated process, with lots of conflicting constraints, redundancy, nonlinearity, and noise in the data; that is, things may not always be what they appear to be. Some optimization methods are better able than others to deal with such messy problems; we assert that social forms of optimization are at least as powerful and are more realistic for cognitive simulations than the inside-the-head methods that have been proposed.

   A good thing about network models of cognition is that the kinds of judgments they make are similar to those that humans make. For instance, symbol-processing methods require *linear separability* of elements for categorization. This mouthful means that if you graphed the attributes of some category members on a chart, you should be able to draw a straight line or surface that separates two categories of things from one another. A simple rule or combination of rules should distinguish them. In fact, though, humans do not categorize by linear separation; neither do neural networks. Categories are not sets. Another advantage is that, since networks work by a kind of statistical analysis of data, they can be more robust in the face of noise and measurement error. They don't assume that symbols or the rules that connect them are hard-and-fast discrete entities, with information flowing through one logical branch or another.

Neural networks and the closely related fuzzy logic models have been called "universal function approximators" because they can reproduce the outputs of any arbitrarily complex mathematical function. Their power as psychological models has grown as well because of their ability to simulate human processes such as perception, categorization, learning, memory, and attention, including the errors that humans make.

These errors seem to be at the heart of the Turing test of machine intelligence. Imagine that a judge in a Turing test situation asked the subject on the other side of the wall to answer a mathematical question, say, what is 43,657 times 87,698? If the perfectly correct answer came back after several milliseconds, there would be no question about whether a human or mechanical mind had answered. Even a highly trained mathematician or a savant would have to take a few seconds at least to answer such a question. The "most human" response would be an estimate, and it would take some time. People make errors in crisp calculations, but at the same time exceed any known computational models in "simple" tasks such as recognizing faces.

So far we have considered a few issues regarding the adaptiveness of living things, and in particular adaptation as it applies in the mental world. By perusing a few kinds of structures that have been used to simulate human thinking processes, we have inched our way closer to the goal of writing computer programs that embed such structures in social contexts. As we said, these models will tend to be complicated—in fact their complexity is part of what makes them work—and it will be difficult to find the right combinations of weights and activations. The next sections examine some methods for finding solutions to complex problems.

Problem Solving and Optimization

Both evolution and mind are concerned with a certain kind of business, and that is the business of finding patterns that satisfy a complex set of constraints. In one case the pattern is a set of features that best fit a dynamic ecological niche, and in the other case it is a pattern of beliefs, attitudes, and behaviors that minimize conflict with personal, social, and physical constraints. Some propensities to acquire particular mental qualities seem to be inherited, though the topic of individual differences in the various propensities (traits such as intelligence, introversion, creativity, etc.) is still the subject of lively controversy. While a tendency to learn, or to learn in a particular way, may be inherited, learning can only take place within a lifetime; there is no genetic transfer of learned knowledge from one generation to another. As we have seen, the two great stochastic systems have different ways of going about their business, though both rely heavily, if not exclusively, on some version of *trial and error*. In this section we discuss some aspects of problem solving where the goal is to fit a number of constraints as well as possible.

It may seem that we use the word "problem" in an unusual way. Indeed, the word has specialized meanings to specific groups who use it. We use it to describe situations—not necessarily mathematical ones (they might be psychological, ecological, economical, or states of any kind of phenomena)—where some facts exist and there is a need to find other facts that are consistent with them. A changing environment might be a problem to an evolving species. An ethical dilemma is a problem to a moral person. Making money is a problem for most people. An equation with some unknowns presents a problem to a mathematician. In some cases, the facts that are sought might be higher-level facts about how to arrange or connect the facts we already have.

Optimization is another term that has different connotations to different people. Generally this term refers to a process of adjusting a system to get the best possible outcome. Sometimes a "good" outcome is good enough, and the search for the very best outcome is hopeless or unnecessary. As we have defined a problem as a situation where some facts are sought, we will define optimization as a process of searching for the missing facts by adjusting the system. In the swarm intelligence view of the present volume, we argue that social interactions among individuals enable the optimization of complex patterns of attitudes, behaviors, and cognitions.

A Super-Simple Optimization Problem

A problem has some characteristics that allow the goodness of a solution to be estimated. This measurement often starts contrarily with an estimate of the error of a solution. For example, a simple arithmetic problem might have some unknowns in it, as $4 + x = 10$, where the problem is to find a value (a pattern consisting of only one element in this trivial example) for x that results in the best performance or fitness. The error of a proposed solution in this case is the difference between the actual and desired results; where we want $4 + x$ to equal 10, it might be that we take a guess at what x should be and find that $4 + x$ actually equals 20, for instance, if we tried $x = 16$. So we could say that the error, when $x = 16$, is 10.

Error is a factor that decreases when goodness increases. An optimization problem may be given in terms of minimizing error or maximizing goodness, with the same result. Sometimes, though, it is preferable to speak in terms of the goodness or fitness of a problem solution—note the tie-in to evolution. Fitness is the measure of the goodness of a genetic or phenotypic pattern. Converting error to goodness is not always straightforward; there are a couple of standard tricks, but neither is universally ideal. One estimate of goodness is the reciprocal of error (e.g., $1/e$). This particular measure of goodness approaches infinity as error approaches zero and is undefined if the denominator equals zero, but that's not necessarily a problem, as we know that if the denominator is zero we have actually solved the problem. Besides, in floating-point precision we will probably never get so close to an absolutely perfect answer that the computer will think it's a zero and crash. Another obvious way to measure the goodness of a potential solution is to take the negative of the error (multiply e by -1); then the highest values are the best.

We can use the super-simple arithmetic problem given above to demonstrate some basic trial-and-error optimization concepts. Our general approach is to try some solutions, that is, values for x, and choose the one that fits best. As we try answers, we will see that the search process itself provides us with some clues about what to try next. First, if $4 + x$ does not equal 10, then can we look at how far it is from 10 and use that knowledge—the error—to guide us in looking for a better number. If the error is very big, then maybe we should take a big jump to try the next potential solution; if the error is very small, then we are probably close to an answer and should take little steps.

There is another kind of useful information provided by the search process as well. If we tried a number selected at random, say, 20, for x, we would see that the answer was not very good (error is $(|10 - 24|) = 14$). If we tried another number, we could find out if performance improved or got worse. Trying 12, we see that $4 + 12$ is still wrong, but the result is nearer to 10 than $4 + 20$ was, with error $= 6$. If we went past $x = 6$, say, we tried $x = 1$, we would discover that, though error has improved to 5, the sign of the difference has flipped, and we have to change direction and go up again. Thus various kinds of facts are available to help solve a problem by trial and error: the goodness or error of a potential solution gives us a clue about how far we might be from the *optimum* (which may be a *minimum* or a *maximum*). Comparing the fitness of two or more points and looking at the sign of their difference gives us *gradient* information about the likely direction toward an optimum so we can improve our guess about which way to go. A gradient is a kind of multidimensional slope. A computer program that can detect a gradient might be able to move in the direction that leads toward the peak. This knowledge can be helpful *if* the gradient indicates the slope of a peak that is good enough; sometimes, though, it is only the slope of a low-lying hill. Thus an algorithm that relies only on gradient information can get stuck on a mediocre solution.

Three Spaces of Optimization

Optimization can be thought to occur in three interrelated number spaces. The *parameter space* contains the legal values of all the elements—called parameters—that can be entered into the function to be tested. In the exceedingly simple arithmetic problem above, x is the only parameter; thus the parameter space is one-dimensional and can be represented as a number line extending from negative to positive infinity: the legal values of x. Most interesting optimization problems have higher parameter dimensionality, and the challenge might be to juggle the values of a lot of numbers. Sometimes there are infeasible regions in the parameter space, patterns of input values that are paradoxical, inconsistent, or meaningless.

A function is a set of operations on the parameters, and the *function space* contains the results of those operations. The usual one-dimensional function space is a special case, as multidimensional outputs can be considered, for instance, in cases of multiobjective optimization; this might be thought of as the evaluation of a number of functions at once, for example, if you were to rate a new car simultaneously in terms of its price, its appearance, its power, and its safety. Each of these measures is the result of the combination of some parameters; for instance, the assessment of appearance might combine color, aerodynamic styling, the amount of chrome, and so on.

The *fitness space* is one-dimensional; it contains the degrees of success with which patterns of parameters optimize the values in the function space, measured as goodness or error. To continue the analogy, the fitness is the value that determines whether you will decide to buy the car. Having estimated price, appearance, power, and so on, you will need to combine these various functions into one decision-supporting quantity; if the quantity is big, you are more likely to buy the car. Each point in the parameter space maps to a point in the function space, which in turn maps to a point in the fitness space. In many cases it is possible to map directly from the parameter space to the fitness space, that is, to directly calculate the degree of fitness associated with each pattern of parameters. When the goal is to maximize a single function result, the fitness space and the function space might be the same; in function minimization they may be the inverse of one another. In many common cases the fitness and function spaces are treated as if they were the same, though it is often helpful to keep the distinction in mind. The point of optimization is to find the parameters that maximize fitness.

Of course, for a simple arithmetic problem such as $4 + x = 10$ we don't have to optimize—someone a long time ago did the work for us, and we only have to memorize their answers. But other problems, especially those that human minds and evolving species have to solve, definitely require some work, even if they are not obviously numeric. It may seem unfamiliar or odd to think of people engaging in day-to-day optimization, until the link is made between mathematics and the structures of real human situations. While we are talking about the rather academic topic of optimizing mathematical functions, what is being said also applies to the dynamics of evolving species and thinking minds.

Fitness Landscapes

It is common to talk about an optimization problem in terms of a *fitness landscape*. In the simple arithmetic example above, as we adjusted our parameter up and down, the fitness of the solution changed by degree. As we move nearer to and farther from the optimum $x = 6$, the goodness of the solution rises and falls. Conceptually, this equation with one unknown is depicted as a fitness landscape plotted in two dimensions, with one dimension for the parameter being adjusted and the second dimension plotting fitness (see Figure 2.6). When goodness is the negative of error, the fitness landscape is linear, but when goodness = $1/\text{abs}(10 - (4 + x))$, it stretches nonlinearly to infinity at the peak. The goal is to find the highest point on a *hill* plotted on the *y*-axis, with the *peak* indicating the point in the parameter space where the value of x results in maximum fitness. We find, as is typical in most nonrandom situations, that solutions in the *region* of a global optimum are pretty good, relative to points

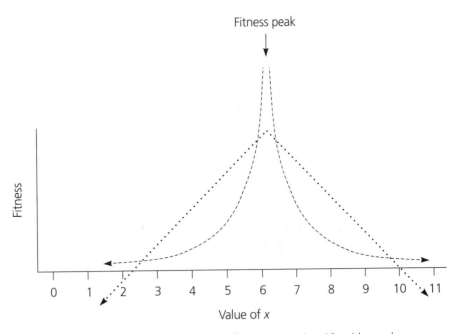

Figure 2.6 Fitness landscapes for the one-dimensional function $x + 4 = 10$, with goodness defined two different ways.

in other regions of the landscape. Where the function is not random—and therefore impossible to optimize—a good optimization algorithm should be able to capitalize on regularities of the fitness landscape.

Multimodal functions have more than one optimum. A simple one-dimensional example is the equation $x^2 = 100$. The fitness landscape has a peak at $x = 10$ and another at $x = -10$. Note that there is a kind of bridge or "saddle" between the two optima, where fitness drops until it gets to $x = 0$, then whichever way we go it increases again. The fitness of $x = 0$ is not nearly so bad as, say, the fitness of $x = 10,000$. No matter where you started a trial-and-error search, you would end up finding one of the optima if you simply followed the gradient.

John Holland has used a term that well describes the strategic goal in finding a solution to a hard problem in a short time; the issue, he said, is "the optimal allocation of trials." We can't look everywhere for the answer to a problem; we need to limit our search somehow. An ideal algorithm finds the optimum relatively efficiently.

Two basic approaches can be taken in searching for optima on a fitness landscape. *Exploration* is a term that describes the broad search for a relatively good region on a landscape. If a problem has more than one peak, and some are better than others, the lesser peaks are known as *local optima.* Normally we prefer to find the global optimum, or at least the highest peak we can find in some reasonable amount of time. Exploration then is a strategic approach that samples widely around the landscape, so we don't miss an Everest while searching on a hillside. The more focused way to search is known as *exploitation.* Having found a good region on the landscape, we wish to ascend to the very best point in it, to the tip-top of the peak. Generally, exploitation requires smaller steps across the landscape, in fact they should often decrease as the top of a peak is neared. The most common exploitational method is *hill climbing,* in which search proceeds from a position that is updated when a better position is found. Then the search can continue around that new point, and so on. There are very many variations on the hill-climbing scheme. All are guaranteed to find hilltops, but none can guarantee that the hill is a high one. The trade-off between exploration and exploitation is central to the topic of finding a good algorithm for optimization: the optimal allocation of trials.

The examples given above are one-dimensional, as the independent variable can be represented on a single number line, using the y-axis to plot fitness. A more complex landscape exists when two parameters affect the fitness of the system; this is usually plotted using a three-dimensional coordinate system, with the z-axis representing fitness; that is, the parameters or independent variables are plotted on a plane with the fitness function depicted as a surface of hills and valleys above the plane. Systems of more than two dimensions present a perceptual difficulty to us. Though they are not problematic mathematically, there is no way to graph them using the Cartesian method. We can only imagine them, and we cannot do that very well.

The concepts that apply to one-dimensional problems also hold in the multidimensional case, though things can quickly get much more complicated. In the superficial case where the parameters are independent of one another, for example, where the goal is to minimize the sphere function $f(xi) = \sum x_i^2$, then the problem is really just multiple (and simultaneous) instances of a one-dimensional problem. The solution is found by moving the values of all the x_i's toward zero, and reducing any of them will move the solution equally well toward the global optimum. In this kind of case the fitness landscape looks like a volcano sloping gradually upward from all directions toward a single global optimum (see Figure 2.7). On the other hand, where independent variables interact with one another, for example, when searching for a set of neural network weights, it is very often the case that what seems a good position on one dimension or subset of dimensions deteriorates the optimality of

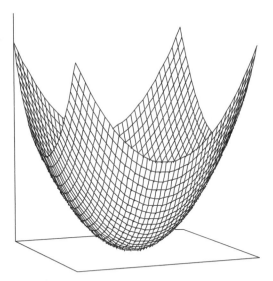

Figure 2.7 The sphere function in two dimensions.

values on other dimensions. Decreasing one of the x_i's might improve performance if and only if you simultaneously increase another one and deteriorates goodness otherwise. In this more common case the fitness landscape looks like a real landscape, with hills and valleys and sometimes cliffs.

High-Dimensional Cognitive Space and Word Meanings

It may seem odd that we have wandered away from the interesting subject of minds and cultures into a discussion of mathematics, of all things. Before digging any deeper into this topic, let us mention one example of why we think that optimization of complex functions has anything to do with minds.

One of the most exciting findings in cognitive science in the past decade has to do with the discovery of computer algorithms that can begin to find the meanings of words. Two groups of scientists have independently developed approaches for statistically extracting the meaning of a word from its context.

The story goes back to the 1950s, when Osgood, Suci, and Tannenbaum (1957) investigated the placement of words and concepts in a multidimensional space. Their approach, called the *semantic differential,* presented people with a word on the top of a page and asked them to rate that word on a lot of scales. For instance, the word might be "radio," and people were asked to rate it on scales such as cold to hot, cruel to kind, and so on (see Figure 2.8). (The first surprise was that people generally did not have much trouble doing this.)

The well-known *halo effect* occurs when a person or object with some positive characteristics is assumed to have other positive characteristics as well. In other words, ratings along different dimensions tend to correlate with one another. These multiple correlations can be identified using principal components or factor analytical techniques. With these statistical methods, Osgood and his colleagues were able to show that three major dimensions most affected the words and concepts they looked at in their studies. By far the most important dimension was one they labeled "evaluation." People rate things in terms of good versus bad, liked versus disliked, and their opinions about other aspects of the thing tend to go along with these evaluative ratings. Two other orthogonal factors, which they labeled "potency" and "activity," were also seen to be important, but not as important as evaluation.

In the 1990s, researchers at the University of Colorado and at the University of California, Riverside, started to revisit and extend this work. First they noted that there is an inherent weakness in picking target words and rating scales out of the blue; the researchers' choices of rating scales might be affected by their own preconceptions, for instance, and not really say anything at all about how ordinary people think of the words and concepts. The investigators decided to do a simple-sounding thing. They took a large body of writing, called a corpus, and analyzed the co-occurrences of the words in it.

If you have ever read Usenet news, then you know that there is a lot of text there, a lot of words. Usenet is a set of discussion groups on the Internet; nobody knows how many there are really, but there are easily more than 30,000 different ones, many of which are very busy with participants writing or "posting" to the groups, replying to one another's postings, and replying to the replies. Again, if you have ever visited Usenet, you will know that this is not a place for the King's English; there is a lot of slang and, yes, obscenity, with flame wars and other tangents mixed in with level-headed talk of love and software and philosophy and pet care and other things. Riverside psychologists Curt Burgess and Kevin Lund (e.g., Burgess, Livesay, and Lund, 1998; Burgess, 1998; Lund, Burgess, and Atchley, 1995) downloaded 300 million words of Usenet discussions for analysis by their program, which they call HAL, for "Hyperspace Analogue to Language."

Rate how you feel about this object on each of the scales below.

> **Radio**

Evaluation

Bad	1 2 3 4 5 6 7	Good
Cruel	1 2 3 4 5 6 7	Kind
Ugly	1 2 3 4 5 6 7	Beautiful
Sad	1 2 3 4 5 6 7	Happy
Negative	1 2 3 4 5 6 7	Positive
Unpleasant	1 2 3 4 5 6 7	Pleasant
Worthless	1 2 3 4 5 6 7	Valuable

Potency

Weak	1 2 3 4 5 6 7	Strong
Small	1 2 3 4 5 6 7	Large
Soft	1 2 3 4 5 6 7	Hard
Light	1 2 3 4 5 6 7	Heavy
Shallow	1 2 3 4 5 6 7	Deep
Submissive	1 2 3 4 5 6 7	Assertive
Simple	1 2 3 4 5 6 7	Complex

Activity

Passive	1 2 3 4 5 6 7	Active
Relaxed	1 2 3 4 5 6 7	Tense
Slow	1 2 3 4 5 6 7	Fast
Cold	1 2 3 4 5 6 7	Hot
Quiet	1 2 3 4 5 6 7	Noisy
Dim	1 2 3 4 5 6 7	Bright
Rounded	1 2 3 4 5 6 7	Angular

Figure 2.8 An example of the semantic differential scale, where items that measure the three most important dimensions—evaluation, potency, and activity—are separated.

Their program went through the entire corpus, looking at each word through a "window" of its 10 nearest neighbors on each side. A matrix was prepared, containing every word (there were about 170,000 different words) in the entire corpus as row and as column headings. When a word occurred directly before the target word, a 10 was added to the target word's row under the neighbor's column. A word appearing two positions before the target had a 9 added to its column in the row corresponding to the target word, and so on. Words appearing after the target word were counted up in that word's columns. Thus the matrix contained bigger numbers in the cells defined by the rows and columns of words that appeared together most frequently and smaller numbers for pairs of words that did not co-occur often in the Usenet discussions.

Words that occurred very rarely were removed. The result of this process was a 70,000 × 70,000 matrix—still too big to use. Burgess and Lund concatenated the words' column vectors to their row vectors, so that each word was followed by the entire set of the association strengths of words that preceded it and ones that followed it in the corpus. Some of the words were not really words, at least they did not turn up in a standard computer dictionary; also, some of the words had very little variance and did not contribute much to the analysis. By removing the columns that contributed little, the researchers found that they could trim the matrix down to the most important 200 columns for each of the 70,000 words.

The effect of this is that each word has 200 numbers associated with it—a vector of dimension 200, each number representing the degree to which a word appears in the proximity of some other word. A first measure that can be taken then is *association,* which is simply shown by the size of the numbers. Where target word *A* is often associated with word *B,* we will find a big number in *B*'s column of *A*'s row. A second and more interesting measure, though, is possible, and that is the Euclidean distance between two words in the 200-dimensional space:

$$D_{AB} = \sqrt{\sum (a_i - b_i)^2}$$

where a_i is the *i*th number in *A*'s row vector and b_i is the *i*th number in *B*'s row vector. A small distance between two words means that they are often associated with the same other words: they are used in similar contexts. When Burgess and his colleagues plotted subsets of words in two dimensions using a technique called multidimensional scaling, they found that words that were similar in meaning were near one another in the *semantic space* (see Figure 2.9).

Similar work was going on simultaneously at the University of Colorado, Boulder, where Thomas Landauer was working with Susan Dumais, a researcher for Bellcore, in New Jersey (Landauer and Dumais, 1997). Their paradigm, which they call Latent Semantic Analysis (LSA), is implemented somewhat differently, in that word clusters in high-dimensional space are found using a statistical technique called singular value decomposition, but the result is essentially identical to Burgess and Lund's—a fact acknowledged by both groups of researchers.

The view that depicts meaning as a position in a hyperspace is very different from the usual word definitions we are familiar with. A

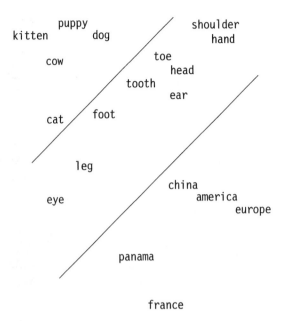

Figure 2.9 Burgess and Lund's multidimensional scaling analyses find that words with similar meanings are associated with the same regions of semantic space, here collapsed to two dimensions. (From Burgess, 1998, p. 191.)

dictionary, for instance, purports to explain the meanings of words by considering them as discrete symbols and providing denotative lists or descriptions of their attributes. Plotting words in a high-dimensional contextual space, though, depicts the words' meanings connotatively, from the inside out. Landauer and Dumais point out that the average seventh-grade child learns approximately 10 to 15 new words every day. These are not typically learned by having their definitions explained, and who ever saw a seventh-grader look up anything in a dictionary? It seems clear that humans learn word meanings from contexts, just as HAL and LSA do. The view of high-dimensional semantic space is a *bottom-up* view of language, as opposed to the *top-down* imposition of definitional rules in the usual dictionary reference.

Landauer and Dumais have suggested that the LSA/HAL model could be presented as a neural network. In fact the connectionist paradigm in general supposes that many aspects of human cognition can be represented as structure in a high-dimensional space. Human cognition and development then can be seen as a gradual process of optimization within a complex space; our thesis is that this process is collaborative in nature.

People do not consciously compute with matrices of thousands and thousands of words all at once, yet somehow the interrelationships of many thousands of words are understood, and meaningful verbal communication is possible. A child's vocabulary is small, and errors are common; learning to map concepts into the appropriate region of the cognitive space occurs gradually and takes a long time. So how do people learn the meanings of words? How do they learn to form connections between concepts found in different regions of the semantic hyperspace? The easy and obvious answer (and a central theme of this book) is that they learn from other people. The symbols themselves are arbitrary, and so different cultures can use different sounds and glyphs to mean the same thing. Yet within a culture meanings are agreed upon. The "swarm" model we are building toward considers the movements of individuals within a commonly held high-dimensional space. Thus the psychological issue is one of adaptation and optimization, of finding our way through labyrinthine landscapes of meaning and mapping related regions appropriately.

Two Factors of Complexity: *NK* Landscapes

In cases where variables are independent of one another, optimization is a relatively simple case of figuring out which way to adjust each one of them. But in most real situations variables are interdependent, and adjusting one might make another one less effective. Language is possibly the penultimate example of this; as we have seen, the meaning of a word depends on its context. For instance, what do the words "bat" and "diamond" mean? If we have been talking about caves, they might refer to flying mammals and found treasures, but if we have been talking about baseball, these words probably mean a sturdy stick and a playing field. The interrelatedness of linguistic symbols is a necessary part of their usefulness; we note that most people learn to use language fluently, but despite decades of striving no programmer has succeeded at writing a program that can converse in fluent natural language.

Stuart Kauffman (1993, 1995) has argued that two factors affect the complexity of a landscape, making a problem hard. These factors are *N*, the size of a problem, and *K*, the amount of interconnectedness of the elements that make it up. Consider a small network of five binary nodes. We start with an ultimately dull network where there are no connections between nodes (see Figure 2.10). In such a simple network, each node has

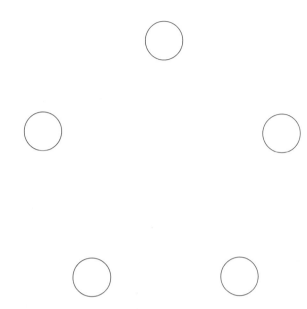

Figure 2.10 *NK* network with *K* = 0 and *N* = 5.

a preferred value; that is, it is "better" for it to take on the value of one or of zero. In Kauffman's research, a table of fitness values is determined by random numbers. For instance, if node 1 has a state value of 1, its fitness is 0.123, and if it is zero, its fitness is 0.987. We want to maximize fitness, so we see in this case that it is better, or more optimal, for this node to have a value of zero. A fitness table for this overly simple graph is shown in Table 2.1.

Since these nodes are binary, we can write the state of the network as a bitstring; for instance, 10101 means that the first node has a state value

Table 2.1 Fitness table for simple *NK* network where *K* = 0.

Node	Value if 0	Value if 1
1	0.987	0.123
2	0.333	0.777
3	0.864	0.923
4	0.001	0.004
5	0.789	0.321

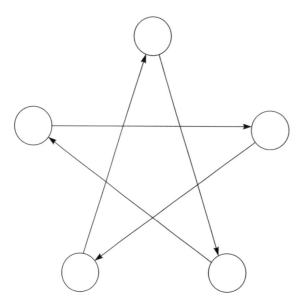

Figure 2.11 When $K = 1$, the better state of each node depends on the state of the node that depends on it and the state of the node that it depends on.

of one, the second is zero, the third is one, the fourth is zero, and the fifth is one. Looking at Table 2.1, we note that the best possible state of the network is 01110. This pattern of node states results in the highest fitness at each site in the graph and thus produces the highest sum over the entire graph.

Now that is not much of a graph, we agree. While there are five nodes, which we will write as $N = 5$, they don't interact; none has any effect on the other. Kauffman says that such a system has $K = 0$, where K stands for the average number of inputs that each node receives from other nodes. Obviously, with $K = 0$ it is very easy to find the global optimum of the network. We can simply pick a site at random and flip its sign from zero to one or vice versa; if the network's total score increases, we leave it in the new state, otherwise we return it to its original state (this is called a *greedy* algorithm). We only need to perform N operations; once we have found the best state for each node, the job is done.

If we increase K to 1, each node will receive input from one other node, and the size of the table doubles. When $K = 1$, the fitness of a node depends on its own state and the state of the node at the sending end of an arrow pointing to it. A typical network is shown in Figure 2.11.

In this connected situation, we have to conduct more than N operations to find the optimal pattern. It is entirely possible that reversing the state of a node will increase its performance but decrease the

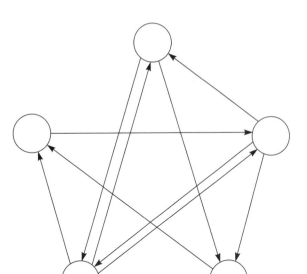

Figure 2.12 As *K* increases, the better state of each node depends on the state of the nodes that receive its inputs, and the states of the nodes that provide inputs to it. Here, *N* = 5 and *K* = 2.

performance of the node it is connected to, or the opposite, its performance will decrease while the other's increases. Further, while we might find a combination that optimizes the pair, the state of the receiving node affects the node *it* is connected to, too, and it is entirely likely that the state of that node is best when this node is in the opposite state—and so on.

When *K* = 2, the fitness of each node depends on its own state (zero or one) and the states of two other nodes whose arrows point to it (see Figure 2.12). The size of the lookup table increases exponentially as *K* increases; its size is $N2^{K+1}$. Reversing the state of a node directly affects the fitness of two other nodes, perhaps changing their optimal state, and their states affect two others, and so on. *K* can be any number up to *N* − 1, at which point every node is connected to every other node in the whole system.

It is easy to see why Kauffman has theorized that the two parameters *N* and *K* describe the *complexity* of any system. First, as *N*, the dimensionality of the system, increases, the number of possible states of the system increases exponentially: remember that there are 2^N arrangements of *N* binary elements. This increase is known as *combinatorial explosion*, and even if it seems obvious, this is a significant factor in determining how hard it will be to find an optimal configuration of elements.

Each new binary element that is added doubles the patterns of node activations. It quickly becomes impossible to test all possible combinations, and so it is necessary to find a way to reduce the size of the search; we need a good algorithm for the optimal allocation of trials.

The second factor, K, is also known as *epistasis*. This term has been borrowed, perhaps inaccurately, from genetics, where it is often seen that the effect of a gene at one site on the chromosome depends on the states of genes at other sites. Kauffman has shown that when K becomes higher relative to N, landscapes become irregular and eventually random. When K is high, the highest peaks are poorer, due to conflicting constraints, and the paths to peaks on the landscape are shorter. When $K = 0$, there is one peak, and as we saw, that peak is found when each node takes on its better value. The path to it is simple and direct: a Fujiyama landscape.

Combinatorial Optimization

Optimizing problems with qualitative variables, that is, variables with discrete attributes, states, or values, is inherently different from optimization of quantitative or numerical problems. Here the problem is one of arranging the elements to minimize or maximize a result. In some cases we might want to eliminate some elements as well, so that the number of things to be rearranged is itself part of the problem. Combinatorial optimization is particularly addressed within the field of traditional artificial intelligence, especially often where the task is to arrange some propositions in such a way that they conform to some rules of logic. "Production rules," if-then statements such as "if A is true then do C," are evaluated and executed in series, to obtain a valid logical conclusion or to determine the sequence of steps that should be followed in order to accomplish some task. Arranging the order of the propositions is a combinatorial optimization problem.

The statement of the goal of an optimization problem is the *objective function,* sometimes called a "cost function." For example, consider a classic combinatorial problem called the traveling salesman problem (TSP). In this problem, a map is given with some number of cities on it, and the objective function to minimize is the distance of a route that goes through all the cities exactly once, ending up where it started.

The *simple permutations* of some elements are all the possible orderings of an entire set. In a very small traveling salesman problem, we might have three cities, called A, B, and C, and their simple permutations would be ABC, ACD, BAC, BCA, CAB, and CBA. Many combinatorial

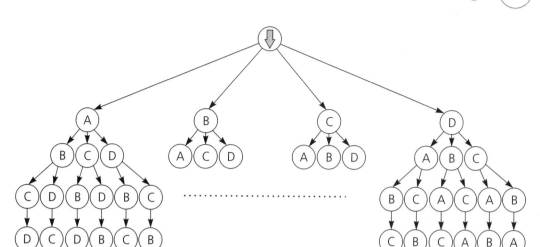

Figure 2.13 Simple permutations of four elements depicted as a tree. The dotted line indicates where branches were omitted due to lack of space.

optimization problems require finding the best simple permutation of some objects. There are *n* factorial (written as *n*!) simple permutations of *n* elements. Thus with our example of three things, there are $1 \times 2 \times 3 = 6$ permutations. The point to realize here is that this number, which represents the size of the search space or set of possible solutions to the problem, gets big quickly. For instance, with only 10 elements the search space contains 3,628,800 possible simple permutations. Any problem that is likely to exist in the real world, for instance, a complicated scheduling or routing problem or cognitive model, will almost certainly have more elements in it than that; we need to find ways to search for an answer without trying every possible solution. Optimal allocation of trials, again.

Simple permutation problems are often represented, conceptually at least, as trees. The three-element problem is just too trivial, so let's consider a simple permutation problem with four elements (see Figure 2.13). The first decision is where to start, and it has four options, A, B, C, or D. Once we've selected one of these, say, A, then we are faced with the next decision: B, C, or D? Each step of the process reduces the number of options by one. Say we choose B, now the decision is between C and D. Selecting C, we have no real choice, but end at D. We see that there are indeed *n*! paths through the decision tree—and so of course it is impossible to actually draw a graph when the problem is big. But the conceptualization is often useful to keep in mind. In an optimization situation the task is to find the path through the decision tree that gives the optimal evaluation or result, whether this is the shortest path from a start to a finish, or any feasible path from a premise to a conclusion, or some other kind of problem.

There are two general ways to search through such a tree data structure. *Breadth-first search* involves going into the tree one layer at a time, through all branches, and evaluating the result of each partial search. In our four-element example, if we are searching for the shortest path from start to end, we might ask, for instance, starting on A, what is the distance to B, to C, to D? If we started on B, what is the distance to A, to C, to D? and so on. It may be that we can eliminate some routes at this point, just because a step is so large that we are sure not to have the shortest route. *Depth-first* search comprises going from the start all the way to the end of a path, evaluating one entire proposed solution. Once the end is reached, depth-first search goes back to the first decision node and then searches to the end of the next path, and so on.

Heuristics are shortcuts in the search strategy that reduce the size of the space that needs to be examined. We proposed a simple heuristic for breadth-first search in the previous paragraph: abandoning a search as soon as there is reason to expect failure. If we were using a depth-first strategy, we might establish a rule where we abandon a path when its distance becomes longer than the shortest we have found so far. Maybe after a few steps into the tree we will be able to eliminate a proposed path or subtree by this method. There are many clever general heuristics and very many ingenious heuristics that apply only to specific problems.

While "order matters" in permutations, the term *combination* refers to the situation where some number of elements is selected from a universe of possibilities, without regard to their order. Thus, a universe of {A, B, C, D} contains these combinations: {A}, {B}, {C}, {D}, {A, B}, {A, C}, {A, D}, {B, C}, {B, D}, {C, D}, {A, B, C}, {A, B, D}, {A, C, D}, {B, C, D}, {A, B, C, D}. Since order doesn't matter in combinations, the combination {A, B, C} is identical to {B, C, A}, {C, A, B}, and so on.

In many situations the problem is defined in such a way that some permutations or combinations are inadmissible; for instance, nodes are connected only to certain other nodes in the graph (there is no road connecting city B with city F). In these cases the size of the search space is of course reduced, though the problem may be more difficult than a simple permutation problem if at each step the appropriateness or correctness of a branch in the tree needs to be validated.

We can muse upon combinatorial aspects of the great stochastic systems. Even though it is a simple-sounding problem of ordering genetic proteins, natural evolution could not have gone from bacterium directly to human (which is not to suggest that we are the ultimate achievement of evolution—we just make a good reference point for the present). The search of the space of possibilities that led down to human intelligence needed to discover RNA, DNA, nucleated cell structures, multicellularity, breathing out of water, spines and centralized nervous systems, and so forth in order to reach the social, linguistic primate called *Homo sapiens*. At the same time the environment had to change and evolve, affecting and being affected by the new forms that inhabited it. It may have been "technically" possible to jump from bacterium to human in one huge mutational leap, but the probability of that happening was miniscule. Evolution can be seen as a kind of search of the space of possible life-forms, where various adaptations are tried, continuing into the future. An important aspect of this search is that it is conducted in parallel; that is, a number of mutations and genetic combinations are proposed and tested simultaneously throughout the populations of species. Since traits cannot be passed from one species to another, evolutionary search at the macro level can be conceptualized as a branching tree.

Minds as well search their space of ideas in parallel, but the search is not well represented as a tree since the nodes at the ends of the branches are connected to one another. Various cultures explore their own regions of a cognitive parameter space, and within those cultures particular individuals search particular regions of the vast search space of beliefs, attitudes, and behaviors. Here and there a culture hits a dead end, and cultures interact, passing heuristical information back and forth to one another, and out on the little branches individuals convey information to one another that shapes their searches.

A human born into this world alone, to make sense of what William James called the "buzzing, blooming confusion" of the world, would not be able in a lifetime to understand very much of it at all—if indeed the word "understand" even has meaning for an asocial hominid. Information sharing is our main heuristic for searching the large space of possible explanations for the world.

Binary Optimization

A frequently used kind of combinatorial optimization, which we have already seen in *NK* landscapes and some Hopfield networks, occurs when elements are represented as binary variables. The binary encoding scheme is very useful, for a number of reasons. First of all, as the versatility of the digital computer indicates, almost anything can be represented to any degree of precision using zeroes and ones. A bitstring can represent a base-two number; for instance, the number 10011 represents—going from right to left and increasing the multiplier by powers of two—$(1 \times 1) + (1 \times 2) + (0 \times 4) + (0 \times 8) + (1 \times 16) = 19$. A problem may be set up so that the 19 indicates the 19th letter of the alphabet (e.g., S) or the 19th thing in some list, for instance, the 19th city in a tour. The 19 could be divided by 10 in the evaluation function to represent 1.9, and in fact it can go to any specified number of decimal places. The segment can be embedded in a longer bitstring, for instance, 10010010111011*100110*110, where positions or sets of positions are evaluated in specific ways.

Because zero and one are discrete states, they can be used to encode qualitative, nonnumeric variables as well as numeric ones. The sites on the bitstring can represent discrete aspects of a problem, for instance, the presence or absence of some quality or item. Thus the same bitstring 10011 can mean 19, or it can be used to summarize attendance at a meeting where Andy was present, Beth was absent, Carl was absent, Denise was present, and Everett was present. While this flexibility gives binary coding great power, its greatest disadvantage has to do with failure to distinguish between these two kinds of uses of bitstrings, numeric and qualitative. As we will see, real-valued encodings are usually more appropriate for numeric problems, but binary encoding offers many advantages in a wide range of situations.

The size of a binary search space doubles with each element added to the bitstring. Thus there are 2^n points in the search space of a bitstring of n elements. A bitstring of more than three dimensions is conceptualized as a *hypercube*. To understand this concept, start with a one-dimensional bitstring, that is, one element that is in either the zero or the one state: a bit. This can be depicted as a line segment with a zero at one end and a one at the other (see Figure 2.14). The state of this one-dimensional "system" can be summarized by a point at one end of the line segment. Now to add a dimension, we will place another line segment with its zero end at the same point as the first dimension's zero, rising perpendicular to the first line segment. Now our system can take on 2^2 possible states: (0,0), (0,1), (1,0), or (1,1). We can plot the first of these states at the origin, and the other points will be seen to mark the corners of a square whose sides are the (arbitrary) length of the line segments we used. The (1,1) point is diagonally opposed to the origin, which is meaningful since it has nothing in common with the (0,0) corner and thus should not be near it. Corners that do share a value are adjacent to one another; those that don't are opposed.

To add a third dimension, draw a cube extending above the page into space. Now the 2^n points are three-dimensional: (0,0,0), (0,0,1), (0,1,0), (0,1,1), (1,0,0), (1,0,1), (1,1,0), and (1,1,1), and each state of the system can be represented as a corner of a cube. As with the two-dimensional bitstring, points with elements in common are connected to one another by an edge of the cube, and points with none in common, for instance, (1,0,1) and (0,1,0), are not connected. The number of positions in which two bitstrings differ is called the *Hamming distance* between them, and it is a measure of their similarity/difference as well as the distance between them on the cube. If two points have a Hamming distance of h between them, it will take a minimum of h steps to travel from one of the corners of the hypercube to the other.

A hypercube has the characteristics of a cube, that is, bitstrings are conceived as corners of the hypercube. It is possible to depict a low-dimensional hypercube (see Figure 2.14), and some insights can be gained from inspecting the graph, but with more dimensions it quickly becomes incomprehensible.

Binary optimization searches for the best corner of the hypercube. Even though the size of binary problems only increases by powers of two, they can get *intractable;* it is not practical or possible to consider every corner of a high-dimensional hypercube. For instance, a 25-bit prob-lem—which is not an especially big problem—represents a hypercube with 33,554,432 corners. There is obviously a need for methods to reduce the size of the search space, so that we can find a good answer without evaluating every possibility, so we can optimally allocate trials.

A couple of years ago, a hamburger restaurant advertised that they could prepare their hamburgers "more than 1,023 different ways." Any-one with experience in binary arithmetic would immediately recognize that 1,024 is 2^{10}, thus indicating how "more than 1,023" came to be im-portant. If you could order your hamburger with or without (1) cheese, (2) lettuce, (3) mayonnaise, (4) pickles, (5) ketchup, (6) onions, (7) mus-tard, (8) relish, (9) dressing, and (10) sesame seed bun, you could have it more than 1,023 different ways. Each combination of the presence or absence of each item represents a corner on a 10-dimensional hyper-cube, which could be evaluated. For instance, you might consider a 1110000011, which is a hamburger with cheese, lettuce, mayonnaise, dressing, and sesame seed bun. For me, I prefer corners of the hypercube where pickles = 0, cheese = 1, and mustard = 0. Some notational sys-tems use the "#" sign to mean "don't care": you might say I prefer my hamburgers prepared as 1##0##0###. The preferred state (absent or pres-ent) of the other ingredients might depend on my mood, how other peo-ple's food looks, the smell of the place, and so on, but mainly they are

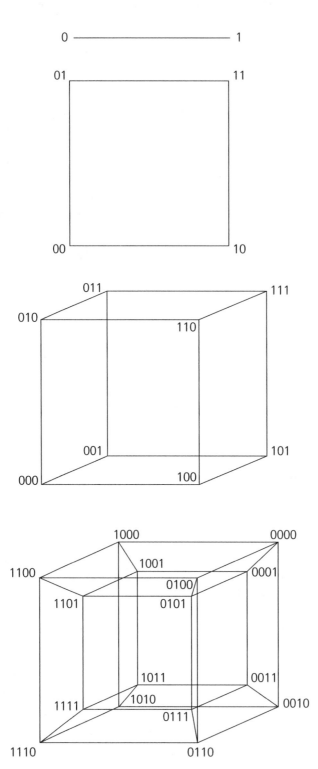

Figure 2.14 One-, two-, three-, and four-dimensional binary spaces: lines, squares, cubes, and hypercubes.

not very important. At least I can focus my search in the region of the hypercube that has those values.

Random and Greedy Searches

What is a rational strategy for searching for a good point in the binary space? We will assume that the space is large, that we don't want to consider every possible corner on the hypercube. So how could we reduce the size of our search? Let's start by making up a naive algorithm.

Common sense tells us we need to start somewhere. Our natural sense of order suggests that we start at the extremes, say—assuming a 10-bit problem—0000000000 or 1111111111. Now, which one of these is a better place to start? We don't have any way to know. Is it better to picture a bare hamburger and imagine adding things to it or to start with the "everything" burger and remove stuff? We don't know. If we started at one extreme (say, all zeroes) and the optimum was at the other end, we would end up having to go the entire distance before we found it. The best answer is probably somewhere in the middle. One commonsense strategy is just to pick a point randomly, say, 1001011010.

Since we don't know anything about our problem, we could try a *random search* strategy, also called "generate and test." Random bitstrings are generated, one after the other, each one is evaluated, and the best one is saved. After some prespecified number of iterations, the algorithm stops, and the best bitstring found is taken as the solution. There is no reason to expect the random search strategy to find a very good answer, and in fact it really doesn't produce very impressive results on big problems.

Another method that won't work especially well is to take a *random walk* on the landscape. We could generate a bitstring solution, flip one bit chosen at random, evaluate the new bitstring, flip another bit, evaluate that, and so on. Such a search can be imagined as movement around the hypercube, going from corner to adjacent corner, one step at a time. If we were looking for an optimum, we could just store the best solution found so far and use it after the walk was finished. As far as optimization goes, a random walk is just an exploitational random search, focusing in one region of the search space.

One thing that sometimes works is the so-called *greedy* approach that we tried with *NK* landscapes. Start by generating a bitstring and evaluating it. Then select the first bit on the bitstring, flip it, and evaluate the new bitstring. If flipping the bit resulted in improved performance, the "flipped" value is kept, or else it is set back to what it was. The greedy algorithm goes through the entire bitstring in this way, flipping, evaluating, and choosing, once for every position, 10 times if $N = 10$. This algorithm will result in improvement unless the search happened to randomly start on a really good corner. What the greedy algorithm doesn't do is to capitalize on patterns of bits that work together. The difficulty of this can be gleaned from the discussion of NK landscapes; when K is high, the effect of a site depends on the states of many other sites. The greedy algorithm may be a reasonable, quick first-guess approach to optimizing a small or simple system and will probably work fine where epistasis, or interaction among elements, is weak.

Hill Climbing

A *hill-climbing* search strategy is a kind of random walk that modifies a pattern, accepts the changes if they result in improvement, and then tries changes to the new pattern (Kauffman calls this an "adaptive random walk"). A good way to do this on a binary landscape is to set a probability of flipping a bit, say, 0.10; this will be called a *probability threshold.* Then at every position on the bitstring a random number between 0.0 and 1.0 is generated; if it is less than 0.10, that bit is flipped, and if it is not, it stays as it was. Once we have gone through the entire bitstring, we evaluate the new pattern. If its fitness is better than the best so far, then we keep this bitstring. On the next iteration, we stochastically flip the bits on this new bitstring. If this leads to improvement, we keep it, otherwise we retain the pattern, so fitness is always improving, as long as it can.

It is easy to see why this is called hill climbing and also why it is not a perfect method. If the *mutation rate,* or proportion of bits flipped, is small, for instance, where an average of one bit per iteration or less is flipped, then a hill climber will climb to the top of whatever hill it started on—it will improve until it reaches the top of that hill. Then it's stuck. This is called finding a local optimum, and while a hill climber is guaranteed to find the top of the hill eventually, we really want to find the global optimum—the highest peak on the entire landscape. (Hill-climbing terminology assumes we are maximizing something, but the same concepts hold for minimization problems.)

Increasing the mutation rate would enable the hill climber to jump from one hill to another—maybe higher—one, but it probably would not climb that hill very well; it would be as likely to jump to another hill as to climb the one it's on. When the Hamming distance between states of a hill climber on successive iterations is large, the algorithm can cover large areas of the landscape and might find better regions than it would have if it had only climbed its original hill. Here is the trade-off between exploration and exploitation. At one extreme, the random search algorithm explores perfectly well, testing one region of the landscape after the other, but fails miserably to zero in on the peaks in those regions. At the opposite extreme, the hill climber is excellent for zeroing in on the best point in its area, but fails entirely to explore new regions.

In fact random search and hill climbing can be seen as two variations on the same theme. Random search, where a new bitstring is randomly generated at every time step, can be variously considered as new individuals being created every time, or it can be considered as successive generations or states of the same individual. In this view, if each bit is determined randomly, it will appear that about half the individual's bits change state each time, on the average. In comparison, if you took a hill climber with a low mutation rate, say, $1/n$, where n is the number of bits in the string—guaranteed to find a local optimum—and increased the mutation rate to 0.5, you would have exactly the search algorithm we have been calling random search. Thus, from at least this point of view, the difference between exploration and exploitation is the mutation rate, or size of steps through the search space.

Simulated Annealing

One of the more powerful approaches to binary optimization is known as *simulated annealing*. Simulated annealing is based on the metaphor of molecules cooling into a crystalline pattern after being heated. In a molten metal the molecules move chaotically, and as the metal cools they begin to find patterns of connectivity with neighboring molecules, until they cool into a nice orderly pattern—an optimum. Simulated annealing takes the basic idea of hill climbing and adds to it a stochastic decision and a *cooling schedule*. A bitstring is modified by flipping randomly selected bits, and if the modified bitstring performs better, it replaces the original. If the modified bitstring performs worse, though, it can still be accepted if a probability test is passed. The probability threshold is a function of the system's "temperature," which decreases over time. Thus the probability of accepting a poorer problem solution decreases as the system cools; the effect of this is that the algorithm roams over wide areas of the search space in the early iterations, bouncing into and out of locally optimal regions. Later in the search the algorithm will be focused on the more promising regions of the problem space.

The salient lesson to be learned from simulated annealing is that sometimes you have to choose to do worse in the short run to do better in the long run. It may seem counterproductive to accept a bad solution, but it is necessary in order to make sure that the search moves beyond local optima, increasing the chances of finding global optima, or at least better local optima. Simulated annealing explores early in the experiment and exploits later—an approach that has been shown to be quite successful for many kinds of problems.

There is another, subtler, lesson to be learned here, too, having to do with the usefulness of metaphor for understanding abstract mathematical systems. There is nothing about the procedure of the simulated annealing algorithm that ties it inextricably to the cooling of molecules in a crystal lattice. The algorithm can be written in abstract algebraic symbols, just like any other algorithmic process. But it was the metaphor of annealing that allowed the conceptualization and creation of the method in the first place, and it is the metaphor that enables people to understand how it works: it would be much more difficult to convey the process through purely abstract symbols. In discussions of some of the paradigms in this book, there may be ambiguity and even confusion between computer programs that are intended to simulate real processes and computer programs that are explained in terms of real processes. Metaphor is the other side of simulation, and considering the two together gives insights into both. There is a computer program, and there is something in the world, and we perceive some degree of resemblance between them. A simulation is a program that helps us understand something about the world, but sometimes the world simply offers a template that helps us understand a program. The world and the program are two systems whose contours are parallel in some relevant way. In one view they are separate and different things, with imagined similarities; in another view they are two ways of doing the same thing. On the one hand, we say a computer program is like a process in the physical world, and on the other, it is often insightful to recognize that the physical world is like a computer program. Beings in computers evolve—and computers think.

Binary and Gray Coding

Flipping bits increments a numeric bitstring with a certain lack of subtlety and grace, especially as binary encoding of real numbers introduces

Hamming cliffs into the landscape. Imagine some parameters were being adjusted, getting close to a solution, and perhaps needed to move only one unit—for instance, from 15 to 16—in order to get to the optimum. Though this is a step of only one unit on a number line, in binary code it requires a jump from 01111 to 10000: this is a Hamming distance of 5— the maximum, as all five bits have to be flipped. A binary algorithm searching for a quantity (for instance, a hill climber) would encounter an obstacle at these cliffs and might never make the complete reversal at all positions necessary to improve the bitstring's fitness.

Gray coding overcomes this impediment while retaining the advantages of binary operations. The challenge is to devise a scheme, using zeroes and ones, to encode integers where the Hamming distance between adjacent numbers equals one; this is called the *adjacency property*. There can be many ways to accomplish this for any length bitstring; the most commonly used version is called "binary-reflected Gray code" (Gray, 1953; Gardner, 1986). As shown in Table 2.2, Gray-coded integers that are one unit different in value are also one unit distant in Hamming distance.

The algorithm for generating Gray code from binary is quite simple. The length of the Gray bitstring will be the same length as the binary version. Further, the leftmost bit will be the same. Starting at the second position from the left, then, the formula is

$$G_i = XOR(B_i, B_{i-1})$$

where G_i is the bit in the ith position of the Gray code, B_i is the bit in the ith position of the binary code, counting from the left-hand side, and the function $XOR()$ returns 1 if the adjacent bits are different from one another, 0 if they are the same. In other words, set the most significant bit on the Gray bitstring equal to the same bit on the binary bitstring, and move to the right. Where a bit matches the bit to the left of it on the binary bitstring, place a 0 in the Gray bitstring, otherwise put a 1. Go down the line doing this at each position. With Gray coding, a movement of one unit on the number line is performed by flipping a single bit, allowing a numeric bit-flipping optimizer to climb more gracefully toward optima.

Step Sizes and Granularity

The search of a parameter space of real numbers will likely entail movement where the length of the steps is important. In binary encoding

Table 2.2 Comparison of binary and Gray code. Hamming cliffs, for instance, between 7 and 8 in the binary column, do not exist in Gray coding.

Decimal	Binary	Gray
0	0000	0000
1	0001	0001
2	0010	0011
3	0011	0010
4	0100	0110
5	0101	0111
6	0110	0101
7	0111	0100
8	1000	1100
9	1001	1101
10	1010	1111
11	1011	1110
12	1100	1010
13	1101	1011
14	1110	1001
15	1111	1000

schemes, including Gray coding, it is important to note that bits at the right end of the bitstring represent smaller numbers than bits on the left, increasing exponentially as we move from right to left. For instance, the binary bitstring 1001011 represents the decimal integer 75. Flipping the rightmost bit from 1 to 0 decrements the numeric value to 74. If instead we flipped the leftmost bit from 1 to 0, we would decrement the number to 11—a step size of 64 units in one operation. Thus random bit flipping on a bitstring that encodes true numbers results in a step of unknown distance through the space. A random-mutation algorithm may test consecutive points that are near one another in the number space, or ones that are far apart, and this feature is uncontrolled. It is extremely unlikely that a researcher would intentionally search the space with such uncontrolled step sizes, but this is indeed what can happen when optimization is conducted with real numbers encoded in bitstrings.

A second issue has to do with the floating-point granularity of bitstring representations, how many decimal places a number can have. Of course a digital computer can represent numbers to hundreds of decimal places using an internal binary representation—it can be done. But the other side of that fact is that an investigator can use the computer's built-in encoding scheme to represent floating-point numbers rather than using an explicit binary encoding with division in the user interface. For example, the number 0.852 could be represented as a bitstring 11001010100, and the program would decode and divide by 1,000—or the number can be represented as it is, as the decimal 0.852, using the computer's internal translation and performing optimization operations on the explicit decimal number. In sum, when numbers are considered as quantities, binary coding might introduce weaknesses and complications that could be avoided by searching in a real-numbered space.

Of course we are mentioning this for a reason. Since Holland's influential *Adaptation in Natural and Artificial Systems* (1975), many genetic algorithm researchers have used binary encoding on all kinds of problems. Some lore and some theory have arisen regarding the right and wrong ways to operate on binary-encoded data. Binary encoding can have advantages, but some operations, for instance, indiscriminate mutation of numeric bitstrings, may interfere with problem solving more often than they help. Almost all genetic algorithm operations can be performed with real-number encodings—while acknowledging some advantages of binary encoding, we recommend using floating-point numbers when they are more appropriate.

Optimizing with Real Numbers

The size of a real-numbered parameter space cannot be calculated: it is infinite. Even a bounded single dimension with end-points at zero and one has an infinity of points on it. So in a strict sense, optimization of real-numbered problems should be impossible. But of course it's not, because real-numbered landscapes, like other types, almost always have some kind of order to them.

Many of the methods used to optimize continuous problems are similar to those used for binary and combinatorial problems. Distance, including step sizes, now is not Hamming distance, but (usually) Euclidean distance. If one point A in an n-dimensional space is defined by coordinates $(a_1, a_2, \ldots a_n)$, and another, B, has coordinates $(b_1, b_2, \ldots b_n)$, then the distance between the two points is (the useful formula, again)

$$D_{AB} = \sqrt{\sum (a_i - b_i)^2}$$

Distance is the square root of the sum of the squared differences between pairs of coordinates. In one dimension, the square root of the difference squared is just the difference itself, which seems like a reasonable measure of distance. In two dimensions, it is the familiar Pythagorean theorem, the length of the hypotenuse of a triangle, and in more dimensions this conceptualization is known as the square root of a *sum of squares*.

If we imagine a random problem, where there is no systematic relationship between points in the parameter space and points on the fitness surface, then the distance between two points is not a useful piece of information in looking for an optimum. In fact, there would be no shortcut for looking for an optimum on a random landscape; any guess would be as good as any other. At the other extreme, if there were a problem with a smooth landscape and a single global optimum, then you could reasonably make the assumption that points that were near each other would have correlated fitness. Moving in a direction that improved fitness on the previous step would probably improve fitness on the current step, too.

This might seem trivial, even obvious—and it is, but it points to an important property of high-dimensional landscapes, which can be exploited by an optimizer. That property is sometimes called *landscape correlation:* the distance between two points in a multidimensional space is seen to be correlated with the difference in their fitness values. This property can be exploited in the search for an optimal point in a continuous space where the landscape is not random. If an algorithm can find a "good" point, for instance, a point with a good fitness value, then it has a clue that there may be a better point nearby. At least it might be worthwhile to search in the area for other good points.

Summary

In this chapter we have been leaning heavily on the assumption that minds and species engage in something like optimization. This is quite a safe assumption really, if we are not too strict with the definition. Nature doesn't necessarily look for the globally optimal solution to a problem (nature isn't *looking* for anything at all!); all that is required is that a solution be good enough, what Herbert Simon (1979) has called a *satisficing* solution. Both evolution and cognition require search of enormous spaces. As we have suggested, there can be a great number of search strategies. The social strategies of humans and evolutionary search are both population based, meaning that many solutions can be tested in parallel; in both cases, interactions among population members result in problem-solving intensity greater than the sum of individuals' solitary efforts.

chapter
three

On Our Nonexistence as Entities: The Social Organism

While human phenomenology by its nature tends to aggrandize the individual's contribution, cognition and behavior can be viewed through multiple "zoom angles" or frames of reference that can provide insights for understanding and explaining mental life and intelligence. These perspectives range from the microscopic to the cosmic, and include consideration of the origin of life, the dynamics of planetary ecology, and the evolution of social behaviors. Modern thinkers exploring the implications of these different zoom angles have frequently turned to computer simulations, robotics, and other creative projects to test ideas, and we report some of these experiments. The history of these models provides fundamental context for the swarm intelligence paradigm described later in the volume. ∎

Views of Evolution

Incredibly, as the second millennium drew to a close, the State of Kansas Board of Education voted to eliminate evolutionary theory from the state's elementary school curriculum. Polls showed that only about a tenth of Americans could wholeheartedly accept the premises of Darwinism, that life has evolved to its present state through natural selection. "Creation scientists," whose conclusions are biblically predetermined and whose arguments are based on the absence of some "missing links," ignoring the tremendous body of positive evidence that does exist, are cited in the media as if they were real scientists. The Creation Science Association for Mid-America helped write Kansas' curriculum proposal, over the objections of biologists and educators, who now must attempt to teach religion disguised as science.

Science strives to remain skeptical and cannot refer to faith to judge the validity of theoretical constructs. The evidence for evolution of species through the mechanism of natural selection is widespread and profound; it is not an "unproven theory," but is the very backbone of modern biology. Religion is an important part of human life, but when religious beliefs based on faith are contradicted by empirical evidence that can be held in the hand and seen with the eyes, the evidence has to win. It is hard to see how anything can be gained by clinging to unsubstantiated beliefs that contradict reality.

The disagreement between religious advocates and evolutionary scientists comes down to this: the creationists know how life began on earth, and the evolutionists don't. The biologists are sure it didn't arrive fully formed during one important week a few thousand years ago, but there is still plenty of debate in the scientific communities about the mechanisms that might have produced the first terrestrial organisms.

Wherever they came from, the first molecules of life appeared on this planet when it was very young. The earth formed about four and a half billion years ago, and the earliest known signs of life date back to 3.45 billion years ago—just 300 million years after the surface of the earth had cooled enough to allow the survival of living things. Those first living molecules were most likely suspended in water with other chemicals. As the earth's atmosphere, in what has been called for obvious reasons the Hadean Eon, was very different from what it became after the evolution of oxygen-emitting flora, we can expect that the surface of the planet was constantly scorched with solar radiation that tore at fragile molecular strands. Volcanoes belched poisonous sulphurous gases into the atmosphere. Uranium 235 was 50 times more abundant than it is now and bombarded the environment with deadly radioactivity. Corrosive compounds would have torn at those primal organisms; lightning and lava and other volatile forces must have made the environment very hard to endure. Somehow in that tough world the self-replicating molecules continued to increase in number, surviving, mutating, changing, adapting to conditions in the liquid, and on the solid land, and in the atmosphere, such as it was then.

What did it take to survive in that environment? Some microorganisms mutated in ways that doomed them, that caused them to fail to reproduce, and some formerly secure molecules (the first living things must not have been much more than that) found themselves in environments that destroyed them. On the other hand, some mutations improved the molecules' chances of reproducing, and occasionally mutations improved their ability to adjust to conditions that were previously forbidding, so the simple life-forms could extend their territory to new kinds of surroundings. The environment presented a kind of challenge, to which the evolving molecules needed to be able to respond; if the environment was a problem, then emerging life needed to evolve a solution to it. Where dangers prevailed, either reproduction ceased or mutation resulted in adaptive features.

The early environment was of course very different from what we have today. We could even say that life today *is* its own environment, or at least has evolved its own world. The adaptations of the first life-forms would not be adaptive today. Evolution is a constantly adjusting process, with organisms changing to meet the requirements of an environment that contains themselves and other organisms. One of the products of that process is the thinking ape, *Homo sapiens*.

Gaia: The Living Earth

Widening the zoom, it is possible to view the entire planet earth as one large organism, with various plant and animal species carrying out their functions much as cells do in our bodies. The atmosphere, the oceans, dry land, and the species that inhabit them conspire (literally, "to breathe together") to maintain the health of the biosphere. It is also possible to view a species as a single distributed organism whose component parts are the individuals who carry on and perpetuate the genetic heritage, and it is possible to see individual organisms as vehicles for even smaller viable units, and so on ad infinitum. There is no "correct" zoom angle for looking at nature, but it is interesting, perhaps even enlightening, to try looking through different-sized scopes.

In the late 18th century James Hutton, who is now known as the father of geology, wrote that he considered the earth to be a superorganism whose characteristics should be studied by a kind of science resembling physiology. His argument, which he called uniformitarianism, was that the physical earth had undergone many changes as a result of life's presence. Two hundred years later, British chemist James Lovelock (1979, 1988) revived the idea that the earth is a living organism, calling his view the *Gaia hypothesis.* Lovelock (1972) wrote his famous controversial statement in the journal *Atmospheric Environment:* "Life, or the biosphere, regulates or maintains the climate and the atmospheric composition at an optimum for itself."

As evidence for the Gaia hypothesis, Lovelock noted that the atmospheres of Mars and Venus are stable with 95 percent carbon dioxide, though there is only a trace of it in the earth's atmosphere. His argument was that early life-forms, especially algae and bacteria, took the carbon dioxide out of the air and replaced it with oxygen. Oxygen is not usually found plentifully (it makes up more than one-fifth of the earth's atmosphere), as it is a very volatile element; it easily combines with other chemicals in the phenomenon we call combustion, or fire. The earth's atmosphere is stable, though the combination of gases that comprise it is extremely unstable. The earth exists in a state of deep chemical disequilibrium, balanced, as it were, at the edge of chaos.

While Lovelock's hypothesis was eagerly accepted by whole-earth environmentalists and New Age philosophers, more hard-nosed natural scientists were reluctant to accept it. The main point of contention had to do with the implication that the earth-organism had a sense of purpose, as suggested by the idea that the planet is "trying" somehow to maintain an environmental state that is beneficial for itself. In consequent discussions, Lovelock admitted he had confused the situation by introducing *teleological* language, that is, language that implied purpose. He revised his statement to reduce this effect, removing implication of the "will" of the superorganism from the hypothesis. Whether the details of the theory are ultimately confirmed or rejected, it is relevant for our discussion simply to consider human life and thought from this level, as a force or factor in dynamics on a global scale.

Also relevant for our discussion, Lovelock demonstrates Gaia theory dynamics in a computer simulation he calls Daisyworld. In one version, Daisyworld is a virtual planet upon which three species of daisies, a dark species, a gray species, and a light species, grow. In the beginning, the planet is cold, as its young sun has not yet reached full strength; the dark

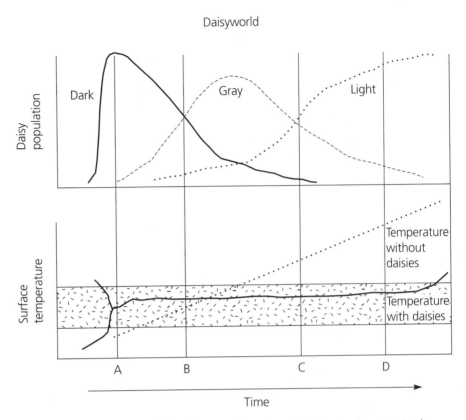

Figure 3.1 Lovelock's Daisyworld simulations let several species of daisies—in this case, Dark, Gray, and Light—grow in their optimal climate. (Adapted from Lovelock, 1988.)

daisies are better suited to that kind of climate, as they are able to hold the heat from the world's sunlike star. They would reproduce, taking over the world, but their heat absorption ends up heating the surface of the Daisyworld, until it becomes inhospitable for the dark daisies, too warm for them (A in Figure 3.1), and their numbers decrease quickly. By warming the environment, however, the dark daisies have made it a good place for white and gray daisies to thrive; these lighter daisies cool their environment by reflecting heat (B). As the young sun reaches its full strength, the lighter daisies have even more advantage, and as their numbers grow, their effect is to bring the planet's temperature down (C). The prevalence of the white daisies keeps the planet's climate much milder than it would have been without life (D). Thus with no knowledge of the environment, and with no intent to do so, the biosphere is able to make adjustments that keep it relatively optimal. The Daisyworld simulation can be run with various parameters such as starting temperature, number of species, heat of the planet's sun, and so on, with a consistent result: through selecting the prevalence of each variety of daisy, the system maintains the temperature at the planet's surface at a viable level.

Gaia theory holds that the entire planet functions as one single integrated organism, including "inorganic" systems such as the climate and the oceans as well as the biosphere. It is a literally global view that places individual organisms and even species in the roles of cells and subsystems, supporting the larger dynamical system through their breathing, eating, breeding, dying. In yet another blow to the human ego, this perspective seems to trivialize the individual—in fact, from the Gaia perspective our species appears to be more of a blight or a cancer than a beneficent organ of the planetary body. Our contribution seems to be to introduce disturbance, imbalance, disruption. So far, seen from the planetary perspective, our human intelligence has simply empowered the destruction of a global system that preceded us by billions of years.

Differential Selection

In 1962, the Scottish ecologist V. C. Wynne-Edwards theorized that evolution selected against animals that reproduced too much, to prevent overpopulation and decimation of a species. He stated:

> Experiment generally shows that . . . many if not all the higher animals can limit their population-densities by intrinsic means. Most important of all, we shall find that self-limiting homeostatic methods of density-regulation are in practically universal operation not only in experiments, but under "wild" conditions also (Wynne-Edwards, 1962, p. 11).

According to him, many behaviors that had previously been unexplained, for instance, noisy group vocalizations, communal displays, and winter roosting aggregations, might serve the function of informing species members about the size of their population. With this knowledge, species could adjust their reproduction rate to maintain equilibrium, avoiding the disaster they could cause by overextending themselves in their limited environment. Acknowledging that many species live in social groups, it may be that a trait resulting in reproductive restraint would survive if it provides advantage for the group as a whole.

The problem with Wynne-Edwards' model, called *group selection*, is that the theory must assume the existence of a gene that carries an automatic selection against itself. If a gene resulted in a reduction of reproduction (e.g., through delayed fertility or some instinctive form of infanticide), then its frequency would decrease as competing genes—ones that reproduced at a higher rate—dominated the population. The idea has been controversial since it was first proposed; if members of a population risk fatal sacrifice, then their genes would be less likely to be passed on to succeeding generations. Though the concept is not subscribed to by most biologists today, argument regarding group selection did focus interesting discussions on the question of altruism, which is seen widely through nature and presents some fascinating puzzles to evolutionary theorists.

In 1964, W. D. Hamilton proposed an explanation for behaviors that don't make sense under the assumption that individuals seek self-gain. *Inclusive fitness* is the concept that organisms strive to ensure the survival of others whose genes most resemble their own. He observed that altruistic behaviors frequently increase the probability of the altruistic individual's genes being passed on, even when the individual will not be the actual ancestor. Animals will sometimes risk or sacrifice their own lives to guarantee the survival of their genetic relatives—noting the confounding fact that a local group of conspeciates is likely to contain numerous near kin. A mother animal who charges a predator to save her offspring risks paying the ultimate cost to herself individually, yet it happens every day, in very many species. The same goes for animals who give a warning call in the presence of a predator, thereby calling attention to themselves. Inclusive fitness suggests that there is a general tendency in nature for genes to do what they can to improve their own chances, even at the cost of individual lives. The majority of self-sacrificing altruistic behaviors seen in nature are performed in the service of siblings, offspring, and cousins.

Think of the cost of reproduction itself. To reproduce means having to have sexual intercourse, which puts an animal in a vulnerable situation, besides forcing it to associate with its potentially dangerous—sometimes cannibalistic—conspeciates. Being pregnant, as we know, means eating for the number in the litter plus one, besides slowing the mother down and making her easier prey or a less adept predator. The cost of nurturing a litter is extremely high; the mother, and sometimes the father, must keep the helpless little ones safe and well fed, and parental sacrifices are high. If the system were oriented around the individual's needs, would we choose to take on all these troubles? Animals' lives, including our own, are organized around the phenomenon of reproduction, from courtship to sexual jealousy to altruistic protection of kin.

Darwin was puzzled by marked differences between the sexes in many species. For example, male elephant seals weigh four times as much as females (4,000 pounds to 1,000)—how could this have happened, if both sexes faced the same challenges from the environment? Males of some species have developed heavy antlers or plumage that obviously imperils their ability to escape predators. In response to these paradoxes, Darwin created a kind of second theory of evolution, having to do with the evolution of sexual characteristics. Though it was not widely accepted at first, the theory of *sexual selection* is now considered to be an important part of evolutionary theory. If we consider that evolution works via a process of differential reproduction, it seems obvious that features that promote reproduction per se—and not necessarily adaptation to environmental challenges—will be selected along with adaptive features.

Further, it is apparent that the investment in reproduction is different for males and females, especially in species such as humans where the pregnancy is relatively long and the newborn offspring are helpless. According to sociobiological theory, "sexier" individuals have more opportunities to mate, for almost all species (though standards of sexiness admittedly differ from species to species!). Where females have an innate preference for mates who can provide shelter and protection, males have an interest in choosing mates who show signs of being fertile. Further, because maternity is certain but paternity is not, it is in the male's best interest to elicit a long-term, exclusive sexual commitment from the female; otherwise he may invest much in perpetuating somebody else's genes. On the other hand, as the male is capable of impregnating multiple females, there is less pressure for males to be sexually exclusive, and in fact evidence is accumulating that human males tend to be more promiscuous than females across cultures.

This is illustrated by a phenomenon known as the Coolidge Effect, named after former U.S. President Calvin Coolidge. The story goes that President Coolidge and his wife were touring a farm. When the First Lady came through the chicken house, the farmer mentioned that roosters are capable of copulating many times in a single day. "Please mention that to the President," she requested. When the President came to the chicken house, the dutiful farmer repeated the fact to him. Silent Cal contemplated this for a few seconds, then asked, "Same hen every time?" "Why, no," said the farmer. "Please mention that to Mrs. Coolidge," said the President.

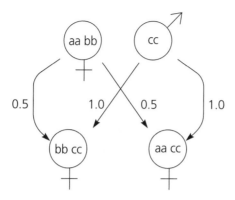

Figure 3.2 Haplodiploid sexual reproduction.

Hölldobler and Wilson (1990) argued that a colony of ants should be thought of as a single superorganism. In order to understand ant behavior it is necessary to look at the behavior of the colony as a whole, rather than the individual. In fact some species of ants die at a rate of 6 percent of the population per day; thus it would seem that the individual's contribution cannot be very great.

The slavish subjugation of the individual to the needs of colonies of social insects of the order *Hymenoptera,* which includes ants, termites, bees, and wasps, can be explained uniquely in terms of kin selection. Hamilton noted that species in this order reproduce by a method known as *haplodiploidy* (see Figure 3.2). Females inherit half their alleles from the mother. Assuming a 50 percent chance of receiving either of two copies of a gene (one from each of a pair of chromosomes), about one-fourth of the genes that two sisters get from the mother will be the same. But the male insects grow from unfertilized eggs, and so all the males receive all their genes from the mother, that is, the queen. Further, father's sperm cells are produced by mitosis, or cell division, so there is no variation among them. Thus all sisters receive the same genes—half their inheritance—from the father's side. Add them up, and it is seen that *Hymenoptera* sisters' genes are on average three-fourths the same, daughters and fathers are 50 percent the same, and mothers and daughters are also about half the same.

This suggests that, rather than seeking to reproduce herself, in order to perpetuate her own genes the sterile female worker would increase the chances of her own genes being passed on by serving the queen, who gives birth to more sisters. The ideal strategy is to produce only enough males to ensure the insemination of young queens, thus passing on the individual's genes to a new colony.

Richard Dawkins' readable and influential books *The Selfish Gene* (1976) and *The Blind Watchmaker* (1985) accomplished several important goals. For one, Dawkins introduced the concept of *memes,* those units of epistemological evolution that correspond to genes in physical evolution. He also suggested, through his Biomorphs, a way for using computers to simulate evolution to better understand the processes of mutation and selection and how they work together. And his accessible prose introduced a large population of literate but nontechnical readers to evolutionary concepts such as kin selection.

Unfortunately, Dawkins' choice of the term "selfish gene" to describe kin selection left some with the impression that genes have a mind of their own, that they have goals and intent, a sense of purpose. This teleological interpretation is incorrect, and it must be understood that the author never stated any such thing—it was only the title of the book that suggested it. When we are discussing trial-and-error evolutionary adaptation, it seems clear that there is no place for purpose; we cannot say that a species "wants to" adapt, or even survive. As for the individual organisms comprising a species, of course the whole point is that their motivations, including their desire to survive, are subservient to the requirements of genetic propagation. An animal might do all it can to sustain its own life, within the constraints of the behavioral repertoire that biology has provided it, but in the long run the behavioral traits that are passed down through the generations are the ones that maximize their own survival as genetic programs. This is a statistical effect and not a teleological one, and the difference is very important.

Ross Buck and Benson Ginsburg (1997) have a somewhat different take on the evolution of social behaviors. They argue that sociality requires communication, which has an innate basis in the displays and preattunements in organisms; displays are visible (i.e., public) behaviors, and conspeciates have evolved the tendency to "tune in" to particular forms of display, to detect and respond to them. They contrast their view with the selfish gene hypothesis; where Dawkins implies that communication is inherently manipulative and exploitative, Buck and Ginsburg argue that social communications create the foundation that allows organisms to support the survival of the species.

Buck and Ginsburg argue that the "general affiliative phenomenon" is a property of protoplasm itself, a fundamental characteristic of all living things. Social—and especially emotional—communication was described by Darwin as a tool for creating and maintaining social order; Buck and Ginsburg elaborate on this theme by noting that spontaneous communication exists in all living things, as a kind of raw social knowledge that is nonpropositional and direct. An animal knows that another animal is angry or sexually excited, and this knowledge is direct; it is not something that has to be "figured out." The *communicative gene* hypothesis suggests that displays and preattunements evolve together through the selection of communicative relationships.

Most recently, Howard Bloom (2000) has argued for a reexamination of group selection as an explanation for much social behavior throughout the animal kingdom. Bloom compares the spread of information in a population of social animals to a neural network. Selection of socially dominant members of the population allows continual strengthening of the information-processing network, improving the adaptive abilities of the species in general. He argues provocatively, "It is time for evolutionists to open their minds and abandon individual selectionism as a rigid creed which cannot coexist with its supposed opposite, group selection. For when one joins the two, one can see that the networked intelligence forecast by computer scientists and physicists as a product of emerging technologies has been around for a very long time" (pp. 12–13).

The question of the unit of selection is a troublesome one for evolution theorists, as we have seen with some awkward and frequently contentious theorizing about group selection, altruism, inclusive fitness, selfish genes, and so on (for an insider's review of the sociobiology debate see Segerstråle, 2000). While the low-level mechanism for the process of evolution is the survival and reproduction of the individual organism, it is obvious that individuals' connections to their conspeciates as well as to other aspects of the environment are critical to survival and largely determine the direction that adaptive evolution will take. Because genetic transmission works across generations and through populations, common sense suggests looking at the phenomenon at the species or population level and in the adaptive relationships that the species develops with elements of its world.

Our Microscopic Masters?

Finally, as we think about different levels of looking at the behaviors and the minds of individuals, let us consider the presence of foreign life-forms within the cells of plants and animals including ourselves. Mitochondria are a kind of *organelle,* like a cell-within-a-cell, found in almost all eukaryotic cells, that is, cells with nuclei. There are usually several or many mitochondria in each cell; they assist the cell by burning sugar to provide energy. The mitochondria are formed of two membranes with a space between. Plants contain another kind of organelle called chloroplasts, which convert light energy to chemical energy in photosynthesis.

The *endosymbiont theory* states that these organelles were originally forms of bacteria that entered plant and animal cells at a very early period of evolution. Both mitochondria and chloroplasts make their own DNA, which is independent of that of the host, and conduct some but not all of their own protein synthesis—chemical life processes. Not only do they move around inside the cell, changing shape, but they control their own replication within the cell. The two layers of membrane suggest that one layer originates in the organelle and the other in the host cell. In sexual reproduction it is seen that the organelle's DNA comes only from the mother's cells, not from the father's, so it descends from a lineage unique from that of the cell in which it resides. Analysis of organelles' DNA suggests that they come from a very primitive and archaic form of bacteria, perhaps cyanobacteria.

In *Lives of a Cell,* Lewis Thomas (1974) argues for our nonentity as individuals. He suggests the perspective that we so-called "organisms" are really just the vehicles for the perpetuation of the master race, the organelles—mitochondria and chloroplasts. As individuals our job is to strive to maintain the viability of nutrients for our masters, and we live to pass their DNA on to future generations, requiring that we adapt to our circumstances, survive, and reproduce.

Looking for the Right Zoom Angle

We present these various views because they seriously challenge the intuitive perception of the individual as a self-interested autonomous being—only time will tell which ones of these theories pass the tests of scientific investigation. Evolution, or life in the broader sense, fails to make sense as a method for fulfillment of the individual's needs. From the evolutionary zoom view, the individual organism, whether tree, insect, or human, is a medium for perpetuation of the species' heritage and for introducing variations to the line. As the theory of evolution has bumped us down a few notches closer to the monkeys as biological entities, further insulting human pride, so these various perspectives on behavior tend to demote the individual from master of his own psyche to a single atomic participant in a transpersonal information-processing effort, or cosmic slave to some bacterial hitchhikers.

One thing we have seen in these various explanations is an assumption that life embodies a dynamic process of optimization. Those first molecules—wherever they came from—are important to us today because they were able to maximize the probability of self-reproduction, as has life ever since. Given an environment that serves exactly as a complex function to optimize—a mixture of gases, solids, liquids, and variations of energy, of nutrients and poisons, of edible cohabitants and predatory ones—the life-system adapts, changing, flowing around obstacles and toward balance and stability, even as the environment is adapting and flowing.

Because we think with biological brains evolved through natural selection to serve an evolutionary end, it may be impossible for humans to *really* extract ourselves from our egoistic perspective, to see ourselves as disposable stepping stones for some bigger or littler entity's progress and survival. It is our nature to attribute causes to individuals, to two-eyed, four-limbed sacks of skin with presumed intentions, feelings, thoughts, and personalities locked up inside them—locked up but visible through strange insightful empathic perceptions. It is science's task to find ways around these limitations, to enable us as well as possible to see the world, including ourselves, as it really exists.

Our role in nature, whether it is to pass on genes or mitochondria, to support our group, to maintain the Gaia-sphere, or even to provide for the survival of the individual organism, involves our ability to minimize error in our adaptations to the environment. Where the environment puts up an obstacle, some species has almost always found a way to overcome it, and our species in particular has found ways to adapt to every environment we have encountered. Konrad Lorenz (1973) even rated us higher than rats and roaches in terms of being able to adapt to diverse environments. Where the environment offers snow and ice, humans live in igloos; where it offers clay, they live in adobe; where palms grow, they thatch their dwellings with fronds, capitalizing at every turn on the opportunities the environment provides, no matter how subtle. Our ability to think—our intelligence—allows us to optimize the probability of our surviving to reproduce. Even if we think of intelligence as a trait of individuals, it seems obvious that bright individuals innovate by building on the existing accomplishments of their social group, often to the benefit of the less-bright members of the group.

Evolution is commonly regarded as an excellent algorithm for the adaptation of organisms. Social behavior, especially culture, is often compared to it, but it has not seemed that an adequate theory has been proposed to explain how culture can result in adaptation. In the following sections we elaborate some of the benefits of sociality generally throughout the animal kingdom, and in humans specifically. While it is interesting to consider some of the varieties of social experience in different animals, we are also interested in the computer programs that implement or simulate social behaviors. Some of these are simulations, where a population of individuals is modeled in a program, and some are programs that interact socially with a user. There is much to learn from both.

Flocks, Herds, Schools, and Swarms: Social Behavior as Optimization

Mutation and sexual genetic recombination, along with selection and perhaps with self-organization (discussed in a later chapter), empower evolution to search for optima on difficult landscapes. The important function of random mutation and recombination is to introduce variation or innovation to a population; selection then ensures that better solutions persist over time. More than one observer of human nature has argued that creativity requires some kind of operation like mutation, some way to randomly generate new solutions. For instance, psychologist Donald Campbell (1960) described creative thinking in terms of "blind variation and selective retention" and suggested that mental creativity might be very much analogous to the processes of biological evolution.

Trial-and-error learning must include some way to create new varieties of problem solutions. The generation of random behaviors is ubiquitous throughout the animal world. The chaotic behaviors of a hooked fish, the head-shaking of a dog playing tug-of-war, the bucking of a bronco or a bull, the zig-zagging of a chased rabbit through a meadow—all these indicate that the ability to generate random, that is, unpredictable, behaviors is adaptive for animals. Random activity is useful for predator avoidance and is also useful for an organism that is searching for something, whether it is food, a mate, a location to build a nest, or a safe hideout. In an insightful chapter entitled "Oscillation and Fluctuation as Cognitive Functions," Konrad Lorenz (1973) discusses the importance of random fluctuations of an organism's movements for search as well as escape from danger. For example (he says), the marine snail waves its long breathing tube from side to side in order to detect the scent of something to eat, as it moves randomly across the bottom of the ocean. The snail is sensitive to the differences in the strength of a scent at the two extremes of the breathing tube's motion. These differences are naturally greatest when the snail is turned at right angles to the goal, so that the food is on one side or the other. But instead of turning at right angles toward the stimulus, the snail makes a sharp reversal that resembles an

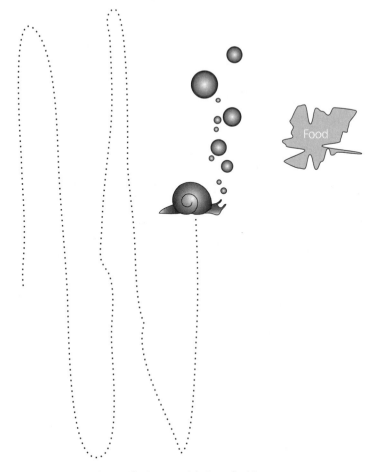

Figure 3.3 The marine snail's oscillations enable it to find its target.

escape response and continues crawling along so that the odor will strike the receptor on the other side. The result is a repetitive zig-zag path that takes the snail toward its goal (see Figure 3.3).

Lorenz compares the kind of search typified by the marine snail to fluctuations in human public opinion. He deserves quoting here:

> As has already been said, the public idea of what is true and real is based on a highly complicated system of social interactions. Since processes having a definite degree of inertia play a part in these, public opinion tends to oscillate. It was Thomas Huxley who said that every new truth starts as a heresy and ends up as an orthodoxy. If one were to interpret the word orthodoxy as signifying a body of rigid and mindless dogma, this would be a depressing statement, but if we take

it as meaning the body of moderate views subscribed to by the majority of people in a given culture, we can see that what Huxley is describing is a characteristic cognitive function of human society (Lorenz, 1973, p. 240).

We quote Lorenz at such length for one reason: he concisely summarizes an important part of the paradigm that we are building here, the view of cognition as a function of collective adaptive search. He barely stops short of giving our particle swarm formula.

Species benefit from sociality in different ways. The bottom-line, bare-necessity requirement for species practicing sexual reproduction is interaction during mating, and in many species the strongest bonds are formed between mothers and offspring. Other species of animals form lifetime monogamous pair bonds between mates. The advantage of this appears to be improved rearing of offspring and the ability of a pair to better defend their nesting territory. While living in social groups increases the chances of any individual finding a mate, it also introduces the negative specter of competition for sexual partners, especially for males, since one male can inseminate many females. Thus males of social species tend to show less affiliation with the group.

In many species that live in groups, social networks of kin help take care of young communally. The gregarious life also permits the ability to share food and to provide warning as well as collaborative defense when predators intrude. Groups of prey may be more conspicuous than solitary individuals, but running with a pack reduces the probability of any individual being selected for attack. Foraging efficiency can be increased when many eyes look for food. Also, groups of animals can acquire information about the environment from conspeciates, for instance, by noting which direction a well-fed flockmate just flew in from. For some predators there is additional advantage in cooperative strategic hunting. Groups have the potential for division of labor, which is most clearly seen in the social insects. For the various social animals, some of these benefits are more important and some less; in general, though, we see that social organisms can attain enormous survival advantage.

Doug Hoskins (1995) has shown how the simplest behaviors of the simplest organisms in the animal kingdom can be shown to function as an optimization algorithm. *E. coli* bacteria show what is perhaps the simplest "intelligent" behavior imaginable. (Hoskins defines intelligence as the "ability of the organism to control its long term distribution in the environment," p. 2.) These single-cell organisms are capable of two kinds of locomotion behaviors, called "run" and "tumble." Running is a forward motion implemented by rotating the flagella counterclockwise. Tumbling occurs when the flagella are rotated clockwise. The change causes the bundle to fall apart, so that the next "run" begins in a random direction. The cell tumbles more frequently in the presence of an adverse chemical, and the overall change of direction is enough to increase the bacterium's probability of survival, enabling it to escape from toxins. This is about as simple as an adaptive behavior can be. Hoskins uses the bacterial adaptation as the cornerstone of his iterated function system model of neuronal interactions in the human brain.

The paramecium's behavior repertoire includes a phobic response that is adaptive and just as simple as the run and tumble of *E. coli*. When a paramecium encounters an aversive chemical environment, it reverses its direction briefly. Then the cilia on one side act to spin the body randomly, until it moves forward again.

Eshel Ben-Jacobs, at the University of Tel Aviv, argues that colonies of bacteria are able to communicate and even alter their genetic makeup in response to environmental challenges. Indeed, Ben-Jacobs asserts that the lowly bacteria colony is capable of computing better than the best computers of our time, and attributes to them properties of creativity, intelligence, and even self-awareness. Discoveries of DNA sequences that can move from one site on the chromosome to another, as well as discoveries of plasmids, which are genetic substances found outside the chromosome in many bacterial species, support the theory that bacterial cells can transmit genetic material to one another, outside the bounds of genetic processes as they had been previously understood. Further, a number of studies have shown that bacteria can mutate nonrandomly, producing particular mutations in response to certain selection pressures. From these discoveries and others, Ben-Jacobs theorizes that the genome operates as an adaptive cybernetic unit, able to solve complex problems posed by the environment.

Not only do the cells adapt as solitary units, but Ben-Jacobs argues that, due to effects that have been observed such as direct physical and chemical interactions between cells, indirect communication through release of chemicals into the environment, and long-range chemical sensing, bacteria are able to organize into what is essentially a single, multicellular superorganism. While research is yet necessary to fill in the gaps in Ben-Jacobs' theory, the social phenomena are well known and noncontroversial.

It is surprising to think that single-celled microbes might possess sophisticated intelligence—maybe it is just because they are so small, so removed from our perceptions. Near the beginning of the 20th century, the biologist H. S. Jennings made the interesting comment that if an amoeba were the size of a dog, we would not hesitate to attribute to it the power of subjective experience (quoted in Lorenz, 1973). The single mechanism of mobility for an amoeba is its familiar flowing, effected by alternating softening and hardening of the protoplasm; this most simple form of locomotion is, in the amoeba, the method both for phobia, that is, avoidance of noxious stimuli, and for approaching positive stimuli. Even this simplest semiliquid creature adapts and responds appropriately to its environment.

Perhaps the most extreme example of the intimate relation between the individual and the social group is found in the life of the lowly slime mold. When food is plentiful, this organism exists as an amoeba, a single-celled organism that feeds on bacteria and reproduces by cell division. When food becomes scarce, though, the amoebas gravitate toward one another and actually bond together to form a kind of slug that can crawl around. At this point they have lost their autonomy and exist only as cells in the aggregate organism. When the slug has found a good spot, it sends up a stalk that releases spores into the environment; the spores eventually become amoebas, and the process starts over.

UC San Diego physicist Herbert Levine has shown how slime-mold amoebas are capable of communicating among themselves and organizing in a way that allows the emergence of superorganisms from clusters of microbes (e.g., Levine, 1998; Ben-Jacobs and Levine, 1998). If members of the species *Dictyostelium discoideum* are spread over a surface that provides no nutrients, they will begin to emit a chemical signal; an amoeba of that species that detects the presence of the signal will emit the signal itself, in what becomes an autocatalytic or self-amplifying process. The effect is that spiral waves of the signaling chemical (cyclic adenosine monophosphate) form in the population. Amoebas then begin to move toward the center of the spiral, and when they meet other amoebas they merge with them in what has been described as a "river system" of protists. Eventually the organisms will have aggregated into a single organism, which then produces reproductive spores that can be spread to more favorable environments. In this way the simple amoeba is able to cope socially with adversity and perpetuate itself.

Accomplishments of the Social Insects

The optimization potential of simple behaviors has been most noted in studies of insects, and in particular in the behaviors of the social insects.

An insect may have only a few hundred brain cells, but insect organizations are capable of architectural marvels, elaborate communication systems, and terrific resistance to the threats of nature.

E. O. Wilson began his systematic study of the social behaviors of ants in 1953, when he attended Konrad Lorenz' lectures at Harvard University. Lorenz is probably best known as the scientist who showed the phenomenon of "imprinting" in animals: some kinds of baby birds adopt the first thing they see when they hatch as their parent and follow it everywhere it goes. In some famous photographs, Lorenz is seen leading a line of happy goslings who had adopted him as their mother. Imprinting is a form of instinctive behavior Lorenz called a *fixed action pattern*. A fixed action pattern is a behavior that an organism emits in response to a particular, often very specific, stimulus. For example, in the springtime the male European robin seeks to drive away other males by song, display, and outright attack. It has been shown that the fixed action stimulus for these behaviors is the red breast. If a stuffed male robin with an olive breast is set near the nest, the male will ignore it, but a tuft of red feathers mounted on a wire frame elicits the entire range of territorial threats.

Extrapolating from Lorenz' descriptions of the fixed action patterns, the stereotypical "instinctive" behaviors of organisms, Wilson theorized that the dazzling accomplishments of ant societies could be explained and understood in terms of simple fixed action patterns, which, Wilson discovered, included behavioral responses to *pheromones,* chemicals that possess a kind of odor that can be detected by other ants. Wilson showed that ants emit specific pheromones and identified the chemicals and even the glands that emitted them. He also laboriously identified the fixed action responses to each of the various pheromones. He found that pheromones comprise a medium for communication among the ants, allowing fixed action collaboration, the result of which is a group behavior that is adaptive where the individuals' behaviors are not.

For Wilson, the problem of the construction of mass behaviors from the behaviors of single ants was the central problem in the sociobiology of insects. Given that the behavior of a single ant is *almost* random, with a stochastic tendency to gravitate toward paths that have been trodden by other ants, the achievements of swarms of ants are most incredible. An isolated ant's behavior quickly results in the demise of the individual, but the mass behavior of a colony of ants provides sustenance and defensive protection for the entire population.

The study of complex systems and the rise to prominence of computer simulation models of such systems gave scientists the tools they needed to model the simple behaviors of ants and how they could combine to produce an effect that is much more than the sum of its parts, and these insights have in turn led to more insights about the nature of man and society and about the physical world. Insect sociality is a classic example of the emergence of global effects from local interactions.

There is a long and interesting tradition in artificial intelligence research of using ant examples to explain important points. These examples make a fascinating barometer of their historical eras and demonstrate a fundamental shift in the way we view ourselves and the workings of complex social and cognitive systems, beginning (as far as we know) with the negative rhetoric of Norbert Wiener's 1950 *The Human Use of Human Beings*. In the aftermath of World War II (and before Wilson's research), Wiener thought that the view of human societies as antlike was essentially a fascist view in which rulers rule forever, followers always follow, and everyone knows and accepts their place: "This aspiration of the fascist for a human state based on the model of the ant results from a profound misapprehension both of the nature of the ant and of the nature of man" (Wiener, 1950/1954, p. 51). Wiener theorized that the structure of a machine or organism is an index of the performance it can be expected to produce. Because ants' bodies are rigid and crude, we should not expect much of them: Wiener describes the ant as "an essentially stupid and unlearning individual, cast in a mold which cannot be modified to any great extent" (p. 51). Humans, on the other hand, are viewed as complex, creative, adaptable, able to learn and to control their own behaviors and the outcomes that result from them—and the sophistication of individuals largely explains the accomplishments of human societies.

In his 1969 *The Sciences of the Artificial,* Herbert Simon proposed that the path of an ant across a "wind- and wave-molded beach" is very irregular, nearly random, but with a general sense of direction. Simon used this image to make the strong point that the complexity of its path is not a characteristic of the ant, but of the environment. The ant itself is a very simple being, unable to generate anything very complex at all. Then (ignoring Wiener) he suggests substituting the word "man" for "ant." His conclusion: "A man, viewed as a behaving system, is quite simple. The apparent complexity of his behavior over time is largely a reflection of the complexity of the environment in which he finds himself" (p. 25).

Of course Simon's characterization of humans as antlike in our simplicity is a tough pill to swallow, but it did serve one pragmatic end for early AI researchers: it made the simulation of human behavior and human cognition seem tractable. If people are like ants, then researchers should be able to write computer programs that do what people do, using complicated data structures and relatively simple programs. This gives us a small insight into the roots of some of the problems that artificial intelligence researchers have been unable to overcome. Wiener was right: an ant *by itself* is not only simple but very stupid and unable to accomplish anything. Trying to figure out how a solitary ant's mind works does not sound like much of a challenge: it simply follows a small set of rules. Ants succeed by collaborating and so do people.

As early as 1979, Douglas Hofstadter was suggesting in his inimitable half-cipher prose that maybe the brain was like an ant colony. No single neuron in a brain contains knowledge in itself, it is only through their interaction that thinking can occur. As Hofstadter's Anteater explains:

> There is some degree of communication among the ants, just enough to keep them from wandering off completely at random. By this minimal communication they can remind each other that they are not alone but are cooperating with teammates. It takes a large number of ants, all reinforcing each other this way, to sustain any activity—such as trail building—for any length of time. Now my very hazy understanding of the operation of brains leads me to believe that something similar pertains to the firing of neurons . . . (Hofstadter, 1979, p. 316).

Hofstadter's view of ants is in line with the contemporary appreciation for the emergence of complex dynamical systems from the interactions of simple elements following simple rules. Cellular automaton virtual ants by Christopher Langton and subsequent computational investigations of emergent patterns in antlike populations has resulted in a kind of fascination with the intelligent accomplishments of large numbers of unintelligent agents. Mitchel Resnick has stated that "Indeed, ants have become the unofficial mascots of the ALife community" (Resnick, 1998, p. 59).

The year 1999 saw the release of the Santa Fe Institute volume entitled *Swarm Intelligence: From Natural to Artificial Systems* (Bonabeau, Dorigo, Theraulz, 1999), which builds a half-dozen systems for problem solving based on the behaviors of swarms of insects. Eric Bonabeau, Marco Dorigo, and Guy Theraulz have built upon E.O. Wilson's early research on the effects of pheromone communication in ant swarms, as pioneered in the computational studies of Jean-Louis Deneubourg, to show that the model of the humble ant can be adopted to solve some kinds of combinatorial problems that were previously thought to be too hard to attempt.

In the introduction to that text, the authors note that the term "swarm intelligence" was originally used to describe a particular paradigm in robot research. They continue:

> Using the expression "swarm intelligence" to describe only this work seems unnecessarily restrictive: that is why we extend its definition to include any attempt to design algorithms or distributed problem-solving devices inspired by the collective behavior of insect colonies and other animal societies (p. 7).

In fact, all their published models derive from the activities of the social insects. Our use of the term is even less restrictive than Bonabeau, Dorigo, and Theraulz'. We note, for instance, that the term "swarm intelligence" has been used in the field of semiotics to describe the kind of irrational buzz of ideas in the mind that underlies the communication of signs between two individuals. More to the point, we found this description of the concept of "swarm" in an early version of an FAQ document from Santa Fe Institute about the Swarm simulation system:

> We use the term "swarm" in a general sense to refer to any such loosely structured collection of interacting agents. The classic example of a swarm is a swarm of bees, but the metaphor of a swarm can be extended to other systems with a similar architecture. An ant colony can be thought of as a swarm whose individual agents are ants, a flock of birds is a swarm whose agents are birds, traffic is a swarm of cars, a crowd is a swarm of people, an immune system is a swarm of cells and molecules, and an economy is a swarm of economic agents. Although the notion of a swarm suggests an aspect of collective motion in space, as in the swarm of a flock of birds, we are interested in all types of collective behavior, not just spatial motion.

We agree with the SFI definition (forgiving their casual use of the term "agents") and explicitly add the possibility of escaping from physical space, defining swarms as they might occur in high-dimensional cognitive space, where collision is not a concern.

The shift in scientific perspectives over the past half-century has been from considering the ant as an individual without dignity, locked in a tyrannical web of conformity, to the ant as a simple but powerful information processor, to the metaphor of the individual ant as a component of a brain, to the modern view that the ant colony, composed of many collaborating individuals, is itself a powerful information processor. Contemporary ants are individuals as individuals, rather than parts of individuals, and their accomplishments are credited to their interindividual interactions.

Some tropical termites are able to build elaborate domed structures that are begun as pillars; in the course of building, the pillars are tilted toward one another until their tops touch and they form an arch. Connecting arches results in the typical dome. As it is frequently remarked that the invention of the arch was a major milestone in the development of the architecture of civilized man, we might wonder how in the world a swarm of simple-minded termites could accomplish the feat. If *we* were building an arch, we would start with a plan, that is, a central representation of the goal and the steps leading to it. Then, as the work would probably require more than one person (unless it was a very small arch), a team of workers would be organized, with the architect or someone who understands the plan supervising the laborers, telling them where to put materials, controlling the timing of the ascension of the two pillars and their meeting. We are so dependent on centralized control of complex functions that it is sometimes impossible for us to understand how the same task could be accomplished by a distributed, noncentralized system.

It appears that the termites build a dome by taking some dirt in their mouths, moistening it, and following these rules:

- Move in the direction of the strongest pheromone concentration.
- Deposit what you are carrying where the smell is strongest.

After some random movements searching for a relatively strong pheromone field, the termites will have started a number of small pillars. The pillars signify places where a greater number of termites have recently passed, and thus the pheromone concentration is high there. The pheromone dissipates with time, so in order for it to accumulate, the number of termites must exceed some threshold; they must leave pheromones faster than the chemicals evaporate. This prevents the formation of a great number of pillars, or of a wasteland strewn with little mouthfuls of dirt.

The ascension of the pillars results from an *autocatalytic* or positive feedback cycle. The greater the number of termites depositing their mouthfuls in a place, the more attractive it is to other termites. Autocatalysis is a significant aspect of many complex systems, enabling the amplification of apparently trivial effects into significant ones. As termite pillars ascend and the termites become increasingly involved in depositing their loads, the pheromone concentration near the pillars increases. A termite approaching the area then detects the pheromone, and as there are multiple pillars and the termite is steering toward the highest concentration, it is likely to end up in the area between two pillars. It is attracted toward both, and eventually chooses one or the other. As it is approaching the pillar from the region in between, it is more likely to climb up the side of the pillar that faces the other one. As a consequence, deposits tend to be on the inner face of the pillars, and as each builds up with more substance on the facing side, the higher it goes the more it leans toward the other. The result is an arch.

Termite builders are one kind of *self-organizing* system. There is no central control, the intention of the population is distributed throughout its membership—and the members themselves are unaware of the "plan" they are carrying out. Actors in the system follow simple rules, and improbable structures emerge from lower-level activities, analogous to the way gliders emerge from simple rules in a cellular automaton.

When insects work collaboratively there is no apparent symbolic communication about the task. A social insect may possess a repertoire of pheromones that amount to messages, but the content of the message is hardly symbolic. The message content may be something like "an angry ant was here," or "somebody was building something here." Pheromone communications do not have an intended audience; they are simply the output of gland secretion in response to a stimulus.

Much of the communication among social insects is accomplished indirectly by a method that the entomologist P. P. Grassé (1959) has called *stigmergy*. If members of a species, for instance, have an inbred response to encountering a pheromone-drenched ball of mud, then any bug that deposits such a ball is unintentionally sending a message to any others that might pass by. Stigmergy is communication by altering the state of the environment in a way that will affect the behaviors of others for whom the environment is a stimulus. Holland and Beckers distinguish between cue-based and sign-based stigmergy. In cue-based stigmergy, the change in the environment simply provides a cue for the behavior of other actors, while in sign-based stigmergy the environmental change actually sends a signal to other actors. Termite arch-building contains both kinds of stigmergy, with pheromones providing signals while the growing pillars provide cues.

Ant corpse-piling has been described as a cue-based stigmergic activity. When an ant dies in the nest, the other ants ignore it for the first few days, until the body begins to decompose. The release of chemicals related to oleic acid stimulates a passing ant to pick up the body and carry it out of the nest. Some species of ants actually organize cemeteries where they deposit the corpses of their conspeciates. If dead bodies of these species are scattered randomly over an area, the survivors will hurry around picking up the bodies, moving them, and dropping them again, until soon all the corpses are arranged into a small number of distinct piles. The piles might form at the edges of an area or on a prominence or other heterogeneous feature of the landscape.

Deneubourg and his colleagues (Deneubourg et al., 1991) have shown that ant cemetery formation can be explained in terms of simple rules. The essence of the rule set is that isolated items should be picked up and dropped at some other location where more items of that type are present. A similar algorithm appears to be able to explain larval sorting, in which larvae are stored in the nest according to their size, with smaller larvae near the center and large ones at the periphery, and also the formation of piles of woodchips by termites. In the latter, termites may obey the following rules:

- If you are not carrying a woodchip and you encounter one, pick it up.
- If you are carrying a woodchip and you encounter another one, set yours down.

Thus a woodchip that has been set down by a termite provides a stigmergic cue for succeeding termites to set their woodchips down. If a termite sets a chip down where another chip is, the new pile of two chips becomes probabilistically more likely to be discovered by the next termite that comes past carrying a chip, since it's bigger. Each additional woodchip makes the pile more conspicuous, increasing its growth more and more in an autocatalytic loop.

Optimizing with Simulated Ants: Computational Swarm Intelligence

Marco Dorigo, Vittorio Maniezzo, and Alberto Colorni (1996; Dorigo and Gambardella, 1997) showed how a very simple pheromone-following behavior could be used to optimize the traveling salesman problem. Their "ant colony optimization" is based on the observation that ants will find the shortest path around an obstacle separating their nest from a target such as a piece of candy simmering on a summer sidewalk.

As ants move around they leave pheromone trails, which dissipate over time and distance. The pheromone intensity at a spot, that is, the number of pheromone molecules that a wandering ant might encounter, is higher either when ants have passed over the spot more recently or when a greater number of ants have passed over the spot. Thus ants following pheromone trails will tend to congregate simply from the fact that the pheromone density increases with each additional ant that follows the trail.

Dorigo and his colleagues capitalized on the fact that ants meandering from the nest to the candy and back will return more quickly, and thus will pass the same points more frequently, when following a shorter path. Passing more frequently, they will lay down a denser pheromone trail. As more ants pick up the strengthened trail, it becomes increasingly stronger (see Figure 3.4). In their computer adaptation of these behaviors, Dorigo et al. let a population of "ants" search a traveling salesman map stochastically, increasing the probability of following a connection between two cities as a function of the number of other simulated ants that had already followed that link. By exploitation of the positive feedback effect, that is, the strengthening of the trail with every additional ant, this algorithm is able to solve quite complicated combinatorial problems where the goal is to find a way to accomplish a task in the fewest number of operations.

At the Santa Fe Institute and Los Alamos National Laboratory, Mark Millonas and his colleagues have developed mathematical models of the dynamics of swarms and collective intelligence, based on the example of pheromone-sniffing, simple-minded ants (Millonas, 1993; Chialvo and Millonas, 1995). Research on live ants has shown that when food is placed at some distance from the nest, with two paths of unequal length leading to it, they will end up with the swarm following the shorter path. If a shorter path is introduced, though, for instance, if an obstacle is removed, they are unable to switch to it. If both paths are of equal length, the ants will choose one or the other. If two food sources are offered, with one being a richer source than the other, a swarm of ants will choose the richer source; if a richer source is offered after the choice has been made, most species are unable to switch, but some species are able to change their pattern to the better source. If two equal sources are offered, an ant will choose one or the other arbitrarily.

Like Hofstadter, Millonas compares the communications network within a swarm of ants to the highly interconnected architecture of neurons in a brain. Both cases can be described in terms of three characteristics:

- Their structure comprises a set of nodes and their interconnections.

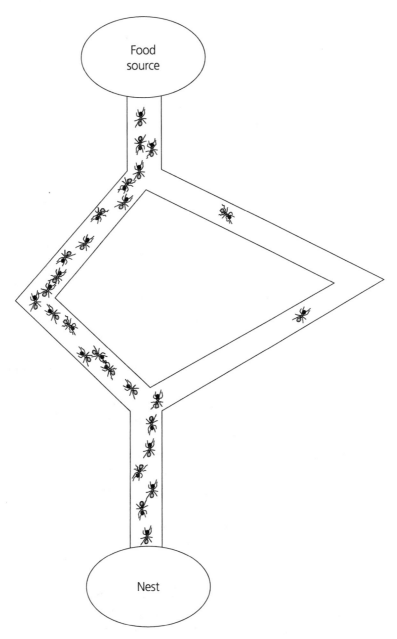

Figure 3.4 Pheromones accumulate on the shorter path because any ant that sets out on that path returns sooner. (Adapted from Goss et al., 1989.)

■ The states of node variables change dynamically over time.

■ There is learning—changes in the strengths of the connections among the nodes.

(This argument is based on a famous paper by Doyne Farmer (1991)—another Santa Fe Institute habitué—depicting "The Rosetta Stone of Connectionism.") Millonas argues that the intelligence of an ant swarm arises during phase transitions—the same transitions that Langton described as defining "the edge of chaos." The movements of ants are essentially random as long as there is no systematic pheromone pattern; activity is a function of two parameters, which are the strength of pheromones and the attractiveness of the pheromone to the ants. If the pheromone distribution is random, or if the attraction of ants to the pheromone is weak, then no pattern will form. On the other hand, if a too-strong pheromone concentration is established, or if the attraction of ants to the pheromone is very intense, then a suboptimal pattern may emerge, as the ants crowd together in a sort of pointless conformity. At the edge, though, at the very edge of chaos where the parameters are tuned correctly, Millonas says, like a pencil balanced on its end that will fall, but we don't know which way it will fall, the ants will explore and follow the pheromone signals, and wander from the swarm, and come back to it, and eventually coalesce into a pattern that is, most of the time, the shortest, most efficient path from here to there.

In these real and simulated examples of insect accomplishments, we see optimization of various types, whether clustering items or finding the shortest path through a landscape, with certain interesting characteristics. None of these instances include global evaluation of the situation: an insect can only detect its immediate environment. Optimization traditionally requires some method for evaluating the fitness of a solution, which seems to require that candidate solutions be compared to some standard, which may be a desired goal state or the fitness of other potential solutions. The bottom-up methods of the insect societies, though, permit no evaluation—no ant knows how well the swarm is doing. In general, the method of pheromone communication means that a more successful path will be somewhat more attractive, with an autocatalytic accumulation of pheromone resulting in the population's convergence on the most-fit behavior—all done at the local level.

In a 1992 presentation to the Workshop on Artificial Life, Mark Millonas cited several "serious" scientific topics that could be illuminated through the study of emergent swarm behaviors, then acknowledged: "In the end perhaps the most pervasive appeal of swarms centers on a kind of emotional attractiveness of the subject . . . More than a paradigm, swarms are almost, at times, an archetype" (Millonas, 1993, p. 418). Our study, too, of swarm intelligence and collection adaptation is motivated in part by the uninformed suspicion there is wisdom to be gained from it, and by the feeling that there is something about the disorderly interactions of dumb actors and their achievements that is just, well, fascinating. It seems that there is something profound and meaningful in these phenomena, something that transcends the compulsive rationality imposed by our intellectual tradition. We are pulled toward the study of these systems with hope that some good result will justify the time spent.

Staying Together but Not Colliding: Flocks, Herds, and Schools

Where ants move more or less randomly around their physical world, some other social animals move about in more orderly ways. Many species of fish, for instance, swim in schools that seem to take on an emergent life of their own. A fish school appears to move as one, with hundreds if not hundreds of thousands of fish changing direction, darting at what appears to be the same exact instant. In 1954, a Smithsonian zoologist named Breder attempted to meet the interesting challenge of contriving a mathematical model to describe the behavior of schooling fishes.

Analyzing empirical biological data, Breder argued that the cohesiveness of a school is a function of the number of fish in the school, the distance between fish, and something he called the "potential" of each individual, which varies with size, among other things. Breder showed that the attraction of a school for a solitary fish (measured by placing a fish in one side of a divided aquarium, with various numbers of fish in the other side where they could be seen, and seeing how much the isolated fish swam on the "social" half of its side) was described by the formula

$$c = kN^t$$

where k and t are constants and N is the number of fish in the other tank.

In one of his examples, $k = 0.355$ and $t = 0.818$. The effect of having an exponent, t, less than 1 is that the attractiveness of the group increases, but the increase becomes less as the group increases in size (Figure 3.5). A larger school is more attractive than a smaller school, but the impact of adding one more individual to the group is more pronounced

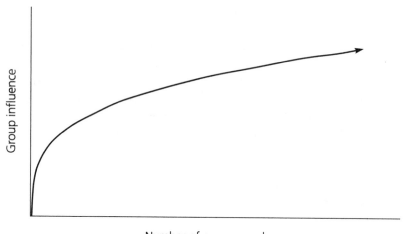

Figure 3.5 As the number of fish—or of people—increases, their effect increases as well, but the rate of increase slows with size.

in a smaller group. The difference between four and five members may be the same, numerically, as the difference between ninety-nine and a hundred, but the effect of adding one to a group of four is much greater in terms of social attractiveness.

Breder's conclusion is interesting indeed in light of a more recent theory of human social behavior. Bibb Latané's social impact theory, conceived in the 1970s and still evolving, resulted from dozens of laboratory and field experiments with human subjects (Latané, 1981). Latané finds that the impact of a group on an individual is a function of the Strength, Immediacy, and Number of sources of influence, that is, other people. Strength is a kind of social influence variable, similar to status or persuasiveness or "potential," Immediacy is the inverse of distance, such that Immediacy increases as distance decreases, and Number is simply the number of people influencing the target individual.

Latané had found, for instance, that the size of the tip left by a group of people in a restaurant is a function of the number of people sitting at the table—the more people in a party, the larger the total amount but the smaller the percentage each pays of the bill. The nervousness of participants in a college talent show was shown to be a function of the number of people in the audience. In these cases and many more, the effect of the number of sources of influence was found to fit a formula:

$$\hat{i} = kN^t$$

where $\hat{i}$ is impact, N is the number of sources of influence, k and t are constants, and $t < 1$. Thus, the impact of a group increases monotonically with the size of the group, but the increase slows with group size—just as it does in fish schools.

It may not be the most flattering self-image for us to view ourselves as flocking, schooling conformists. But the literature of social psychology since the 1930s consistently shows us to be, in spite of ourselves, herding creatures. Whenever people interact they become more similar, as they influence and imitate one another, teach and learn from one another, lead and follow one another. Norms and cultures, and, we would say, minds, are the result. It is not the usual case that humans move in synchronously choreographed swoops and dashes across the landscape, as fish and birds do; human physical behaviors are not flocklike or schoollike, but the trajectories of human thoughts through high-dimensional cognitive space just might be.

We are working toward a model that describes peoples' thinking as a social phenomenon. Thinking differs from the choreographed behaviors of fish and birds in two major ways. First, thinking takes place in a space of many more than three dimensions, as we have seen in our discussions of graphs and matrices, high-dimensional analogues of language, and neural nets. Second, when two minds converge on the same point in cognitive space, we call it "agreement," not "collision."

A very influential simulation of bird flocking was published by Craig Reynolds in 1987. Reynolds assumed that flocking birds were driven by three local forces: collision avoidance, velocity matching, and flock centering. That is, "boids" (Reynolds' name for his simulated birds)

- pull away before they crash into one another
- try to go about the same speed as their neighbors in the flock
- try to move toward the center of the flock as they perceive it

Implementing just these three rules, Reynolds' programs show very realistic flocking behavior, with coherent clusters of boids whirling through the three-dimensional simulated space, splitting to flock around obstacles and rejoining again (see Figure 3.6). Reynolds has gone on to provide animation of herds and flocks for movies, and his simple noncentralized algorithm lies behind many animated cinematic sequences.

The biologist B. L. Partridge showed that fish seem to regulate their schooling behavior using visual information and information from the fish's lateral line. Vision guides the "approach" or flock-centering

Figure 3.6 Reynolds showed that a realistic bird flock could be programmed by implementing three simple rules: match your neighbors' velocity, steer for the perceived center of the flock, and avoid collisions.

tendency; blinded pollock swim farther from their neighbors than do fish with vision. The lateral line, on the other hand, enables the collision-avoiding "repulsive force." Fish with their lateral lines removed swam closer to neighbors than did normal fish. Fish that were either blinded or had their lateral lines removed still schooled, but fish with both sensory systems disabled did not maintain their position in the school at all. Thus there is empirical support for a biological infrastructure corresponding to Reynolds' simple rules, at least in fish.

In a classic 1971 paper called "Geometry for the selfish herd," W. D. Hamilton had proposed one explanation for the *why* of herding, schooling, and flocking behavior, a functional explanation showing what it accomplishes. These social congregations are seen more often in prey than in predators. Hamilton noted that an animal at the edge of the herd is more likely to be picked off by a predator. Thus the "selfish" behavior of any member of the population would be to try to move as close to the center of the herd as possible. The population-level effect of this response, along with some simple rules to minimize collisions and other conflicts, is the organized behavior seen in nature.

In his paper "Flocks, herds, and schools: A distributed behavioral model," Craig Reynolds (1987) had compared his model to a *particle system*. Particle systems are a kind of computer graphics method that comprises large numbers of individual agents or objects, each having its own behavior. Reynolds noted that boid behavior is more complicated than the typical particle system—and less complicated than the behaviors of real birds.

At about the same time, the early 1980s, in less widely cited but more empirical research, University of Rhode Island biologist Frank Heppner observed bird flocks, using a pair of movie cameras placed at 90-degree angles, filming at three frames per second (Heppner and Grenander, 1990). Watching the movies one frame at a time, Heppner could distinguish some features of flocking that had previously not been understood. For instance, he concluded that there was not, in fact, a leader to the flock; any bird could lead a maneuver at any time. The flock maintained a kind of dynamic equilibrium without any central control.

Heppner had also been experimenting with a computer program implementing Conway's Game of Life and noted that the spontaneous behaviors of gliders in that program bore a strong resemblance to the spontaneous order that arose in the bird flocks he was studying. He assembled a small team including himself, applied mathematician Ulf Grenander, and an undergraduate programmer named Daniel Potter, and they began investigating the possibility that complex bird flock maneuvers could emerge from programs of very simple rules. The result was a graphical display of artificial bird flocks whose coordinated movements were startlingly similar to the real thing.

The rules that Heppner and his colleagues implemented were somewhat similar to Reynolds', but with some differences. Heppner too implemented an "attractive force" rule, which attracted birds toward one another unless they were too close, in which case they were repelled. Unlike Reynolds' boids, Heppner's were attracted to a roost; the closer they got to it, the stronger the attraction would become. Heppner's birds had a tendency to maintain a target velocity, and finally, birds were occasionally knocked off course by a random force, similar to a gust of wind. Again, the result was a very believable, simple, decentralized simulation of bird flock choreography.

Some of our own earliest experiments were derived from approximations of Reynolds/Heppner bird flocks; in fact, Heppner's roost idea provided an inspiration that led us deep into the study of swarm intelligence (Kennedy and Eberhart, 1995). In our very first experiments, populations of "birds" flew in orderly flocking patterns. If birds could be programmed to flock toward a roost, then how about having them look for something like birdseed? It seems impossible that birds flying hundreds of feet in the air could see something as tiny as seed on the ground—but they are able to find it. A flock of birds might fly over a neighborhood watching for signs of edibles—and importantly, those signs might come through observing other birds eating or circling in on a target, or in seeing another member of the flock turning back and descending toward something that it might have seen. These are social cues of a different order from seeing the food substance itself, seeing somebody see something.

The flock does not know where the seed is, but responds to social signals by turning back, flying past the target, circling around, spiraling in cautiously until the birds are sure they have found food in a safe place. In one of our first publications on the topic of the particle swarm algorithm, we called the function optimum a "cornfield vector," suggesting that the population members have a motivation to find the hidden point. It is like finding something in a landscape where the searcher gets clues from other searchers.

Most recently, physicists Toner and Tu (1999) published a mathematical analysis in a paper with the familiar-sounding title, "Flocks, herds, and schools: A quantitative theory of flocking." Their model analyzes the collective movements of large numbers of self-propelled agents or organisms. They note that flock simulations have several features in common. First, a flock is made up of a large number of individuals attempting to move in the same direction as their neighbors. Second, individuals (these authors also call them "boids") respond only to their nearer neighbors. Third, stochastic noise—randomness—is introduced into the individuals' movements, so that flocking is never perfect. Finally, "rotational symmetry" is maintained, that is, the flock is equally likely to move in any direction.

Toner and Tu assert that one of the most interesting things about flocks is that they never attain equilibrium. The behavior of a flock resembles a Type 4 cellular automaton, continuing in characteristic but unpredictable patterns that extend for long periods of time. (Our particle swarms described later in the book seem to share this attribute with flocks.)

In flocking simulations the important thing to simulate is coordinated movement of the organisms, whether flocks, herds, or schools. Some motives for studying such a topic include the desire to understand biological aspects of social behavior and the wish to create interesting and lucrative graphical effects. There is another, more pragmatic, reason to learn about coordinating movements with simple decentralized rules, and that is to design and develop robot societies.

Robot Societies

The stereotypical movie robot is an android with a voice-recognition interface and a sardonic sense of humor. In fact the most common type of robot is quite a bit less exotic than that, often built out of Legos, with a couple of motors and some big cables that need to be held up by a graduate student who follows the robot around the room. It will be worthwhile for us to sidetrack for a moment and talk about robot intelligence, since some of the most important work in swarms and swarm intelligence is being conducted in that field.

In the early days of electronic computing, robots were a perfect laboratory for the implementation of artificial intelligence, and the classic robots were based on the paradigm that has come to be known as GOFAI (Good Old-Fashioned Artificial Intelligence), which is practically synonymous with the symbol-processing paradigm. Robot intelligence was presumed to operate—in fact, human intelligence was presumed to operate—through the manipulation of symbols. Presumably, the same rules of logic should operate on any set of symbols, no matter what the subject domain. Generally, a robot or human mind was presumed to contain a *central executive* processor; this was presumed to include a kind of *inference engine* that operates on data that have been prepared by a perceptual system, and the executive's output was sent to effectors, or motor devices, which were able to act upon the world.

Given the assumptions of cognitive science and the profound vanity of humans generally, it is not surprising that scientists would assign the will (central executive) to a central role in an artificial mind and implement a system of information processing that followed directly from the classical academic philosophies of the universities, and before that, from the pronouncements of the Church. The early researchers had no way of knowing that scientific psychology would fail to support their intuitions about human nature and mind—remember that, before the "cognitive revolution" of the 1960s and 1970s, almost all scientific psychology was behavioristic, dismissing everything mental as unworthy of scientific study. Thus there was not a well-developed understanding of things cognitive.

In robotics, it turned out that the symbol-processing approach just didn't work out very well. As the robot's world became more complicated, the robot's mind became more complicated; it had more symbols to retain. Worse, maintenance of complicated chains of logic meant that it had to figure out that when one fact changes, other facts have to change, too. It is not trivial to deduce which ones. For instance, if you are told that someone has just turned out the lights, *you* know immediately that they can't see where they're going. A symbol processor has to figure it out. GOFAI robots might stand in one spot for 10 minutes contemplating their next step, trying to figure out whether they would bump into something.

In a highly readable paper called "Elephants don't play chess," MIT researcher Rodney Brooks (1991) proposed an alternative to GOFAI (a tradition with deep roots at MIT, by the way). Brooks argued that symbols must be *grounded* in a physical reality if they are to have any meaning. Brooks' view solved some computational problems that had made real-world robotics intractable, and the successful results seen in Brooks' laboratory lent support to his conjecture.

Brooks' robots' minds are organized according to a principle called the *subsumption architecture*. A subsumption architecture is built from the bottom up; simple robot behaviors are developed, and then these are fitted loosely together. They remain independent of one another, each module doing its specific part without consulting the other. For instance, Brooks' soda-can-finding robot, Herbert, has a collision avoidance module that sends a signal to turn it when it comes too close to something it might bump into, and a can-pick-up module to pick up the cans when it finds them. Herbert can pick up cans whether it is or is not avoiding anything, and can avoid things without picking up cans, or do neither, or do both. The modules are independent. Further, both modules are grounded in the world; that is, avoidance happens when there is something in the world to avoid, and picking up happens when a can is detected in the world. There is nothing like a central executive control.

The quotable Brooks has said, "The world is its own best model. It is always exactly up to date. It always contains every detail there is to be known." Rather than maintaining a symbolic representation of the world, Brooks' robots simply respond to the world as it presents itself to their sensors. The effect of this is that computation can be reduced to a tiny fraction, perhaps 1 or 2 percent, of what was required by previous robots, programs can be smaller, storage and memory requirements are reduced, and as a result smaller, smarter robots can be built more cheaply.

Brooks' robots are able to navigate messy scenes. For instance, Herbert roams around cluttered laboratory offices looking for empty cans, which it picks up and throws away. No set of rules could prepare a symbol processor for all the kinds of complexities that might exist there, for instance, tipped-over chairs, wadded-up papers—tipped-over graduate students, for that matter. The bottom-up subsumption architecture, though, is able to adapt to whatever conditions it finds.

Tom and Jerry were two identical robots that were programmed to interact with one another. They were given an urge to wander about, that is, to perform random actions, a repulsive force that kept them from bumping into things, and were programmed as well with an attractive force, to make them move toward distant things. They also had a motion detector sense that allowed them to detect moving objects—most likely one another—and follow them: a possible genesis of robot sociality.

Construction of subsumption architectures is bottom up: it is incremental; small parts are constructed and then fitted together in a way that Brooks likens to evolutionary change. Communication among various modules is minimal, with tight coupling between modules and the world. The hallmark of such an intelligence is that it is decentralized, a theme that should be getting familiar by now. In the subsumption architecture, behaviors are separate, each one defined to accomplish one small, distinct goal, yet someone watching the robot feels that there is an intentional unity about the way it moves. Wheels are programmed to stop when the robot approaches a possible soda can; arms are programmed to reach up when the wheels stop turning; a hand is programmed to grasp when the light beam between fingers and thumb is broken—an observer sees the robot go to the table, reach out, and pick up a can. Oddly, goal-directed behavior emerges, or so it seems, from a sequence of fixed action patterns linking perceptions to actions. Pushes look like pulls through the prism of our biological brains.

Brooks' model brings up a new facet of this swarm business. We note that the MIT robots are *autonomous:* they are self-contained and self-sufficient—with an occasional human helping hand. But inside the robot's mind there is something of a swarm of modules, you might say, a multitude of subroutines that perform particular functions.

Marvin Minsky, another MIT veteran, calls such a cognitive modularity a "Society of Mind" (Minsky, 1985). Unfortunately, where Brooks makes it clear that he is talking about robots, man-made minds, and simple ones at that, Minsky takes the model of AI modularity and tries to use it to explain the workings of the human mind. Minsky's coffee-table book, *Society of Mind,* contains 270 essays, each one packaged as a pearl of wisdom or aphoristic reading on a psychological or cognitive topic. The theme of the book is that mind is composed of a "society" of specialized functions or "agents" that operate more or less independently of one another. Unfortunately, the subsumption approach that is so successful with robots has no basis in psychology—though there is evidence of modularity in the organization of the brain, the metaphor of a society of cognitive agents is exactly the kind of reified fiction that scientific psychology rejects. There is no homunculus inside the head, never mind teams of homunculi.

Near the end of his book, Minsky summarizes with the statement: "Minds are simply what brains do." The statement is concise and superficially plausible. It seems to do away with the mind-body problem in one slicing phrase and reduces much of Western philosophy to dry chattering. But the statement is obviously incorrect. There is doubtless a relation between brains and minds, but there are very many brains that don't do minds at all, for instance, bird brains, and bug brains, and sleeping brains, and fetus brains, and damaged brains—dead brains definitely don't do minds. There are also lots of things that brains do that are not mind: having seizures, controlling digestion, growing, metabolizing oxygen, hemorrhaging. Further, there doesn't seem to be any reason that minds require brains—we can imagine aliens having minds, and there is a strong enough case for computers having minds, at least potentially, that Minsky's statement just can't be true in an absolute sense.

Even worse, the statement—and the cognitivistic philosophy that supports it—assumes that a brain by itself, we'll even grant that it is a healthy human brain, would develop into a mind, that it would become conscious and learn to think (which we take to be minimum requirements for something to be called a "mind."). This ignores all social science research for the past hundred years. In order for a brain to become mental, it needs to have interaction with other minds. A fetal brain grown to adulthood in a vat containing nutritious fluids would never acquire the qualities of mind. If it were transplanted into a Frankensteinian body, it would be unable to reason, remember, categorize, communicate, or do any of the other things that a minimal mind should be able to do. Feral humans, that is, humans who have grown up in the wild, have never been known to show signs of having anything like a mind—it is not sufficient to have perceptual stimuli; social ones are necessary. The relation between mind and society is unquestionable, but is ignored by Minsky's oversimplifying, where the "society" is moved into the cavern of the skull.

We would have thought the issue was more or less settled in the research debate between Ernest Hilgard, a cognitivist, and Nicholas Spanos, a social psychologist, in a fascinating series of investigations into the nature of hypnotic responding, multiple personalities, and other forms of psychopathology. Hilgard's neodissociation theory (e.g., 1979, 1977) postulated autonomous cognitive processes inside the mind, which under certain circumstances separated and acted independently from the self. For instance, in hypnosis, cognitive subsystems could come under the control of a hypnotist. Hilgard's mechanistic model resembled Minsky's society of mind, with agentic substructures expressing autonomous volition beyond the reach of the individual's self. Spanos (e.g., 1982, 1986), on the other hand, argued that individuals were simply acting in ways they believed were appropriate for situations labeled "hypnosis," "multiple personality," and so on. Anyone can act like they have multiple personalities, but only a few people convince themselves. According to Spanos' sociocognitive theory, hypnosis and certain forms of pathological responding were manifestations of normal behavior; no internal systems or "agents" were invoked to explain behaviors. Hypnosis, in this view, is something the person does—not something that happens to him. Where Hilgard's experiments would seem to show that a hypnotist could communicate with different autonomous subsystems of the hypnotized person, for instance, a "hidden observer" within the subject, sealed off from the rest of the cognitive system, Spanos would replicate the experiments under slightly different conditions to show that the hypnotized individuals were just doing what they thought they should be doing. The neodissociationists' internal special processes were shown to be fictitious, theoretical constructions with no basis in reality. Again and again Hilgard and his colleagues attempted to prove the existence of cognitive substructures, and repeatedly Spanos' case for strategic role enactment prevailed. Nothing is gained in thinking of the mind as a "society" of interacting agents.

There are some differences between multiagent systems such as the subsumption architecture and swarms such as are being described here. In swarm intelligence the global behavior of the swarm is an emergent effect of the local interactions of swarm members; in a system of multiple autonomous agents, however, the system's behavior is more nearly a *sum* of the agent's contributions. The effect may strike a human observer as being surprisingly well choreographed, as if the robot were pursuing a goal, but that is only a matter of attribution by the observer. In fact the behaviors are exactly what they have been programmed to be, and the cause of the surprise is anthrocentrism, not emergence. Further, individuals in a swarm are usually relatively homogeneous, while "society of mind" or subsumption modules are assigned to specialized tasks. One ant is the same as the next; one bee does pretty much the same thing as the next bee. Grasping is very different from looking, which is different from rolling.

Rodney Brooks has expressed the opinion that robots will continue to get smaller and smaller; indeed the simplicity of the subsumption architecture enables such a miniaturization. For instance, he says, imagine a colony of "gnat robots," living in your television, that come out when the set is off and dust the screen. Indeed, the stage is set for such developments; it appears likely, though, that the miniature robots of the future will probably not be seen as autonomous agents, each one fully empowered to accomplish whole tasks. The microrobots of the future will be swarms.

On an even smaller scale are the nanotechnology robot swarms envisioned and described by Kurzweil (1999). Large assemblies of the constituent nanomachines will be able to create any desired environment. One example of nanorobot swarms is described by Hall (1994, 1995), in which each nanorobot, called a Foglet, is a cell-sized device with 12 arms pointing every which way, each with a gripper. A large assemblage of these devices can not only cooperate to form large structures, but also form distributed intelligence. But let's come back to the present for the moment.

Toshio Fukuda's laboratory in Nagoya, Japan, is a day-and-night bustle of graduate students, postdocs, assistant professors, and engineers, planning, designing, and building robots of many types. For Honda Corporation, Fukuda has built a life-size android robot with two legs, two arms, a trunk, and a head, balancing and walking with a humanlike gait. If you shove this robot backwards, it will flail its arms and catch its balance just like a person would. It's a lucky thing, too, since it weighs nearly 600 pounds. Fukuda's lab also produces robotic gibbons, two mechanical arms and the trunk of a body; a gibbon is able to train itself to swing from limb to limb in an artificial jungle with both overhanded and underhanded brachiation movements. When a gibbon's hand is knocked from its branch, the robot swings incrementally higher and higher until it can reach the branch again—just as a real gibbon or monkey would do. In lectures, Fukuda likes to show a video of a gibbon in a jungle setting and say, "That's not a robot, that's a real monkey." In the next scene a robotic gibbon is depicted, with uncanny resemblance to the live one. Fukuda says, "That's a robot" (Fukuda, 1998).

Fukuda's lab is developing robot swarms. Where Brooks' subsumption methodology took the cognitive executive control out of the picture, decentralizing the individual robot's behavior, Fukuda's robots replace the individual's control with reflex and reactive influence of one robot by another. These decentralized systems have the advantages that the task load can be distributed among a number of workers, the design of each individual can be much simpler than the design of a fully autonomous robot, and the processor required—and the code that runs on it—can be small and inexpensive. Further, individuals are exchangeable. We think back to 1994 when the NASA robot named Dante II ambled several hundred feet down into the inferno of Mt. Spurr, a violent Alaskan volcano, to measure gases that were being released from the bowels of the earth. The 1,700-pound Dante II toppled into those self-same bowels and would now be just an expensive piece of molten litter if helicopters bearing earthlings had not saved it—for instance, if the accident had happened on another planet, which was what Dante was being developed for. If, instead, a swarm of cheap robots had been sent into the crater, perhaps a few would have been lost, but the mission could have continued and succeeded. (Dante II was named after Dante I, whose tether snapped only 21 feet down into Mount Erebus in 1993—another potentially expensive loss.)

Just as knowledge is distributed through the connections of a neural network, so *cellular robots,* as they are called, might be able to encode representations through their population, communicating as they investigate diverse regions of the landscape—a real, three-dimensional landscape this time. The concept of cellular robotics, as Fukuda tells it, is based on the concepts found in E. O. Wilson's research on insect societies, where global intelligence emerges from the local interactions of individually unintelligent beings. Researchers R. G. Smith and R. Davis have described two general kinds of cooperation that might take place when such a population tries to solve a problem. *Task sharing* occurs when individuals share the computational load for performing the subtasks required to solve the problem, while *result sharing* comprises a form of cooperation in which the individuals share the results of their individual tasks. Fukuda and his colleagues argue that both types of cooperation are probably necessary for societies of real robots to solve problems in the real world.

Robots are different from pure software agents in one obvious but important respect: they must follow the laws of physics in the three-dimensional world. They have to deal with collisions somehow; the locomotion of cellular robots in physical space in most implementations is programmed as some variation on Reynolds' three-rule boids algorithm, where individuals are attracted to the group, repulsed by too-near neighbors, and attempt to match velocities with neighbors.

For research purposes, it is certainly less expensive to write a program that displays pictures of a population of robots on a computer screen than to make real robots, especially when there are a great number of them. Some of Fukuda's studies in self-organizing societies of robots are in fact implemented with software robots; by programming appropriate physics into their simulations, researchers can get the bugs out of their robots before the cost gets high.

At Georgia Tech, Jennifer Hodgins and her colleagues experiment with software robots, programmed with full simulated real-world physics (Brogan and Hodgins, 1997; Hodgins and Brogan, 1994) . Their standard simulated robot, for some reason, is a one-legged hopping robot, something like a bowling-ball/pogo-stick combo. The simulated robot moves by leaning and hopping. The steeper it leans the farther it hops.

Hodgins' algorithm for group behavior has two parts: a perception model and a placement algorithm. The perception model lets the robot scan its simulated environment. For instance, if there are, as in one of her reports, 105 robots, and each one needs to avoid colliding with its neighbors, then they must be able to tell who is nearby. In a simulation this means the program has to loop through each of the robots one at a time, and for each one calculate the distance to all of the others, in order to determine which ones are nearest—a total of 105^2, or 11,025, calculations at each time step. This has been called the problem of *N-squared complexity,* and it is computationally very expensive; the movements of the robots will be slowed as the computer is busy calculating neighborhoods. In fact this is a good reason for Brooks to say that the world is its own best model. While a computer has to work very hard to tell individual agents what is around them, an embodied agent only has to sense what looms large, or sounds loud, or produces a fast echo, letting the world do its own computation. Hodgins uses some heuristics to reduce the number of calculations required, and she does report that on a powerful Silicon Graphics computer the simulations go faster than real time—so the problem is not technologically insurmountable.

The placement algorithm calculates a desired velocity, based on the positions and velocities of nearby neighbors. Hodgins' one-legged hoppers are generally able to avoid running into one another, but she reports that they do sometimes collide and are sometimes unable to miss obstacles in their paths. Hodgins has conducted similar experiments with "point masses," that is, populations of particles with no physical properties, and reports that they are more successful at coordinating their movements. This might be because the hopping of one-legged robots introduces somewhat erratic trajectories that are hard to coordinate. Hodgins and her colleagues have also run populations of simulated bicycle riders, including in their programs all the physical aspects of balancing, elasticity of objects, friction, and so on. The human rider is modeled as a rigid body of 15 segments with movable joints, and the bicycle is likewise realistic.

Kerstin Dautenhahn (1998, 1999), a German researcher now in the UK, has been conducting an intensive program of investigation into many aspects of socially intelligent computational entities, including robots, virtual robots, and lifelike software agents. Her robots interact with humans and with one another, often learning and adapting during the course of the social interaction. Dautenhahn concludes that *embodiment* is central to social behavior and to intelligence generally. Her research is centered around a set of hypotheses that really amount to a manifesto (Dautenhahn, 1999):

1. Life and intelligence only develop inside a body,
2. which is adapted to the environment that the agent is living in.
3. Intelligence can only be studied with a complete system, embedded and coupled with its environment.
4. Intelligence is linked to a social context. All intelligent agents are social beings.

She further asserts that social robotics can be characterized by these statements (Dautenhahn, 1999):

1. Agents are embodied.
2. Agents are individuals, part of a heterogeneous group (the members are not identical but have individual features, like different sensors, different shapes and mechanics, etc.).
3. Agents can recognize and interact with each other and engage in social interactions as a prerequisite to developing social relationships.
4. Agents have "histories"; they perceive and interpret the world in terms of their own experiences.
5. Agents can explicitly communicate with each other. Communication is grounded in imitation and interactions between agents;

meaning is transferred between two agents by sharing the same context.

6. The individual agent contributes to the dynamics of the whole group (society) as well as the society contributing to the individual.

Dautenhahn allows that the term "agent" can be used to refer to computational, mechanical, or biological entities—as long as they have some degree of autonomy and can interact with their environment. Robotic and biological agents of course do have a physical body and are embedded in an environment that they can interact with, but what about software agents? Of course it is possible to computationally simulate an environment and some corporeal structure, and if the simulation is thorough and precise, then, as with other simulations, the modeled events might perfectly well mimic real physical events. But isn't this cheating? If we have defined our territory in such a way that having a body is a necessary part of it—well, does a simulated body count? Are function inputs and outputs the same as perceptions and motor activities? Dautenhahn asks the questions and proposes an answer: simulation allows experimentation with alternative embodiments. What aspects of the physics of the real world are necessary for the production of intelligent behavior? What if we stripped out some qualities—for instance, what if collisions were allowed, and beings could interpenetrate? What if movement was allowed in three or four or more dimensions? These kinds of questions can only be answered in experimentation with simulated beings.

Dautenhahn has proposed that social intelligence be defined as "the individual's capability to develop and manage relationships between individualized, autobiographic agents which, by means of communication, build up shared social interaction structures which help to integrate and manage the individual's basic ('selfish') interests in relationship to the interests of the social system at the next higher level." She contrasts this kind of social intelligence with swarm intelligence, in particular the kinds of work described above with insect behavior, in which individuals tend to be more or less the same and interactions among them are primitive. She explains that nature has produced two kinds of social models, which she calls "anonymous" and "individualized" societies. Social insects do not keep track of one another as individuals, but only as group members, while members of individualized societies will search for a missing member.

As researchers in the field of robotics search for ways to create a social milieu among their man-made hardware agents, they are forced to work within the constraints of the three-dimensional physical world. The most important communications have to do with the individual agents' relative positions in space and coordination of movements. When we think of mental life in a social context, though, we find that, while some of our communications and thoughts have to do with coordination of movements in physical space, a preponderance of them have to do with seeking agreement on beliefs and attitudes. Agreement means that two individuals occupy the same point in a cognitive space; obviously collision is not an issue here, but something to be desired.

Shallow Understanding

The cognitivistic view of mind assumes that social interaction is a kind of interface between persons and that the important information processing occurs inside the participants. Though there has been a considerable amount of research on *natural language processing,* this research has mostly focused on development of interfaces that would allow users to communicate with the computer in a natural way, without needing to rely on precise and arbitrary commands and hard-to-comprehend outputs. The "deep" processing is performed through logical operations on symbolic representations in the computer's native mode. A user's inputs are translated into commands to the operating system, and then again into binary machine language: zeroes and ones. A message is interpreted by the receiver according to its own inner "deep" structures and constraints.

An interesting minor tradition in artificial intelligence shows the potential, though, for a "shallow" view of language processing and interaction between humans and computers. *Chatterbots* are programs whose specialty is real-time verbal interaction with human users. Some chatterbots are connected to knowledge sources such as databases on particular topics, but their real strength, what makes them different from any other database-querying program, is their ability to deliver natural-sounding responses to questions and conversational comments. The most commonly seen chatterbot application is implemented in a chat room or MUD environment, where the bot might show newcomers around, answer questions and provide information, strike up conversations with people, or make a general nuisance of itself.

Recall that the Turing test defined computer intelligence in terms of social interaction; a computer that could interact with a human in such a manner that the user could not tell whether it was a machine or a person was considered to be able to *think,* according to Turing. This classic definition conspicuously fails to mention aspects of "deep processing," but rather focuses on lifelike, anthropomorphic communication. We suspect that this was a profoundly insightful move on Turing's part. It is part of our positive self-image, especially for us intellectual types who play with sophisticated toys, to see our minds, our understanding of the world, as something deep, mysterious, profound. Computer programs that process complex databases of information, drawing profound conclusions, proving theorems, finding logical contradictions in convoluted inferential arguments—these seem "deep." Programs that can make small talk and silly jokes do not seem deep; we have heard them referred to as "shallow AI." They are also a lot harder to write.

The original chatterbot was ELIZA, a famous psychoanalysis program described in a 1967 paper by Joseph Weizenbaum (Weizenbaum, 1967, 1976). Named after Eliza Doolittle from the musical *My Fair Lady,* ELIZA was intended to model the behavior of a Rogerian therapist interacting with a client. Carl Rogers' "client-centered" method of therapy worked by forming an empathic bond between client and therapist through unconditional acceptance of the client's views. The therapist was not supposed to offer his or her own opinions, but rather encouraged clients to express their feelings and thoughts freely. An important tactic for the therapist was to reflect or rephrase the client's statements. For instance, if the client says, "I can't stand my kids," the therapist might respond, "You say you can't stand your children?" Occasionally the therapist guides the discussion by asking the client to elaborate, but the whole point is to talk about what the client wants to talk about.

In programming ELIZA, Weizenbaum capitalized on the nondirected nature of Rogerian therapy. ELIZA does not offer much in the way of comment or opinion or information, but does zero in on psychologically relevant words and draws the client to explain further. She might change the wording of a sentence around or turn statements into questions, but there is no attempt anywhere in the program to "understand" what the user is saying—there is no "deep" processing of information, only shallow processing of the surface structure of language.

In fact, most of what ELIZA does is scan the input text string for key words and then look up her canned response to those words. If it is easily possible to invert a sentence, to turn a statement into a question, or a question into a statement, then she will do that. For instance, one LISP version of ELIZA has a rule written as

```
(((?* ?x) my mother (?* ?y))
    (Who else in your family ?y))
```

This rule looks to see if the user's input has the text string "my mother" in it. Whatever is in the sentence before the target string is assigned to the variable *x*, and whatever comes after it is assigned to *y*. ELIZA then replies with a question that begins with "Who else in your family" and ends with whatever was just assigned to *y*. For instance, if the user types, "My mother was a wild-eyed Gypsy dancer," the program will respond with, "Who else in your family was a wild-eyed Gypsy dancer?" Thus the interaction is purely, simply, absolutely, and unapologetically superficial. ELIZA makes no pretense of understanding the user's input, yet her responses are, a great deal of the time, perfectly appropriate. The program itself is only a hundred or so lines of code, with another hundred lines in a data set of responses and targets to match.

The Loebner Prize is a monetary award given annually to the designer of a computer program that best performs in a Turing test. A bronze medal and $2,000 are awarded to the "most human" computer, based on the ratings of a panel of judges. Multimillionaire Hugh Loebner has promised an award of $100,000 and a gold medal to the designer of a computer program whose responses are indistinguishable from a human's.

Judges sit at a computer and interact with the programs through a keyboard, asking it questions, giving it information, and freely probing its ability to respond in a humanlike way to their inputs. Though there is some complaining about the procedure, for instance, that no program could have sufficient knowledge about every kind of thing a judge could ask, it does turn out that some of the programs produce a believable interaction script.

The winner of both the 1998 and 1999 contests was a program called "Albert," produced by a programmer in northern Georgia named Robby Garner. Albert is an impudent character who convinces the user he is human through sheer personality. If there is a lull in the conversation, Albert will throw out something, like, "Have you seen any good movies lately?" Albert can be programmed with an encyclopedia of knowledge that can be used to answer users' questions. He also keeps track of previous information the user has entered; for instance, he never seems to forget a name.

Here is a snippet of conversation about *Star Trek* that Albert had with a judge in the competition:

Judge05: What is the Cage?
Program: The United Federation of Planets.
Judge05: No it isn't.

Program: Yes it is.

Judge05: No it's not.

Program: Oh, okay.

Judge05: Stop arguing.

Program: I'm not arguing.

Judge05: Who is your favorite character?

Program: I think Spock is my favorite.

Judge05: Why do you like Spock so much?

Program: He is just so cool.

Judge05: Do you like Scotty?

Program: Yes. He's always there when you need him.

Judge05: What does Scotty say?

Program: I believe he likes scotch and maybe brandy on occasion.

Judge05: What is Scotty's job?

Program: You're kidding, right?

As you can see, Albert's ability to seem human comes largely from his skill at changing the subject and making disingenuous comments in place of following the train of a conversation that has veered outside his realm of knowledge.

This question of deep understanding versus superficial reacting is a tough one. In trying to measure understanding, we are simply sure to fail. Except under extraordinarily confusing circumstances, any time you ask someone if they understand what is happening they will say yes, they do. Even if it is proven that they had an entirely opposite interpretation of the context from everyone else, they will have had *some* interpretation, and to them that constituted "understanding." Imagine if, to ask whether a computer program understands us, we just programmed it to answer "Yes" to the question, "Do you understand?" That would very well mirror the situation as it stands with humans. (Without throwing in our two bits about Searle's Chinese Room, we can suggest that most people don't *really* know what's going on, most of the time.)

Harvard psychologist Ellen Langer (1989) has shown, in a large number of informative experiments, that a great amount of human behavior results from mindless processing of superficial information—we really can't think deeply about everything that happens to us. And even when we do, we often seem to apply socially sanitized methods of processing information. Psychologists have had a very hard time finding examples of real logic being used in human cognition.

We seem here to be arguing in favor of stupidity. Of course that's not the point. The point is that much thinking might not be best character-ized as "deep" information processing. Perhaps it is better thought of as a relatively shallow process of imagined social interactions. From this view, the power of human intellect will not be found hidden away in the mystical recesses of the brain, but rather should be sought in the open, in the links between people.

Agency

What is an agent? We have seen some examples, for instance, in Minsky's societies of agents, Hodgins' bowling-ball/pogo-stick simulated robots, and chatterbots. Some have argued that software agents are simply func-tions in a computer program, or even computer programs themselves, and that is indeed literally true—they are that. But the idea of an agent, especially an autonomous agent or an intelligent agent, supposes some-thing more; it supposes that there is something agentic about the pro-gram. It seems to have something like a will, or an ability to make judg-ments or decisions; it's able to *do* things—and not just what it's told, either.

Stan Franklin and Art Graesser (1996) of the University of Memphis have provided a taxonomy of autonomous agents as a way of under-standing what is—and what is not—an "agent." They ask how an agent differs from other computer programs, and after sifting through all the discussions on the topic, they conclude, "An autonomous agent is a sys-tem situated within and a part of an environment that senses that envi-ronment and acts on it, over time, in pursuit of its own agenda and so as to effect what it senses in the future" (Franklin and Graesser, 1996). The most interesting thing to us about autonomous agents is the tendency we have to anthropomorphize their behavior, to ascribe such things as "sensation," "action," "understanding," and even "agendas" to them. In writing autonomous agent programs, developers attempt to capture some of the attributes that make a human useful, and in so doing they necessarily build on a foundation of their own preconceptions about what people are, how they work, what motivates them.

Social psychologists have long studied questions about the attribu-tion of causality, especially distinguishing between internal and external causes, that is, whether a behavior was caused by something about the person or about the situation, and between stable and unstable causes, mainly between dispositional traits of a person and short-lived responses to situations. Causal agency, intention, free will, consciousness—these are characteristics that are attributed to persons, they are part of our explanation of the world, but can never be demonstrated to exist in any real sense.

From the social science viewpoint, the self is theorized to arise from the individual's participation in a social milieu (James, 1892; Cooley, 1902; Mead, 1934; Goffman, 1959; Baumeister, 1982; Schlenker, 1982). Constructing an identity requires believing our own explanations for what has caused us to act the way we have, and requires us to construct a model of the world that is believable and that justifies our behavior or casts it in a positive light.

In light of chatterbots and autonomous agents, we can consider human thought and interaction from the perspective of *role theory,* where the individual's objective is positive self-presentation, and the function of covert information processing is to support that presentation. From the role-theoretic view, the individual is concerned *mainly* with the interface between self and other. Roles provide us a way to interpret interpersonal actions. Let's say, for instance, that a cognitive psychologist conducts a memory experiment with a sophomore student. From the second she walks in the door, the student is trying to figure out what the experiment is really about and what she can do to make a good impression. She then grants the experimenter's request that she look at a series of words on the computer monitor, and makes every effort to commit them to memory, assuming that she is expected to do this by repeating them silently to herself, over and over again. When the experimenter asks her to recall the words, she wonders whether she should give them back in order—or would that be too obvious and boring? If she wants to make a good impression on the experimenter she will do her best to tell him what words she saw. Thus, the researcher (who is working on a paper with the preliminary title "Proactive interference effects of abstract and concrete words in implicit memory") is measuring the extent to which the participant will comply with his expectations—the strength of the connection between the two of them. According to the role-playing perspective, cognitive mechanisms are only useful as they serve the individual's self-presentation and self-identity. The role-theoretical perspective would predict that individuals retrieve and report different information depending on who makes the request, and in fact that is what is seen— schemata are primed by context.

According to neuroscientist Leslie Brothers (1997), evidence about the evolution of language suggests that it may have come to us as an extension of grooming behaviors seen in primates. Picking parasites from one another's fur is a very important social activity in most species of primates, allowing them to show affection and respect, apologize for aggression, and strengthen alliances. Though grooming confers some small survival advantage by getting rid of parasites, this advantage clearly does not justify the amount of time spent at it. The same can be said for human conversation: the amount of information conveyed in conversation rarely justifies the amount of time people spend talking with one another. The primary purpose of both grooming and language is to establish and maintain social relationships. We share verbiage with others as a display of respect and affection. Our perspective in this book places interpersonal activities in the center of human affairs, not at the periphery.

Summary

We have seen many ways in which individuals interacting according to simple local rules can produce complex and adaptive social patterns of behaviors. Physical constraints limit certain kinds of coordinated behaviors, in particular those that require collision avoidance, and in many systems it is adaptation to physical constraints that results in the interesting behavior. For instance, flocking and schooling, those remarkable group dynamisms, result when a repulsive force is added to the force that attracts individuals toward the center of the aggregation. The repulsive force (so-called, of course it is not a force but a behavioral disposition) is simply an adjustment to prevent collisions, but its inclusion in the trajectory algorithm results in a surprising emergent display of distributed choreography.

Schools, flocks, and herds benefit from the aggregation of individuals, but the advantage of the coordinated movement is mainly that it minimizes the occurrence of contusions, abrasions, and loss of scales, feathers, and fur. The simple fact of their numbers increases each individual's safety from predators, improves their luck at browsing, and bestows other benefits as well. The choreography, though, only keeps them from bumping into one another.

The behavior of the social insects is, on the face of it, simpler, less sophisticated, random, aimless. Yet there are many examples of impressive accomplishments emerging from the simplest rules imaginable. Swarms of insects are not concerned with collision; they routinely climb on and over one another. The choreography itself is not pretty, but the results, whether arch building or larvae sorting or finding the shortest path to a goal, begin to suggest what we would call the emergence of a collective intelligence.

In the next chapter we explore the paradigms of evolution as they have been adapted for computational intelligence in computer programs. We consider four main paradigms: genetic algorithms, evolutionary programming, evolution strategies, and genetic programming. In each paradigm, populations of problem solutions are allowed to evolve over time, breeding the best together to capitalize on evolutionary selection. We will see that evolution in a computer can solve problems that often prove to be intractable otherwise.

chapter
four

Evolutionary Computation Theory and Paradigms

Previously in this book, we have discussed mind and evolution, which have been described as the two great stochastic systems in nature. As if to support Bateson's comment, these two great systems have provided some of the most exciting challenges in the history of computer science. Modeling the information-processing techniques of minds, of course, was the task of the artificial intelligence movement. Modeling the ponderous adaptive processes of natural evolution has been an equal challenge, and the rewards have been tremendous. The several evolutionary computation paradigms have provided insights into the workings of nature as well as a toolbox for engineers and

others who need to solve extremely hard, often poorly specified, problems.

Evolutionary computing paradigms are intimately related to the swarm methods that are the focus of this volume and have been mentioned in numerous contexts in these discussions. This chapter goes into some detail in reviewing the field of evolutionary computation, which consists of machine learning optimization and classification paradigms that are roughly based on evolution mechanisms such as biological genetics, natural selection, and emergent adaptive behavior. Evolutionary computation paradigms provide tools to build intelligent systems that model intelligent behavior.

This book is designed to appeal to people in engineering and the physical sciences as well as those in the social sciences and intelligent laypersons. In this chapter we go into the history and basics of evolutionary computation in some detail—perhaps too much detail if you already have some familiarity with the subject. So if you are an established practitioner of evolutionary computation, you may want to skim through at least the first parts of this chapter. ∎

Introduction

This chapter is designed to provide basic information needed to utilize evolutionary computation tools to solve practical problems. We present the terminology and key concepts, followed by paradigms that are developed from and illustrate the key concepts. The chapter is written largely from the perspective of an engineer or computer scientist, emphasizing the application potential of evolutionary computation tools, and drawing comparisons with other applied problem-solving techniques.

We begin with a brief history of evolutionary computation, followed by an overview of the evolutionary computation field. Each of the four main areas (as we define them) is discussed in more detail in its own section. Comparisons of evolutionary computation tools (in these four areas) and other processing methods appear in each section. Particle swarm optimization, which is sometimes considered a fifth component area of evolutionary computation, is introduced in a later chapter.

Evolutionary Computation History

This history section focuses on people, rather than theory or technology. We do this for two main reasons. First, it seems to us a more interesting way to look at history. History is, after all, just a record of people doing things. Second, the evolutionary computation field, particularly in the early days, revolved around a few key individuals. These individuals and their followers seemed to us to sometimes resemble minicultures.

Having said that, the selection of individuals is somewhat arbitrary because the intent is to provide a broad sample of people who contributed to current technology, rather than an exhaustive list. Some well-known researchers are mentioned only briefly, and others are omitted. The fact that someone is discussed only briefly, or even omitted altogether, is not meant to reflect the authors' opinion of that person's contribution. The selected people and their contributions are discussed roughly in chronological order.

The Four Areas of Evolutionary Computation

The evolutionary computation field has been considered by many researchers to include four areas:

- Genetic algorithms
- Evolutionary programming
- Evolution strategies
- Genetic programming

There are other ways to look at the field, such as considering genetic programming as a branch of genetic algorithms, but we choose this approach. Bentley (1999) takes a similar approach, stating that there are four main types of evolutionary algorithms in use, then several pages later recognizing that genetic programming is a specialized form of genetic algorithm. (As the father of one of the authors [RE] was fond of saying, "There are two kinds of people in the world. People who think that there are two kinds of people in the world, and people who don't.")

Of these four methodologies, more work has been done with genetic algorithms, and so we focus on that field. (We realize that the emphasis on genetic algorithms is fading somewhat. In fact, hybrids of the four methodologies are becoming increasingly popular.) Contributors to the other three areas are also discussed, but in somewhat less detail. Although it might be argued that work in the early 20th century on Darwinian synthesis by Haldane and others may be the place we should start, what is now known as evolutionary computation really began to take shape about 50 years later.

Genetic Algorithms

The development of genetic algorithms has its roots in work done in the 1950s by biologists using computers to simulate natural genetic systems. Of the people doing these simulations, one of those doing work most closely related to our current concepts of genetic algorithms was an Australian, A. S. Fraser, who began publishing in the field in the late 1950s

(Fraser 1957). Our history of evolutionary computation thus (arbitrarily) begins with him.

Fraser was working in the area of epistasis (suppression of the effect of a gene) and represented each of three parameters of an epistatic function as 5 bits in a 15-bit string. He then based his selection of "parents" by choosing those strings whose variable values produced function values between -1 and $+1$. Fraser was working with natural systems, and while his work somewhat resembles function optimization as currently done by genetic algorithms, he apparently did not consider the possibilities of applying his methodology to artificial systems (Fraser 1960, 1962).

Also beginning to publish in the early 1960s was the man who together with his students has probably had more influence on the field of genetic algorithms than any others: John H. Holland of the University of Michigan. Holland attended MIT as an undergraduate, where he was influenced by such luminaries as Norbert Weiner and John McCarthy. He was part of a team that programmed the prototype of the IBM 701 to "learn" something about running a maze, prompting Holland to regard the computer as a sort of "simulated lab rat." After working at IBM, Holland went to the University of Michigan, where, under Arthur Burks, he obtained the first Ph.D. in the United States in computer science (Levy 1992).

Davis (1991) stated:

> John Holland . . . created the genetic algorithm field. The field would not exist if he had not decided to harness the power inherent in genetic processes in the early 1970s and functioned as the technical and political leader of the genetic algorithm field from its inception to the present time. Our understanding of the unique features of genetic algorithms has been shaped by the careful and insightful work of Holland and his students from the field's critical first years to the present time (p. vi).

Holland's interest was in machine intelligence, and he and his students developed and applied the capabilities of genetic algorithms to artificial systems. He taught courses in *adaptive systems* in the early 1960s while laying the groundwork for applications to artificial systems with his publications on adaptive systems theory (Holland 1962). Holland's systems were adaptive because of their robustness in spite of changes and uncertainty in the environment. Further, they were *self-adaptive* in that they could make adjustments based on their interaction with the environment over time.

The GA metaphor is genetic inheritance at the level of the individual. A problem solution is considered as an individual's chromosome, or pattern of genetic alleles, and low-level operations such as those that are seen in the nuclei of cells are proposed for developing new solutions.

One of Holland's many contributions was his use of a population of individuals, conceptualized as chromosomes, in the search process, rather than the use of single individuals as was common at the time. (Fraser used populations, but, as stated previously, didn't apply his methodology to artificial systems.) He also derived the schema theorem, which shows that schema (fundamental building blocks of individual chromosomes) that are more "fit" with respect to a defined fitness function are more likely to reproduce in successive generations of the population of chromosomes. We go into more detail about the schema theorem later in this chapter.

Chromosomes in nature are formed of twisted strands of DNA, composed of the four proteins adenine, cytosine, guanine, and thymine. These strands are presently understood as a kind of computer program that gives instructions to the cells that comprise the organism; the DNA sequence contains instructions about how to develop and what to do. While our digital computers use the base-two or binary number system to encode program instructions and data, chromosomes use a base-four method, encoded in the ordering of the four proteins. Genetic algorithms usually use base-two chromosomes, though the methods developed by Holland and his followers can be applied to any base number system, including floating-point decimals.

Beginning in the 1960s Holland's students routinely used selection, crossover, and mutation in their applications. Several of Holland's students made significant contributions to the genetic algorithm field, often starting with their Ph.D. dissertations. We mention only a few.

The term "genetic algorithm" was used first by Bagley (1967) in his dissertation, which utilized genetic algorithms to find parameter sets in evaluation functions for playing the game of Hexapawn, which is played on a 3 × 3 chessboard in which each player starts with three pawns. Bagley's genetic algorithm resembled many used today, with selection, crossover, and mutation.

For the genetic algorithm field, 1975 was an important year. It was in this year that Holland published one of the field's most important books, entitled *Adaptation in Natural and Artificial Systems*. In the first five years after it was published, the book sold 100–200 copies per year and seemed to be fading into oblivion. Instead, between 1985 and 1990, the number of people working on genetic algorithms—and interest in his book—

increased sufficiently to persuade Holland to update and reissue it (Holland 1992).

Also in 1975, K. A. De Jong, one of Holland's students, published his Ph.D. dissertation entitled "An analysis of the behavior of a class of genetic adaptive systems." As part of his dissertation, De Jong put forward a set of five test functions designed to measure the performance of any genetic algorithm. Two metrics were devised, one to measure the convergence of the algorithm, the other to measure the ongoing performance. De Jong examined the effects of varying four parameters (population size, crossover probability, mutation probability, and generation gap) on the performance of six main kinds of genetic algorithm paradigms (De Jong, 1975). Although a number of other benchmark functions have emerged, De Jong's five-function test bed and two performance metrics are still among frequently referenced criteria for genetic algorithm performance.

De Jong went to the University of Pittsburgh from Michigan, where he taught genetic algorithms to a number of students, among them Steve Smith and John Grefenstette. Smith published a significant dissertation on machine learning involving a classifier system that became known as "Smith's Poker Player" (Smith, 1980). Upon graduation, Grefenstette began teaching yet another generation of students at Vanderbilt University, including J. David Schaffer, who was the first to develop a multiobjective algorithm (Schaffer, 1984), work that has enjoyed a revival in popularity. For a comprehensive survey of multiobjective evolutionary algorithms, see Coello (1999).

Grefenstette developed a genetic algorithm implementation called GENESIS that, in its various incarnations and reincarnations, became perhaps the most widely used genetic algorithm implementation in the late 1980s (Grefenstette, 1984a, 1984b). He also was instrumental in founding and editing the proceedings of the first International Conference on Genetic Algorithms, a premier conference in the field (Grefenstette, 1985).

David E. Goldberg, another of Holland's students, has concentrated on engineering applications of genetic algorithms. He is a former gas pipeline worker; his Ph.D. dissertation considered a 10-compressor, 10-pipe, steady-state, serial gas pipeline problem (Goldberg, 1983). The goal was to provide a strategy that minimizes the power consumed in the pumping stations, subject to pressure-related constraints. He summarized the power the genetic algorithm brought to the pipeline problem when he wrote, "If we were, for example, to search for the best person among the world's 4.5 billion people as rapidly as the GA, we would only need to talk to four or five people before making our near optimal selection" (Goldberg, 1987). Goldberg's 1989 volume is one of the most influential books on genetic algorithms: *Genetic Algorithms in Search, Optimization, and Machine Learning.* He continues to be an important contributor to the field.

The author of another significant genetic algorithm book is self-taught in genetic algorithms. Lawrence (Dave) Davis got interested in them while at Texas Instruments, where he obtained support to evaluate genetic algorithms for 2D bin packing in a chip layout application. He published the *Handbook of Genetic Algorithms,* after moving to the Boston area, where he worked for BBN. The book comprises two main parts. The first is a tutorial on genetic algorithms, and the second is a collection of case studies contributed by a number of researchers (Davis, 1991). In the mid-1990s, two of the most widely read books for people wanting to learn about genetic algorithms were those by Goldberg and Davis.

At approximately the same time that Holland and his students were developing genetic algorithms, two groups were working on opposite sides of the Atlantic on a different approach that did not utilize *crossover,* a feature of the genetic algorithm implementations.

Evolutionary Programming

In the United States, Larry J. Fogel and his colleagues developed what they named *evolutionary programming.* Evolutionary programming uses the selection of the fittest, but the only structure-modifying operation allowed is mutation—there is no crossover. Fogel and his colleagues mainly worked with finite state machines and were interested in machine intelligence, and they were able to solve some problems that were quite difficult for genetic algorithms. Fogel (1994) described evolutionary programming as taking a fundamentally different approach than genetic algorithms:

> The procedure abstracts evolution as a top-down process of adaptive behavior, rather than a bottom-up process of adaptive genetics. It is argued that this approach is more appropriate because natural selection does not act on individual components in isolation, but rather on the complete set of expressed behaviors of an organism in light of its interaction with its environment.

Philosophically, then, evolutionary programming researchers consider each point in the population to represent an entire species, with species competing to fill environmental niches.

Since the original development of evolutionary programming, two of the main extensions have been adding the ability to evolve continuous parameters and the capability for *self-adaptation*. (The original version of evolutionary programming worked only with discrete parameters.) The inclusion of self-adaptation allows the strategy parameters that guide the mutation to be evolved.

Fogel summarizes evolutionary programming as implementing "survival of the more skillful" rather than the "survival of the fittest" emphasized by genetic algorithm developers. In the mid-1960s a book documenting this approach proved to be quite controversial (Fogel, Owens, and Walsh, 1966). Misunderstandings and misinterpretations related to the book have been identified as a contributing factor to problems experienced by researchers in obtaining funding for evolutionary computation in the late 1960s and 1970s (Goldberg, 1989). It is, however, probable that another significant factor was the well-known symbolics versus numerics controversy (temporarily won by Minsky and the symbolics researchers). One of the leading evolutionary programming researchers during the 1970s was at New Mexico State University. Don Dearholt and his students were responsible for a significant number of publications on evolutionary programming during this decade.

Evolution Strategies

At the same time, across the ocean, Ingo Rechenberg and Hans-Paul Schwefel were experimenting with mutation in their attempts to find optimal physical configurations for a series of hinged plates in a wind tunnel and a tube that delivered liquid. The usual gradient-descent techniques were unable to solve the sets of equations for reducing wind resistance. They began experimenting with mutation, slightly perturbing their best problem solutions to search randomly in the nearby regions of the problem space.

Rechenberg and Schwefel used the first computer available at the Technical University of Berlin to simulate various versions of the approach that became known as *evolution strategies* (Rechenberg, 1965; Schwefel, 1965). In the early 1970s, Rechenberg published a book that is considered the foundation for this approach (Rechenberg, 1973), and evolution strategy continues to experience significant activity, especially in Europe. Research developments in Germany and the United States continued in parallel, with each group unaware of the other's findings until the 1980s, although they may have known about each other (Fogel, 2000).

Genetic Programming

The fourth major area of evolutionary computation (although, as we said earlier, some consider it a subset of genetic algorithms) is genetic programming. In genetic programming, computer programs are evolved directly. Some of the earliest related work done by Friedberg (1958) and Friedberg, Dunham, and North (1959) worked with fixed-length computer programs that were coded by another program designed to optimize the performance of the fixed-length program. Their programs, dubbed "Herman" and "Ramsey," each comprised a set of 64 instructions, with each instruction being 14 bits long. The programs were defined such that every arrangement of the 14 bits was a valid instruction and each set of 64 instructions was a valid program. Unfortunately, the results of the efforts did not live up to expectations. In retrospect, there were probably three main reasons for this. First, the programs were limited in length to 64 instructions: a "failure" was tallied if the program did not terminate successfully by the end of the 64th instruction (even if there was a loop). Second, there was only one program; thus, there was a population of just one that evolved. Third, it is not clear that the fitness function used was appropriate.

These limitations were successfully dealt with by Stanford's John Koza (yet another former student of Holland), who developed genetic programming in its current form in the late 1980s. Whereas the other three evolutionary computation approaches use string-shaped chromosomes, Koza evolved computer programs in a population of tree-shaped ones. The units used for crossover were LISP symbolic expressions that are essentially subroutines. Koza has been a prolific producer of documentation, including books (Koza, 1992; Koza et al., 1999) and videotapes related to genetic programming, and it is one of the fastest-growing and most fascinating areas of evolutionary computation. The idea of evolving computer programs has been around for decades and is now becoming a reality.

Toward Unification

As the 1980s came to a close, the four areas of evolutionary computation continued to develop relatively independently, with little cooperation or communication among them. In 1994, however, an important meeting was held that brought together researchers from all four evolutionary computation areas: the IEEE World Congress on Computational Intelligence, held at Walt Disney World, Florida. The World Congress comprised a minisymposium on computational intelligence and three conferences: The International Conference on Neural Networks; the fuzzy logic conference FUZZ/IEEE 1994; and the first IEEE Conference on Evolutionary Computation (ICEC), chaired by Professor Zbigniew Michalewicz of the University of North Carolina at Charlotte.

A total of 96 papers were presented orally in ICEC and 63 in poster sessions, representing authors from 23 countries worldwide. The two volumes of proceedings from this evolutionary computation conference are a landmark in the field (Michalewicz et al., 1994). The field of computational intelligence seems to have come of age with the second IEEE World Congress on Computational Intelligence held in Anchorage, Alaska, in May 1998. Researchers in these four areas of evolutionary computation are now communicating and working significantly more with each other.

Evolutionary Computation Overview

Evolutionary computation (EC) paradigms generally differ from traditional search and optimization paradigms in three main ways by

1. utilizing a population of points (potential solutions) in their search.
2. using direct "fitness" information instead of function derivatives or other related knowledge.
3. using probabilistic, rather than deterministic, transition rules.

In addition, EC implementations sometimes encode the parameters in binary or other symbols, rather than working with the parameters themselves. We now examine these differences in more detail, followed by a quick look at how to put evolutionary algorithms to work.

EC Paradigm Attributes

Most traditional optimization paradigms move from one point in the decision hyperspace to another, using some deterministic rule. One of the drawbacks of this approach is the likelihood of getting stuck at a local optimum. EC paradigms, on the other hand, start with a *population* of points (hyperspace vectors). They typically generate a new population with the same number of members each epoch, or *generation*. Thus, many maxima or minima can be explored simultaneously, lowering the probability of getting stuck. Operators such as crossover and mutation effectively enhance this parallel search capability, allowing the search to directly "tunnel through" from one promising hyperspace region to another.

EC paradigms do not require information that is auxiliary to the problem, such as function derivatives. Many hill-climbing search paradigms, for example, require the calculation of derivatives in order to explore the local maximum. In EC optimization paradigms the fitness of each member of the population is calculated from the value of the function being optimized, and it is common to use the function output as the measure of fitness. Fitness is a direct metric of the performance of the individual population member on the function being optimized.

The fact that EC paradigms use probabilistic transition rules certainly does not mean that a strictly random search is being carried out. Rather, stochastic operators are applied to operations that direct the search toward regions of the hyperspace that are likely to have higher values of fitness. Thus, for example, reproduction (selection) is often carried out with a probability that is proportional to the individual's fitness value.

Some EC paradigms, and especially canonical genetic algorithms, use special encodings for the parameters of the problem being solved. In genetic algorithms, the parameters are often encoded as binary strings, but any finite alphabet can be used. These strings are almost always of fixed length, with a fixed total number of ones and zeroes, in the case of a binary string, being assigned to each parameter. By "fixed length" it is meant that the string length does not vary during the running of the EC paradigm. The string length (number of bits for a binary string) assigned to each parameter depends on its maximum range for the problem being solved and on the precision required.

Implementation

Regardless of the paradigm implemented, evolutionary computation tools often follow a similar procedure:

1. Initialize the population.
2. Calculate the fitness for each individual in the population.

3. Reproduce selected individuals to form a new population.

4. Perform evolutionary operations, such as crossover and mutation, on the population.

5. Loop to step 2 until some condition is met.

Initialization is most commonly done by seeding the population with random values. When the parameters are represented by binary strings, this simply means generating random strings of ones and zeroes (with a uniform probability for each value) of the fixed length described earlier. (While most evolutionary algorithms use population members of fixed length, we should note here that genetic programming also evolves the population members' structure, including each member's length.) It is sometimes feasible to seed the population with "promising" values, known to be in the hyperspace region relatively close to the optimum. The total number of individuals chosen to make up the population is both problem and paradigm dependent, but is often in the range of a few dozen to a few hundred.

The fitness value is often proportional to the output value of the function being optimized, though it may also be derived from some combination of a number of function outputs. The fitness (evaluation) function takes as its inputs the outputs of one or more functions, and outputs some probability of reproduction. It is sometimes necessary to transform the function outputs to produce an appropriate fitness metric; sometimes it is not. For some evolutionary algorithms, only a small percentage of the computational time is required for the algorithm itself; most of the computational time is required for evaluating fitnesses.

Selection of individuals for reproduction to constitute a new population (often called a new generation) is usually based upon fitness values. The higher the fitness, the more likely it is that the individual will be selected for the new generation. Some paradigms that are considered evolutionary, however, such as particle swarm optimization, can retain all population members from epoch to epoch. In fact, Bentley (1999) states that "the preferential selection of some parents instead of others is not essential to evolution." (If there is no selection operator, then all population members produce offspring with equal probability.)

Termination of the algorithm is usually based either on achieving a population member with some specified fitness or on having run the algorithm for a given number of generations.

In many, if not most, cases, a global optimum exists at one point in the decision hyperspace. Furthermore, there may be stochastic or chaotic noise present. Sometimes the global optimum changes dynamically because of external influences; frequently there are very good local optima as well. For these and other reasons, the bottom line is that it is often unreasonable to expect any optimization method to find a global optimum (even if it exists) within a finite time. The best that can be hoped for is to find near-optimum solutions and to hope that the time it takes to find them increases less than exponentially with the number of variables. One leading EC researcher (Schwefel, 1994) suggests that the focus should be on "meliorization" (improvement) rather than optimization. We agree. Put another way, evolutionary computation is often the *second-best* way to solve a problem. Classical methods such as linear programming should often be tried first, as should customized approaches that take full advantage of knowledge about the problem.

Why should we be satisfied with second best? Well, for one thing, classical and customized approaches will frequently not be feasible, and EC paradigms will be usable in a vast number of situations. For another, a real strength of EC paradigms is that they are generally quite *robust*. In this field, robustness means that an algorithm can be used to solve many problems, and even many kinds of problems, with a minimum amount of special adjustments to account for special qualities of a particular problem. Typically an evolutionary algorithm requires specification of the length of the problem solution vectors, some details of their encoding, and an evaluation function—the rest of the program does not need to be changed. Finally, robust methodologies are generally fast and easy to implement.

This brings us to our *Law of Sufficiency:* If a solution is *good* enough, and it is *fast* enough, and it is *cheap* enough, then it is *sufficient*. In almost all real-world applications, we are looking for, and satisfied with, sufficient solutions. (Note that "good enough" means that the solution meets specifications.)

In the next sections we review four areas of evolutionary computation: genetic algorithms, evolutionary programming, evolution strategies, and genetic programming. Genetic algorithms, discussed in the next section, receive a majority of the attention, as they currently account for most of the successful applications in the literature (although this is changing). It is also important to note that hybrids of these approaches, and hybrids of these approaches with other computational intelligence tools such as neural networks, are becoming more prevalent.

Genetic Algorithms

It seems that every technology has its jargon; genetic algorithms are no exception. Therefore, we begin by reviewing some of the basic terminology that is needed to understand the genetic algorithm (GA) literature. A sample problem is then presented to illustrate how GAs work; a step-by-step analysis illustrates a GA application, with options discussed for some of the individual operations. The section concludes with a more detailed look at the fundamental schema theorem and at approaches for improving GA performance in some situations.

Details on implementing GAs are discussed in Appendix B, in which a specific GA implementation is summarized. Executable code for this implementation is on the Internet site for the book.

An Overview

Genetic algorithms are search algorithms that reflect in a primitive way some of the processes of natural evolution. (As such, they are analogous to artificial neural networks' status as primitive approximations to biological neural processing.) Engineers and computer scientists generally do not care as much about the biological foundations of GAs as their utility as analysis tools (another parallel with neural networks). GAs often provide very effective search mechanisms that can be used in optimization or classification applications.

EC paradigms work with a *population* of points, rather than a single point; each "point" is actually a vector in hyperspace representing one potential, or candidate, solution to the optimization or classification problem. A population is thus just an ensemble, or set, of hyperspace vectors. Each vector is called an *individual* in the population; sometimes an individual in a GA is referred to as a *chromosome* because of the analogy to genetic evolution of organisms.

Because real numbers are often encoded in GAs using binary numbers, the dimensionality of the problem vector might be different from the dimensionality of the bitstring chromosome. The number of elements in each vector (individual) equals the number of real parameters in the optimization problem. A vector *element* generally corresponds to one parameter, or dimension, of the numeric vector. Each element can be encoded in any number of bits, depending on the representation of each parameter. The total number of bits defines the dimension of hyperspace being searched. If a GA is being used to find "optimum" (sufficient, in terms of the Law of Sufficiency described previously) weights for a neural network, for example, the number of vector elements equals the number of weights in the network. If there are w weights, and it is desired to calculate each weight to a precision of b bits, then each individual will consist of $w \times b$ bits, and the dimension of binary hyperspace being searched is 2^{wb}.

The series of operations carried out when implementing a "plain vanilla" GA paradigm is the following:

1. Initialize the population.
2. Calculate fitness for each individual in the population.
3. Reproduce selected individuals to form a new population.
4. Perform crossover and mutation on the population.
5. Loop to step 2 until some condition is met.

In some GA implementations, operations other than crossover and mutation are carried out in step 4. Crossover, however, is considered by many to be an essential operation of all GAs.

A Simple GA Example Problem

Because implementing a "plain vanilla" GA paradigm is so simple, a sample problem (also simple) seems to be the best way to introduce most of the basic GA concepts and methods. As will be seen, implementing a simple GA involves only copying strings, exchanging portions of strings, and flipping bits in strings.

Our sample problem is to find the value of x that maximizes the function $f(x) = \sin(\pi x/256)$ over the range $0 \le x \le 255$, where values of x are restricted to integers. This is just the sine function from zero to π radians, as illustrated in Figure 4.1. Its maximum value of 1 occurs at $\pi/2$, or $x = 128$. The function value and the fitness value are thus defined to be identical for the sample problem.

There is only one variable in our sample problem: x. We assume for the sample problem that the GA paradigm uses a binary alphabet. The first decision to be made is how to represent the variable. This has been made easy in this case since the variable can only take on integer values between 0 and 255. It is therefore logical to represent each

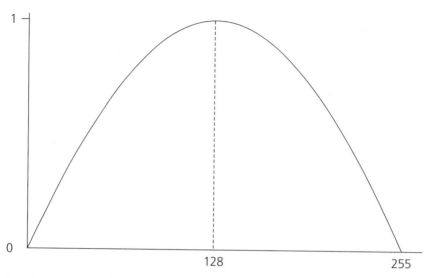

Figure 4.1 Function to be optimized in example problem.

individual in our population with an eight-bit binary string. The binary string 00000000 will evaluate to 0, and 11111111 to 255.

We next decide how many individuals will make up the population. In an actual application, it would be common to have somewhere between a few dozen and a few hundred individuals. For the purposes of this illustrative example, however, the population consists of eight individuals.

The next step is to initialize the population. This is usually done randomly. We therefore use a random number generator to assign a 1 or 0 to each of the eight positions in each of the eight individuals, resulting in the initial population in Figure 4.2. Also shown in the figure are the values of x and $f(x)$ for each binary string.

After fitness calculation, the next step is reproduction. Reproduction consists of forming a new population with the same total number of individuals by selecting from members of the current population with a stochastic process that is weighted by each of their fitness values. In the example problem, the sum of all fitness values for the initial population is 5.083. Dividing each fitness value by 5.083, then, yields a *normalized fitness* value f_{norm} for each individual. The sum of the normalized values is 1.

These normalized fitness values are used in a process called "roulette wheel" selection, where the size of the roulette wheel wedge for each

Individuals	x	$f(x)$	f_{norm}	Cumulative f_{norm}
1 0 1 1 1 1 0 1	189	0.733	0.144	0.144
1 1 0 1 1 0 0 0	216	0.471	0.093	0.237
0 1 1 0 0 0 1 1	99	0.937	0.184	0.421
1 1 1 0 1 1 0 0	236	0.243	0.048	0.469
1 0 1 0 1 1 1 0	174	0.845	0.166	0.635
0 1 0 0 1 0 1 0	74	0.788	0.155	0.790
0 0 1 0 0 0 1 1	35	0.416	0.082	0.872
0 0 1 1 0 1 0 1	53	0.650	0.128	1.000

Figure 4.2 Initial population and $f(x)$ values for GA example.

population member, which reflects the probability of that individual being selected, is proportional to its normalized fitness value.

We spin the roulette wheel by generating eight random numbers between 0 and 1. If a random number is between 0 and 0.144, the first individual in the existing population is selected for the next population. If the number is between 0.144 and (0.144 + 0.093) = 0.237, the second individual is selected, and so on. Finally, if the random number is between (1 − 0.128) = 0.872 and 1.0, the last individual is selected. The probability that an individual is selected is thus proportional to its fitness value. It is possible, though highly improbable, that the individual with the lowest fitness value *could* be selected eight times in a row and make up the entire next population. It is more likely that individuals with high fitness values are picked more than once for the new population.

The eight random numbers generated are 0.293, 0.971, 0.160, 0.469, 0.664, 0.568, 0.371, and 0.109. This results in initial population member numbers 3, 8, 2, 5, 6, 5, 3, and 1 being chosen to make up the population after reproduction, as shown in Figure 4.3.

The next operation is crossover. To many evolutionary computation practitioners, crossover of binary encoded substrings is what makes a genetic algorithm a genetic algorithm. Crossover is the process of exchanging portions of the strings of two "parent" individuals. An overall probability is assigned to the crossover process, which is the probability that, given two parents, the crossover process will occur. This *crossover rate* is often in the range of 0.65 to 0.80; we select a value of 0.75 for the sample problem.

01100011

00110101

11011000

10101110

01001010

10101110

01100011

10111101

Figure 4.3 Population after reproduction.

First, the population is paired off randomly into pairs of parents. Since the order of the population after reproduction in Figure 4.3 is already randomized, parents will be paired as they appear there. For each pair, a random number is generated to determine whether crossover will occur. It is thereby determined that three of the four pairs will undergo crossover.

Next, for the pairs undergoing crossover, two crossover points are selected at random. (Other crossover techniques are discussed later in this chapter.) The portions of the strings between the first and second crossover points (moving from left to right in the string) will be exchanged. The paired population, with the first and second crossover points labeled for the three pairs of individuals undergoing crossover, is illustrated in Figure 4.4(a) prior to the crossover operation. The portions of the strings to be exchanged are in bold. Figure 4.4(b) illustrates the population after crossover is performed.

Note that, for the third pair from the top, the first crossover point is to the right of the second. The crossover operation thus "wraps around" the end of the string, exchanging the portion between the first and the second, moving from left to right. For two-point crossover, then, it is as if the head (left end) of each individual string is joined to the tail (right end), thus forming a ring structure. The section exchanged starts at the first crossover point, moving to the right along the binary ring, and ends at the second crossover point. The values of x and $f(x)$ for the population following crossover appear in Figure 4.4(c) and (d), respectively.

The final operation in this plain vanilla genetic algorithm is mutation. Mutation consists of flipping bits at random, generally with a constant probability for each bit in the population. As is the case with the probability of crossover, the probability of mutation can vary widely

1 2	Individuals	x	f(x)
0 1 1|0 0 0|1 1	0 1 1 1 0 1 1 1	119	0.994
0 0 1|1 0 1|0 1	0 0 1 0 0 0 0 1	33	0.394
1 2			
1|**1 0 1 1**|0 0 0	1 0 1 0 1 0 0 0	168	0.882
1|**0 1 0 1**|1 1 0	1 1 0 1 1 1 1 0	222	0.405
2 1			
0 1|0 0 1 0 1|0	1 0 0 0 1 0 1 0	138	0.992
1 0|1 0 1 1 1|0	0 1 1 0 1 1 1 0	110	0.976
0 1 1 0 0 0 1 1	0 1 1 0 0 0 1 1	99	0.937
1 0 1 1 1 1 0 1	1 0 1 1 1 1 0 1	189	0.733
(a)	(b)	(c)	(d)

Figure 4.4 Population before crossover showing crossover points (a), after crossover (b), and values of *x* (c) and *f(x)* (d) after crossover.

according to the application and the preference of the researcher. Values of between 0.001 and 0.01 are not unusual for the mutation probability. This means that the bit at each site on the bitstring is flipped, on average, between 0.1 and 1.0 percent of the time. One fixed value is used for each generation and often is maintained for an entire run.

Since there are 64 bits in the example problem's population (8 bits × 8 individuals), it is quite possible that none would be altered as a result of mutation, so we will consider the population of Figure 4.4(b) as the "final" population after one iteration of the GA procedure. Going through the entire GA procedure one time is said to produce a new *generation*. The population of Figure 4.4(b) therefore represents the first generation of the initial randomized population.

Note that the fitness values now total 6.313, up from 5.083 in the initial random population, and that there are now two members of the population with fitness values higher than 0.99. The average and maximum fitness values have thus both increased.

The population of Figure 4.4(b) and the corresponding fitness values in Figure 4.4(d) are now ready for another round of reproduction, crossover, and mutation, producing yet another generation. More generations are produced until some stopping condition is met. We may simply set a maximum number of generations to let the algorithm search, let it run until a performance criterion has been met, or stop the algorithm after some number of generations with no improvement.

A Review of GA Operations

Now that one iteration of the GA operations (one generation) for the example problem has been completed, we will review each of the operations in more detail. We will examine various approaches and reasons for each.

Representation of Variables

The representation of the values for the variable x was made (perhaps unrealistically) straightforward by choosing a dynamic range of 256; an eight-bit binary number was thus an obvious approach. Standard binary coding, however, is only one approach; others may be more appropriate.

In this example, the nature of the sine function places the optimal value of x at 128, where $f(x)$ is 1. The binary representation of 128 is 10000000; the representation of 127 is 01111111. Thus, the smallest change in fitness value can require a change of every bit in the representation. This situation is an artifact of the encoding scheme and is not desirable—it only makes the GA's search more difficult. Often, a better representation is one in which adjacent integer values have a Hamming distance of one; in other words, adjacent values differ by only a single bit. One such scheme is Gray coding, which is described in Chapter 2.

Some GA software allows the user to specify the dynamic range and resolution for each variable. The program then assigns the correct number of bits and the coding. For example, if a variable has a range from 2.5 to 6.5 (a dynamic range of 4) and it is desired to have a resolution of three decimal places, the product of the dynamic range and the resolution requires a string 12 bits long, where the string of zeroes represents the value 2.5. A major advantage of being able to represent variables in this way is that the user can think of the population individuals as real-valued vectors rather than as bitstrings, thus simplifying the development of GA applications.

The "alphabet" used in the representation can, in theory, be any finite alphabet. Thus, rather than using the binary alphabet of 1 and 0, we could use an alphabet containing more characters or numbers. Most GA implementations, however, use the binary alphabet.

Population Size

De Jong's dissertation (1975) offers guidelines that are still usually observed: start with a relatively high crossover rate, a relatively low mutation rate, and a moderately sized population—though just what constitutes a moderately sized population is unclear. The main trade-off is obvious: a large population will search the space more completely, but at a higher computational cost. We generally have used populations of between 20 and 200 individuals, depending, it seems, primarily on the string length of the individuals. It also seems (in our experience) that the sizes of populations used tend to increase approximately linearly with individual string length, rather than exponentially, but "optimal" population size (if an optimal size exists) depends on the problem as well.

Population Initialization

The initialization of the population is usually done stochastically, though it is sometimes appropriate to start with one or more individuals that are selected heuristically. The GA is thereby initially aimed in promising directions, or given hints. It is not uncommon to seed the population with a few members selected heuristically and to complete the population with randomly chosen members. Regardless of the process used, the population should represent a wide assortment of individuals. The urge to skew the population significantly should generally be avoided, if our limited experience is generalizable.

Fitness Calculation

The calculation of fitness values is conceptually simple, though it can be quite complex to implement in a way that optimizes the efficiency of the GA's search of the problem space. In the example problem, the value of $f(x)$ varies (quite conveniently) from 0 to 1. Lurking within the problem, however, are two drawbacks to using the "raw" function output as a fitness function—one that is common to many implementations, the other arising from the nature of the sample problem.

The first drawback, common to many implementations, is that after the GA has been run for a number of generations it is not unusual for most (if not all) of the individuals' fitness values, after, say, a few dozen generations, to be quite high. In cases where the fitness value can range from 0 to 1, for example (as in the sample problem), most or all of the fitness values may be 0.9 or higher. This lowers the fitness differences between individuals that provide the impetus for effective roulette wheel selection; relatively higher fitness values should have a higher probability of reproduction.

One way around this problem is to equally space the fitness values. For example, in the sample problem, the fitness values used for reproduction could be equally spaced from 0 to 1, assigning a fitness value of 1 to the most fit population member, 0.875 to the second, and 0.125 to the lowest fitness value of the eight. In this case the population members are ranked on the basis of fitness, and then their ranks are divided by the number of individuals to provide a probability threshold for selection. Note that the value of 0 is generally not assigned, since that would result in one population member being made ineligible for reproduction. Also note that $f(x)$, the function result, is now not equal to the fitness, and that in order to evaluate actual performance of the GA, the function value should be monitored as well as the spaced fitness.

Another way around the problem is to use what is called *scaling*. Scaling takes into account the recent history of the population and assigns fitness values on the basis of comparison of individuals' performance to the recent average performance of the population. If the GA optimization is maximizing some function, then scaling involves keeping a record of the minimum fitness value obtained in the last w generations, where w is the size of the scaling window. If, for example, $w = 5$, then the minimum fitness value in the last five generations is kept and used instead of 0 as the "floor" of fitness values. Fitness values can be assigned a value based on their actual distance from the floor value, or they can be equally spaced, as described earlier.

The second drawback is that the example problem exacerbates the "compression of fitness values" situation described earlier because near the global optimum fitness value of 1, $f(x)$ (which is also the fitness) is relatively flat. There is thus relatively little selection advantage for population members near the optimum value $x = 128$. If this situation is known to exist, a different representation scheme might be selected, such as defining a new fitness function that is the function output raised to some power.

Note that the shape of some functions "assists" discrimination near the optimum value. For example, consider maximizing the function $f(x) = x^2$ over the range 0 to 10; there is a higher differential in values of $f(x)$ between adjacent values of x near 10 than near 0. Thus slight change of the independent variable results in great improvement or deterioration of performance—which is equally informative—near the optimum.

In the discussion thus far, we have assumed that optimization implies finding a maximum value. Sometimes, of course, optimization requires finding a minimum value. Some versions of GA implementations allow for this possibility. Often, the user must specify the maximum value f_{max} of the function being optimized, $f(x)$, over the range of the search. The GA then can be programmed to maximize the fitness function $f_{max} - f(x)$. In this case, scaling, described above, keeps track of f_{max} over the past w generations and uses it as a "roof" value from which to calculate fitness.

Roulette Wheel Selection

In genetic algorithms, the expected number of times each individual in the current population is selected for the new population is proportional to the fitness of that individual relative to the average fitness of the entire population. Thus, in the initial population of the example problem, where the average fitness was $5.083/8 = 0.635$, the third population member had a fitness value of 0.937, so it could be expected to appear about 1.5 times in the next population; it actually appeared twice.

The conceptualization is that of a wheel whose surface is subdivided into wedges representing the probabilities for each individual (see Figure 4.5). For instance, one point on the edge is determined to be the zero point, and each arc around the circle corresponds to an area on the number line between zero and one. A random number is generated, between 0.0 and 1.0, and the individual whose wedge contains that number is chosen. In this way, individuals with greater fitness are more likely to be chosen. The selection algorithm can be repeated until the desired number of individuals has been selected.

There are a number of variations to the roulette wheel procedure. A few of them are reviewed below.

One variation on the basic roulette wheel procedure is a process developed by Baker (1987) in which the portion of the roulette wheel is assigned based on each unique string's relative fitness. One spin of the roulette wheel then determines the number of times each string will appear in the next generation. To illustrate how this is done, assume the fitness

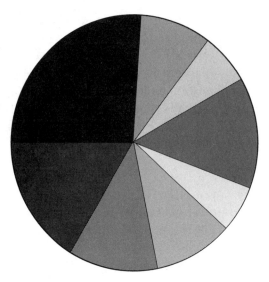

Figure 4.5 In roulette wheel selection, the probability of an individual being selected (the size of its slice of the wheel) is proportional to its fitness (indicated here by shading).

values are normalized (sum of all equals one). Each string is assigned a portion of the roulette wheel proportional to its normalized fitness. Instead of one "pointer" on the roulette wheel spun n times, there are n pointers spaced $1/n$ apart; the n-pointer assembly is spun once. When the wheel comes to rest, the place each of the n pointers points determines a population member in the next generation. If a string has a normalized fitness greater than $1/n$ (corresponding to an expected value greater than 1), it is guaranteed at least one occurrence in the next generation. If the string has a normalized fitness f_n such that $1/n < f_n < 2/n$, it will have either one or two occurrences in the next generation. If $2/n < f_n < 3/n$, it will have either two or three occurrences, and so on.

In the discussion thus far, it has been assumed that all of the population members are replaced each generation. Although this is often the case, it sometimes is desirable to replace only a portion of the population, say, the 80 percent with the worst fitness values. The percentage of the population replaced each generation is sometimes called the *generation gap*.

Unless some provision is made, with standard roulette wheel selection it is possible that the individual with the highest fitness value in a given generation may not survive reproduction, crossover, and mutation to appear unaltered in the new generation. It is frequently helpful to use what is called the *elitist strategy*, which ensures that the individual with the highest fitness is always copied into the next generation. We always include this strategy when using a genetic algorithm.

Crossover

The most important operator in GA is *crossover,* based on the metaphor of sexual combination. (An operator is a rule for changing a proposed problem solution.) If a solution is encoded as a bitstring, then mutation may be implemented by setting a probability threshold and flipping bits when a random number is less than the threshold. As a matter of fact, mutation is not an especially important operator in GA. It is usually set at a very low rate, and sometimes it is omitted altogether. Crossover is more important and adds a new dimension to the discussion of evolution so far.

Other evolutionary algorithms use random mutation plus selection as the primary method for searching the landscape for peaks or niches. One of the greatest and most fundamental search methods that biological life has found is sexual reproduction, which is extremely widespread throughout both the animal and plant kingdoms. Sexual reproduction capitalizes on the differences and similarities among individuals within a species; where one individual may have descended from a line that contained a good solution to one set of environmental constraints, another individual might have evolved to deal better with another aspect of survival. Perhaps one genetic line of rabbits has evolved a winter coloration that protects it through the changing seasons, while another has developed a "freeze" behavior that makes it hard for predators to spot. Mating between these two lines of rabbits might result in offspring lacking both of the advantages, offspring with one or the other characteristic either totally or in some degree, or offspring possessing both of the advantageous traits. Selection will decide, in the long run, which of these possibilities are most adaptable; the ones that adapt better, survive.

Crossover is a term for the recombination of genetic information during sexual reproduction. In GAs, offspring have equal probabilities of receiving any gene from either parent, as the parents' chromosomes are combined randomly. In nature, chromosomal combination leaves sections intact; that is, contiguous sections of chromosomes from one parent are combined with sections from the other, rather than simply shuffling randomly. In GAs there are many ways to implement crossover.

The two main attributes of crossover that can be varied are the probability that it occurs and the type of crossover that is implemented. The following paragraphs examine variations of each.

A crossover probability of 0.75 was used in the sample problem, and two-point crossover was implemented. Two-point crossover with a probability of 0.60–0.80 is a relatively common choice, especially when Gray coding is used.

The most basic crossover type is *one-point crossover,* as described by Holland (1975/1992) and others, for instance, Goldberg (1989) and Davis (1991). It is inspired by natural evolution processes. One-point crossover involves selecting a single crossover point at random and exchanging the portions of the individual strings to the right of the crossover point. Figure 4.6 illustrates one-point crossover; the crossover point is the vertical line, and portions to be exchanged are in bold in Figure 4.6(a).

Another type of crossover that has been found useful is called *uniform crossover,* described by Syswerda (1989). A random decision is made at each bit position in the string as to whether or not to exchange (crossover) bits between the parent strings. If a 0.50 probability at each bit position is implemented, an average of about 50 percent of the bits in the parent strings are exchanged. Note that a 50 percent rate will result in the maximum disruption due to uniform crossover. Higher rates just mirror rates lower than 50 percent. For example, a 0.60 probability uniform crossover rate produces results identical to a 0.40 probability rate. If the rate were 100 percent, the two strings would simply switch places, and if it were 0 percent neither would change.

Values for the probability of crossover vary with the problem. In general, values between 60 and 80 percent are common for one-point and two-point crossover. Uniform crossover sometimes works better with slightly lower crossover probability. It is also common to start out running the GA with a relatively higher value for crossover, then taper off the value linearly to the end of the run, ending with a value of, say, one-half or two-thirds the initial value.

Mutation

In GAs, mutation is the stochastic flipping of bits that occurs each generation. It is done bit-by-bit on the entire population. It is often done with a probability of something like .001, but higher probabilities are not unusual. For example, Liepins and Potter (1991) used a mutation probability of .033 in a multiple-fault diagnosis application.

If the population comprises real-valued parameters, mutation can be implemented in various ways. For instance, in an image classification

1 0 1 1 0 | **0 1 0**   1 0 1 1 0 1 0 0

0 1 0 0 1 | **1 0 0**   0 1 0 0 1 0 1 0

(a)   (b)

Figure 4.6 One-point crossover before crossover (a) and after crossover (b).

application, Montana (1991) used strings of real-valued parameters that represented thresholds of event detection rules as the individuals. Each parameter in the string was range limited and quantized (could take on only a certain finite number of values). If chosen for mutation, a parameter was randomly assigned any allowed value in the range of values valid for that parameter.

The probability of mutation is often held constant for the entire run of the GA, although this approach will not produce optimal results in many cases. It can be varied during the run, and if varied, usually is increased. For example, the mutation rate may start at .001 and end at .01 or so when the specified number of generations has been completed. In the software implementation described in Appendix B, a flag in the run file can be set that increases the mutation rate significantly when the variability in fitness values becomes low, as is often the case late in the run.

Schemata and the Schema Theorem

Exactly *how* do GAs do what they do? How is it possible to develop new population members that, on average, are fitter than the previous generation, while searching new regions of the problem space? Since all that GAs have to work with are (often binary) strings, there must be features related to the fitness inherent in the strings that are utilized.

Schemata

The string features that are relevant to the optimization process are called *schemata* (singular: *schema*). A theorem that describes *why* the canonical GA paradigm is able to efficiently direct an optimization process is the *schema theorem*. (This theorem also applies to other proportional selection methodologies.)

First described for the GA field by Holland (1975/1992), schemata are similarity templates for strings. Each schema defines a subset of strings with identical values at specified string locations. As used here, the word "string" usually refers to substrings of an individual population member string, but can refer to the entire string. Schemata provide a means by which relevant similarities among the individual population members can be described and exploited.

In order to define schemata, the alphabet of the strings is used to define values at specified locations, and an additional character is used as a "don't care" symbol in locations where the value doesn't matter. As is common in the GA literature, the pound symbol (#) is used in this book as the "don't care" symbol. (The asterisk symbol "*" is also commonly used.) Schemata can thus generally be thought of as comprising an alphabet of $a_o + 1$ characters, where a_o is the number of characters in the GA representation. In most cases, as in the example, the GA strings have a binary representation, so the schemata comprise the characters {0, 1, #}.

As an example, consider the schemata of length 4 that may appear in, say, the leftmost four positions of the population strings of the example problem. One such schema is #000, which has two member strings. That is, two strings match the schema: 1000 and 0000. The schema 1##0 has four matching strings: 1000, 1010, 1100, and 1110.

Holland argued that adaptation can be thought of in terms of schemata. Recall the hamburger example of an earlier chapter. Any hamburger that included the schema 1##0##0###—cheese but no mustard or pickles—was fitter than the average. It is possible that fitness could be increased even more by choosing an absolutely perfect combination of other ingredients. In other words, this might not be the very fittest schema in the search space, but as soon as this combination is hit upon it will be found to be quite a bit fitter than most other kinds of hamburgers. In a GA the hamburger that contains 1##0##0### will be likely to be selected to reproduce in the next generation, relative to hamburgers that have mustard or pickles or don't have cheese.

Genetic optimization increases the likelihood that the schemata that most improve the species' fitness will persist to the next generation. Holland argues that crossover among the fittest members of a population will result in the discovery and survival of better schemata. It should be noted, though, that some researchers have recently found errors in Holland's argument, and the issue is currently controversial. Even if the proof is shaky, it can be seen empirically, simply by running GA programs, that crossover is quite effective, if not always fast, for finding good solutions to highly complex problems.

How many schemata are possible for a string length of l and an alphabet of a_o characters? In the previous example, for $a_o = 2$, there can be a 0, 1, or # at each string position, resulting in a total possible number of schemata of $3 \times 3 \times 3 \times 3 = 81$. Generalizing, there are $(a_o + 1)^l$ total possible schemata for any representation of length l.

Another informative measure is the total possible number of unique schemata in a population. Consider a specific string of length 8, taken from the example problem: 01110111. Since each string position can assume the value it has, or the wild-card value, the string belongs to $2^8 = 256$ schemata. Any binary string of length l thus belongs to 2^l schemata. In a population of n individuals, then, there are between 2^l (if all members are identical) and $n2^l$ (if no two schemata are the same) schemata. Populations with higher diversity have a greater number of schemata.

Schemata that are part of an individual with high fitness have a higher than average probability of reproducing. Therefore, highly fit schemata benefit from differential reproduction relative to fitness. If selection were the only operator used, though, no new regions of the search hyperspace would ever be explored. Crossover and mutation provide new schemata to guide the search into new regions.

Crossover is a slightly more complicated matter than reproduction. Consider two schemata: ##1####0 and ###10###. If both are parts of strings of equal fitness, which is more likely to be passed on to the new population? Either one- or two-point crossover is more likely to disrupt the first, since it is quite likely that a crossover point will occur between the two string end-points. The second is more compact in nature and is relatively unlikely to be disrupted by a one- or two-point crossover operation.

Mutation is not likely to disrupt either schema, since it typically occurs at a very low rate. And since it is considered on a bit-by-bit basis, if it does occur it is just as likely to disrupt one as the other.

While crossover and mutation are potentially disruptive, they facilitate an efficient search by introducing innovations. Furthermore, compact (short) schemata that are part of highly fit individuals will, with high probability, appear in ever-increasing numbers in future generations. The schemata are the elements of which future generations are built; Holland (1992) named them "building blocks." The schema theorem sums up all of this and provides a quantitative estimation of one aspect of GA performance.

The Schema Theorem

The schema theorem predicts the number of times a specific schema will appear in the next generation of a GA, given the fitness of the population member(s) containing the schema, the average fitness of the population, and other parameters. The GA can be thought of as effectively working with a large number of schemata simultaneously, ranging from very short schemata to schemata as long as the individual population members. This has been named "intrinsic parallelism" by Holland. The schema theorem provides a quantitative prediction for all schemata, regardless of length. It should be noted that the theorem applies only to "plain vanilla" GAs. As soon as you do anything special, including something as simple as implementing elitism, where the fittest population member is automatically copied into the next generation, the schema theorem no longer applies.

The derivation of the theorem is beyond the scope of this book. The reader is referred to the derivation in Goldberg (1989). The schema theorem is

$$n_{t+1}(S) \geq n_t(S) \frac{f(S)}{f_{avg}} \left[1 - p_c \frac{\delta(S)}{l-1} - o(S)p_m \right] \tag{4.1}$$

In Equation 4.1, n is the total number of examples of a particular schema S. The subscripts $t + 1$ and t refer to time steps, or generations. The parameter $f(S)$ is the average fitness of the individual population members that contain the schema S, while f_{avg} is the average fitness of the entire population. The probabilities of crossover and mutation are p_c and p_m, respectively.

The parameter $\delta(S)$ is called the "defining length" of the schema; it is the distance between the first and last specific string positions. For example, for the schema #01#11#, the defining length is 4. The total length of the string is l, while $o(S)$ is the "order" of the schema, or the number of fixed positions (ones and zeroes) in the schema. In the preceding example, the order of the schema is 4. The order of a schema is just the number of potential "cut" points within the schema that could be affected by crossover.

Summarized, Equation 4.1 states that the expected number of occurrences of schema S in generation $t + 1$ is the number in the current generation multiplied by the average schema fitness divided by the average population fitness, less the disruptive effects caused by crossover and mutation. Schemata with above-average fitness values will be represented an increasing number of times as generations proceed. Those with below-average values will be represented less and less; they will "die out," just as happens in nature.

The schemata with small values for defining length are disrupted least by crossover, so the most rapidly increasing representation in any population will be of highly fit, short schemata, called *building blocks,* which will experience exponential growth. Building blocks illustrate that it is often beneficial to keep some parts of a solution intact. This is the most important consequence of the schema theorem.

Note that the schema theorem, by itself, does not specify how well a GA will solve a particular problem. It should also be noted that there is controversy in the EC community with respect to the usefulness and validity of the theorem. We include it, as have other recent books dealing with GAs such as Mitchell (1996), Pedrycz (1998), and Haupt and Haupt (1998), because we believe it provides useful insights into GA processes.

Final Comments on Genetic Algorithms

In sum, the genetic algorithm operates by evaluating a population of bitstrings (there are real-numbered GAs, but binary implementations are more common) and selecting survivors stochastically based on their fitness, so fitter members of the population are more likely to survive. Survivors are paired for crossover, and often some mutation is performed on chromosomes. Other operations might be performed as well, but crossover and mutation are the most important ones. Sexual recombination of genetic material is a powerful method for adaptation.

The material on genetic algorithms in this chapter has provided only an introduction to the subject. We suggest that you explore GAs further by sampling the references cited in this section. With further study and application, it will become apparent why GAs have such a devoted following. In the words of Davis (1991):

> There is something profoundly moving about linking a genetic algorithm to a difficult problem and returning later to find that the algorithm has evolved a solution that is better than the one a human found. With genetic algorithms we are not optimizing; we are creating conditions in which optimization occurs, as it may have occurred in the natural world. One feels a kind of resonance at such times that is uncommon and profound.

This feeling, of course, is not unique to experiences with GAs; using other evolutionary algorithms can result in similar feelings. An implementation of a genetic algorithm is presented in Appendix B. The software for the GA implementation is on the book's Internet site.

Evolutionary Programming

Evolutionary programming is the second of the four evolutionary computation paradigms we examine in this chapter. It is similar to genetic algorithms in its use of a population of candidate solutions to evolve an answer to a specific problem, and differs in its concentration on "top-down" processes of adaptive behavior. The emphasis in evolutionary programming is on developing behavioral models, that is, models of observable system interactions with the environment. Theories of natural evolution heavily influence the development of evolutionary programming concepts and paradigms.

Evolutionary programming is derived from the simulation of adaptive behavior in evolution: GAs are derived from the simulation of genetics. The difference is perhaps subtle, but important. Genetic algorithms work in the *genotype space* of the information codings, while evolutionary programming (EP) emphasizes the *phenotype space* of observable behaviors (Fogel, 1990). EP therefore is directed at evolving "behavior" that solves the problem at hand; it mimics "phenotypic evolution."

Evolutionary programming is a more flexible approach to evolution than some of the other paradigms. Operators are freely adapted to fit the problem at hand. Generally the paradigm relies on mutation—and not sexual recombination—to produce offspring. Whereas evolution strategies systems usually generate many more offspring than parents (a ratio of seven to one is common, as we will see in the next section), EP usually generates the same number of children as parents. Parents are selected to reproduce using a tournament method; their features are mutated to produce children who are added to the population. When the population has doubled, the members—parents and offspring together—are ranked, and the best half are kept for the next generation.

A significant addition to the basic evolutionary programming methodology is self-adaptation, which provides the capability of strategy parameters to evolve themselves, thus directing mutation into more promising search space. The three main types of evolutionary programming are called *standard EP, meta-EP,* and *Rmeta-EP,* which are distinguished by different levels of self-adaptation (Bentley, 1999).

The process of implementing EP in an application is presented next. Examples of specific application areas follow.

The Evolutionary Programming Procedure

The procedure that is generally followed when implementing an EP appears in the following listing. Following a brief description of the procedure, we review two types of applications: modeling of a predictive system and building a function optimization tool.

The EP procedure is

1. Initialize the population.
2. Expose the population to the environment.
3. Calculate fitness for each member.
4. Randomly mutate each "parent" population member.
5. Evaluate parents and children.
6. Select members of new population.
7. Go to step 2 until some condition is met.

The population is randomly initialized. For problems in real (computable) space, each component variable of each individual's vector is generally a real value that is constrained to some dynamic range. In the two EP examples that follow, the variables (vector elements) represent finite state machine parameters and function variables, respectively. The number of population members is problem dependent, but is often a few dozen to a few hundred, similar to GA populations.

In order to better understand the remaining steps in the EP procedure, we consider two examples. These examples are representative of two main types of problems to which EP paradigms are often applied. The first involves time series prediction using a finite state machine. The second is the optimization of a mathematical function.

Finite State Machine Evolution

Evolutionary programming paradigms are sometimes used for problems involving prediction. One way to represent prediction of the environment is with a sequence of symbols. As with GAs, the symbols must be members of a finite alphabet. We can use a system comprising a finite state machine, for example, to analyze a symbol sequence and to generate an output that optimizes a fitness function, which often involves predicting the next symbol in the sequence. In other words, a prediction is used to calculate a system response that seeks to achieve some specified goal.

A *finite state machine* is defined as "a transducer that can be stimulated by a finite alphabet of input symbols, can respond in a finite alphabet of output signals, and possesses some finite number of different internal states" (Fogel, 1991). The input and output symbol alphabets need not be identical. We must specify the initial state of the machine. We must also specify, for each state and input symbol combination, the output symbol and next state. Table 4.1 specifies a three-state finite state machine with an input alphabet of two characters and three possible output symbols.

Finite state machines are essentially a subset of Turing machines, developed by the English mathematician and computer science pioneer Alan Turing (1937). Turing machines are capable, in principle, of solving all mathematical problems (of a defined general class) in sequence. Finite state machines, as used in EP, can model, or represent, an organism or a system.

Unlike GAs, where crossover is an important component of producing a new generation, mutation is the only operator used in EP systems. Each member of the current population typically undergoes mutation to produce a "child." Given the specification of the finite state machine and its operation, there are five main types of mutation that can occur: As long as more than one state exists, the initial state can be changed, and/or a state can be deleted. A state can be added. A state transition can be changed. Finally, an output symbol for a given state-input symbol can be changed.

Although the number of children produced by each parent is a system parameter, each "parent" typically produces one "child," and the population becomes twice its original size after mutation. After measuring the fitness of each structure, the best one-half are kept, maintaining the population size at a constant value from generation to generation. At some point in some applications, it is necessary to make a prediction of the

Table 4.1 Specification table for a three-state finite state machine (after Fogel, 1991).

Existing state	A	A	B	B	C	C
Input symbol	1	0	1	0	1	0
Output symbol	Y	Y	X	Z	Z	Y
Next state	A	B	C	B	A	B

next symbol in a sequence. The structure with the highest fitness is chosen to generate this new symbol, which is then added to the sequence. (It is also possible to specify the problem so that the symbol predicted is further in the future than one time step.)

Unlike other evolutionary paradigms, in EP systems mutation can change the size of structures (states can be added and deleted). This fact and the potential for changing state transitions lead to another consideration: the specification table for a finite state machine can have unfilled blanks in it. There can be mutations that add states that are never utilized in a given problem; Fogel (1991) calls these "neutral mutations." It is also possible to create the situation via mutation where a state transition that is specified is not possible because the new state has been deleted. Mutations such as this and others, such as changing output symbols, tend to have less effect the more states the machine has, but can still cause fatal errors in the finite state machine if they are not handled properly.

Although Fogel (1995) usually allows a variable-length structure, it is also possible to evolve a finite state machine with EP using a fixed structure. First, the maximum number of states must be determined. For purposes of illustration, using the three-state machine defined earlier as an example, we will assume that no more than four states are allowed.

Each state could then be represented by a fixed six-bit binary element as follows. The first bit could represent the "activation" of the state: if it is 1, the state is active, if 0, the state is inactive (i.e., it does not exist). The next bit could represent the input symbol: 1 or 0. The next two bits could represent the output symbol: X, Y, or Z. (Note that our example above has only three output symbols. We have to either allow four or handle a nonexistent symbol the way nonexistent states are handled.) The final two bits then designate one of four output states.

The population is thus initialized with individuals 24 bits long. For the example it may be a good idea to specify that only individuals with at least two active states can be allowed in the initial population.

A child is now generated for each parent. Given the five possible kinds of mutation outlined earlier, one possible mutation procedure is the following:

1. For each individual, generate a random number from 0 to 1.

2. If the number is between 0.0 and 0.2, change the initial state; if between 0.2 and 0.4, delete a state; and so on.

3. The mutation selected in step 2 is done with a flat probability across all possibilities. For example, if the initial state is to be changed and there are a active states, then one active state is selected to be the initial state; each active state has the probability of $1/a$ of being selected.

4. Infeasible state transitions are modified to be feasible. If a state transition to an inactive state has been specified, one of the active states is selected to be the object of the transition. As above, each active state has the probability of $1/a$ of being selected.

5. Evaluate fitnesses and keep the best 50 percent, resulting in a new population of the same size.

The scenario outlined above is only one of many possibilities. For example, it might be desirable to lower the probability ranges (the ranges between 0 and 1 in step 2) for adding and deleting states, and correspondingly increase the mutation probability ranges for changing input symbols and/or output symbols. It is also possible to *evolve* the ranges, number of states, and so on.

So how do finite state machines relate to what we've been discussing in earlier chapters? One example is the development by Fogel (1995) using evolutionary programming of finite state machines that do very well at playing the prisoners' dilemma game. The payoff function is that used by Axelrod (1980): if both cooperate, each player gets 3 points; if both defect, each player gets 1 point; if one defects and one cooperates, the cooperating player gets no points while the defecting player gets 5 points.

Fogel allowed the finite state machines to have up to eight states. This doesn't represent all possible behaviors *à la* Axelrod, but does allow a dependence on sequences of greater than third order. Fogel was able to evolve finite state machines that had average scores slightly greater than 3.0, which is the score that is achieved through mutual cooperation alone.

Figure 4.7 is the diagram for a seven-state finite state machine (one of many evolved by Fogel) to play prisoners' dilemma. The start state is state 6, and play is begun by cooperating. In the figure, "C" denotes co-operate and "D" denotes defect. The input alphabet comprises [(C,C), (C,D), (D,C), (D,D)], where the first letter represents the finite state machine's previous move and the second the opponent's. So, for example, a label of C,D/C on the arrow leading from state X to state Y means that if the system is in state X and on the previous move the finite state machine cooperated and the opponent defected, then cooperate and transition to state Y. Sometimes, more than one situation can result in the same state transition. For example, in Figure 4.7, assume the machine is in state 6, in which case if the machine and opponent both defected on the previous move, the machine defects (D,D/D) and transitions to state 2. Likewise, a transition from state 6 to state 2 occurs if the machine co-operated and the opponent defected on the previous move; the machine cooperates in this case (C,D/C) as it moves into state 2.

Function Optimization

The second example of a type of problem to which EP paradigms are applied is function optimization. The following example features the modification of each component of the evolving individual structures with a Gaussian random function.

Consider, for the example, optimizing a function with two variables such as $F(x,y) = x^2 + y^2$. The extremum in this case is a minimum at $x = y = 0$. The first step is to establish a random initial population and to specify the dynamic range of the two variables. One plausible approach might be to start with an initial population of 50 individuals, each variable of which is initialized randomly over the range $[-5, 5]$. The fitness value of each of the individuals is then calculated. The inverse of the Euclidean distance from the origin is one reasonable fitness measure.

Each "parent" individual is mutated to create one "child." The mutation method used by Fogel (1991) is to add a Gaussian random variable with zero mean and variance equal to the parent's error value (the Euclidean distance from the origin in this example) to each parent vector component. The fitness of each of the children is then evaluated the same way as the parents.

The process of mutation is illustrated by the following equation:

$$p_{i+k,j} = p_i + N(0, \beta_j \phi_{p_i} + z_j), \quad \forall_j = 1, \ldots, n \tag{4.2}$$

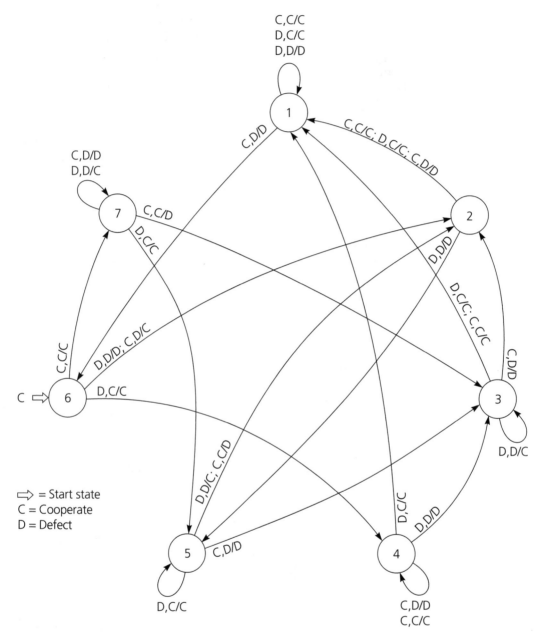

Figure 4.7 A seven-state finite state machine to play prisoners' dilemma. (From Fogel, 1995.)

where $p_{i,j}$ is the jth element of the ith organism, $N(\mu, \sigma^2)$ is a Gaussian random variable with mean μ and variance σ^2, ϕ_{pi} is the fitness score for p_i, β_j is a constant of proportionality to scale ϕ_{pi}, and z_j represents an offset. For the function used in the example, it has been shown that the optimum rate of convergence is represented by

$$\sigma = \frac{1.224\sqrt{f(x)}}{n}$$

where n is the number of dimensions (Baeck and Schwefel, 1993).

Another way to perform mutation involves the process of self-adaptation, mentioned earlier. In this variation, the standard deviations (and rotation angles, if used) are modified based on their current values. As a result, the search adapts to the error surface contours (Fogel, 1995).

Fitness, however, is sometimes not used directly by itself to decide which half of the augmented population will survive to the next generation. Tournament selection is used, with each individual competing with a number, say, 10, of other individuals in the following way.

For each of the 10 competitions with other individuals, a probability of "scoring a point" is set equal to the error score of the opponent divided by the sum of the individual and opponent errors. For instance, if the error of the individual is 2 and that of the opponent (one of 10 opponents) is 3, the probability of scoring a point is 3/5, or 60 percent. The total score is tallied over the 10 competitions for each individual, and the half of the population with the highest total scores is selected for the next generation.

Final Comments

The implementation of evolutionary programming concepts seems to vary more from application to application than GA implementations. A number of factors contribute to the differences in approach, but the most important factor seems to be the "top-down" emphasis of EP. Another is the fact that selection is a probabilistic function of fitness, rather than being tied directly to it. One developer of EP (Fogel, 1991) stated that EP is at its best when it is used to optimize overall system behavior.

Evolution Strategies

As a biological analogy, evolution strategies models problem solutions as species rather than as they have been described earlier, as populations of normally distributed multivariate points scattered around a fitness landscape. The aspect of these populations that permits them to adapt to their environment (in research this is often simulated by a *test function* or hard optimization problem) is their ability to evolve their own evolvability.

If evolutionary programming is based on evolution, then, reasons Rechenberg (1994), the field of evolution strategies is based upon the evolution of evolution. Since biological processes have been optimized by evolution, and evolution is a biological process, then evolution must have optimized itself. Evolution strategies, although utilizing forms of both mutation and crossover (usually called *recombination* in the evolution strategies literature), have a slightly different view of both operations than either evolutionary programming or genetic algorithms.

There are many similarities between evolution strategies and evolutionary programming, and in fact the two paradigms are moving closer together as researchers exchange techniques across the Atlantic. Evolution strategies, like evolutionary programming, take a top-down view. They also stress the phenotypic behavior as opposed to the genotypic. This means, for example, that the phenotypic behavior ramifications of recombination are of importance, rather than what happens to the genotypes. ES paradigms also usually use real values for the variables, rather than the binary coding favored in genetic algorithm implementations.

Mutation

In evolution strategies the goal is to move the mass of the population toward the best region of the landscape. Through application of the simple rule, "survival of the fittest," the best individuals in any generation are allowed to reproduce; their offspring resemble them but with some differences introduced through mutation. An individual is a potential problem solution characterized by a vector of numbers representing phenotypic features. Mutation is performed by adding normally distributed random numbers to the parents' phenotypic coordinates, their position in the search space, so that the next generation of children explores around the area in the landscape that has proven good for their parents.

The amount of mutation—the evolvability of the population—is controlled in an interesting way in ES. An individual is typified by a set of features and by a corresponding set of *strategy parameters*. These are usually variances or standard deviations (the square root of the variance), though other statistics are sometimes used. The strategy parameters are used to mutate the feature vectors for the individual's offspring; for instance, standard deviations can be used to define the variability of the normal distribution used to perturb the parent's features. Random numbers can be generated from a probability distribution with a mean of zero and a standard deviation defined by the strategy parameters; adding these random numbers to the values in the parent's feature vector simulates mutation in the offspring. They resemble the parents but differ from them to some controlled extent. Since the evolutionary process is applied to the strategy parameters themselves, the range of mutation, or the variability of the changes introduced in the next generation, evolves along with the features that are being optimized. This is analogous to the self-adaptation process in evolutionary programming that we previously discussed.

Intuitively it can be seen that increasing the variance is like increasing the step size taken by population members on the landscape. High variance equals exploration, wide-ranging search for good regions of the landscape, and corresponds to a high rate of mutation, while low variance is exploitation, focused search within regions. The strategy parameters stochastically determine the size of the steps taken when generating offspring of the individual; a large variance means that large steps are likely to be taken, that the children are likely to differ greatly from their parents. As the children are randomly generated from a normal distribution, though, a large variance *can* produce a small step size and vice versa. It is known that 68.26 percent of random normal numbers generated will fall within one standard deviation, 95 percent fall within 1.96 standard deviations of the mean, and so on. So widening the standard deviation widens the dispersion of randomly generated points.

ES's unique view of mutation includes the concept of an *evolution window*. The theory behind the concept is that mutation operations result in fitness improvement only if they land within a defined step size band, or window (Rechenberg, 1994). Recombination (crossover) and mutation operations that land outside the evolution window are not helpful. A theoretical derivation of Rechenberg states that if mutations are carried out with an optimal standard deviation, the probability of a "successful" (helpful) mutation is about one-fifth. Evolution strategies carry the idea of the evolution window still further. They assert that dynamic adjustment of the mutation size to a dynamic evolution window can provide benefits called *meta-evolution,* or evolution of the second kind (Rechenberg, 1994).

Like evolutionary programming, ES employs Gaussian noise functions with zero mean to determine mutation magnitudes for the variables. For the strategy parameters, log normal distributions are sometimes used as mutation standard deviations.

ES theory states that mutation rates should be inversely proportional to the number of variables in the individual population member and should be proportional to the distance from the function optimum. In real-world applications, of course, the exact value of the optimum is usually unknown. However, some knowledge often exists about the optimum. It is often known within an order of magnitude; sometimes to within a factor of two or three. Even limited knowledge such as this can be helpful in guiding an evolution strategy search.

Recombination

In ES, recombination manipulates entire variable values. This is usually done using one of two methods. The first and more common method (the *local* method) involves forming one new individual using components (variables) from two randomly selected parents. The second method, the *global* method, uses the entire population of individuals as potential sources from which individual components for the new individual can be obtained.

Each of the two methods, local and global, is generally implemented in one of two ways. The first is called *discrete recombination,* which consists of selecting the parameter value from either of the two parents. In other words, the parameter value in the child equals the value of one of the parents. The second way, called *intermediate recombination,* involves setting each parameter value for a child at a point between the values for the two parents; typically, the value is set midway between those values. If the parents are denoted by *A* and *B,* and the *i*th parameter is being determined, then the value established using intermediate recombination is

$$x_i^{new} = x_{A,i} + C(x_{B,i} - x_{A,i})$$

where C is a constant, usually set to 0.5 to yield the midpoint between the two parent values.

Thus we see that evolutionary strategies contain a component representing sexual combination of features. In intermediate recombination, for instance, the children's features are computed as a kind of average of the two parents' features, while in discrete recombination, individual features may come intact or mutated from one parent or the other.

In the experience of ES practitioners, the best results often seem to be obtained by using the local version of discrete recombination for the parameter values and the local version of intermediate recombination for the strategy parameter(s). In fact, Baeck and Schwefel (1993) report that implementation of strategy parameter recombination is mandatory for success of any ES paradigm.

Selection

In evolution strategies, as in all Darwinian models, an individual's fitness determines the probability that it will reproduce in the next generation. There can be many ways to decide this; for instance, we could rank all the individuals from best to worst, chop off the bottom of the list, and save only the proportion that we want to survive. This proportion depends on how many offspring they shall have, assuming the population size remains constant from one generation to the next. In nature, of course, there is no ranking of individuals; the survival of each depends on the environment and that individual's chance encounters. Imagine a snowshoe hare that has a mutation that makes its fur turn black in the winter. In the snow this hare is more visible than its camouflaged cousins. It might just happen, though, that no predators come into the area where this hare lives, so they don't see it, and subsequently it reproduces, passing on the mutation. It *can* happen—it is just that the likelihood is reduced relative to the alternative, which is that a predator that comes into the area immediately notices this contrastive morsel and eats him, rather than his harder-to-see littermates. In nature, the measure of fitness has a great amount of error in it; possible improvements are commonly lost.

This suggests that selection needs to be probabilistic—you can't just propagate the best so-many individuals to the next generation. The lesson we learned from simulated annealing was that sometimes a step backward is productive in the long run. In the same way, natural evolution lets some less-fit individuals reproduce, and it is quite likely that eventual improvement is transmitted through the less obvious route. Evolutionary computation researchers have come up with a number of techniques for stochastically selecting survivors for the next generation.

In order to better model the stochastic aspect of natural selection—what could be called survival of the luckiest—several computational methods of selection have been devised. Common methods include ranking, roulette wheel selection, and tournament selection.

Ranking is the simplest procedure, though it does not have the advantage of allowing selection of less-fit individuals. The population is sorted from best to worse, and individuals above the cutoff in the list are chosen. One salient objection to this method is that it requires global information. Knowledge of all fitness values is needed in order to determine the rank of any individual. Obviously nature does not work this way; only local information is used in natural selection, and errors in ranking—occasions where more-fit members fail to reproduce or less-fit members succeed—contribute to the adaptation of the population. This might be a weaker argument than it seems; there are plenty of times, though, where a computer needs to use global information in order to accomplish things that nature does without it. For instance, to detect collisions in virtual worlds requires computation of the relative positions of all objects in the world, while in the physical world things behave appropriately without any such computations. Running into a brick wall stops you, period. So evolution in a computer program might be acceptable even if it requires global information as a way to accomplish an end.

Roulette wheel selection was discussed in the previous section on genetic algorithms. Recall that, in roulette-wheel selection, each individual is given a probability of selection proportional to its fitness.

In our discussion of optimization we discussed three number spaces: the parameter space, the function space, and the fitness space. Almost all studies in artificial evolution treat the function space as if it is identical to the fitness space; that is, the function output provides a number that indicates how close to the global optimum the search algorithm is. We note, though, that there are dangerous ambiguities in the confusion of these two quantities. Recall the two measures of fitness that could be used with the simple problem $4 + x = 10$ in the earlier chapter. The landscape was very different when fitness was defined as a multiplicative inverse from when it was simply the negative of additive error. The ideal fitness measure should generally be scaled between zero and one, representing the probability of a population member's survival.

Outside of Eden there is always some selection pressure; a limited number of individuals can survive, so fitness should remain constant on the average. When population size is limited—as it always is—an individual's fitness must be considered to be relative to that of other individuals in the population, as well as relative to the function result. Selection based on simple ranking is consistent with the concept of competition, where one contestant wins and one loses, though an individual that is a little bit better gets the same advantage as one that is very much better. We see in sum that fitness contains two components that are correlated but not necessarily identical: the first is simple fitness, having to do with the ability of the individual to meet the demands of the environment, and the second derives from the inability of the environment to support all the individuals that meet the first criterion. The first should ideally be reported in floating-point quantities, and the second requires rank information only.

Tournament selection uses local competitions to determine survivors. In its simplest form, individuals are paired at random, and the best member of each pair is selected to reproduce. This can be repeated until the next generation is sufficiently populated. Other tournament methods pair up individuals in some number of competitions, adding a point to their score each time they win, and then keep individuals with more than a critical number of points; other methods select subgroups at random from the population and allow the one with the highest fitness to survive to the next generation (see Figure 4.8).

The results of tournament selection correlate with the results of ranking; that is, fitter individuals survive in general. One-on-one, winner-take-all tournaments allow the most error in terms of less-fit individuals being selected; while the very best individual is guaranteed to survive and the very worst is guaranteed not to, it is entirely possible that the next-to-worse individual is paired with the worst one and thus is selected. Repetitive and subgroup tournaments decrease the amount of error while increasing the correlation with ranking results, until an algorithm where each individual engages in $n - 1$ unique tournaments, where n is the population size, is exactly equivalent to ranking.

Differences exist between evolution strategies and other paradigms of evolutionary computation with respect to selection. ESs generally operate with a surplus of descendants. Schwefel (1994) describes the most common versions of ES selection, known as the (μ, λ) and $(\mu + \lambda)$-ES. In both versions, the number of children generated from μ parents is $\lambda > \mu$. Commonly used is a λ/μ ratio of 7.

In the original (1+1)-ES, one parent produces one offspring, with only the fitter of the two surviving. Although it is called "two-member ES," this is an evolutionary algorithm with a population of one (Bentley, 1999). This version is now seldom used.

The difference between the "plus" and "comma" versions comes in the next step. In the (μ, λ) version, the μ individuals with the highest

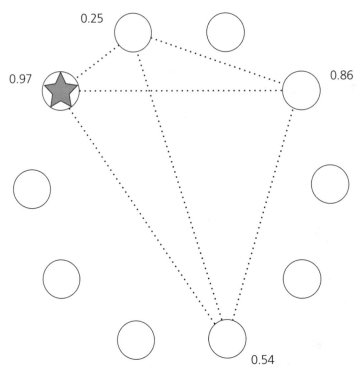

Figure 4.8 The winner of this tournament, with a fitness of 0.97, will be one of the members of the next generation.

fitness values out of the λ children are selected. Note that the μ parents are not eligible for selection in this scheme, only the children. In the (μ + λ) version, the best μ individuals are selected from a pool of candidates that includes both the μ parents and the λ children, that is, the union of the two groups of individuals. Whichever of the methods is used, the μ individuals that are left have thus been selected completely deterministically and have equal probabilities to mate and have descendants in the next generation.

In the discussion of genetic algorithms, the elitist strategy was mentioned, in which the individual with the highest fitness in each generation is guaranteed to survive to the next generation. This individual may be carried over from the previous generation or appear as a result of operations in the current one. As can be seen from the preceding discussion, the (μ + λ) version implements elitism, as the most-fit parent will be retained, while the (μ, λ) version does not. Elitism is generally considered to be helpful in GA applications. With evolution strategies, however, the

(μ, λ) version is generally observed to yield better performance (Baeck and Schwefel, 1993).

The following list summarizes the procedure used in most evolution strategies:

1. Initialize population.
2. Perform recombination using the μ parents to form λ children.
3. Perform mutation on all children.
4. Evaluate λ or $\mu + \lambda$ population members.
5. Select μ individuals for the new population.
6. If the termination criterion is not met, go to step 2; otherwise, terminate.

In sum, in evolution strategies mutation is applied to the parent's features to generate children that resemble the parent but differ stochastically from it. Each of the survivor's positional coordinates is entered as the mean of a normal distribution, the corresponding strategy parameter is entered as the variance or standard deviation, and a child vector of numbers is generated for both positions and strategy parameters. These children are evaluated, selection is applied, and the cycle repeats. The evolution of strategy parameters suggests the evolution of evolvability, adaptation of the mutability of a species as it searches for, then settles into, a niche.

Genetic Programming

The three areas of evolutionary computation discussed thus far have all involved individual structures that are defined as strings. Some are strings of binary values, some include real-valued variables, but all are strings, or vectors. The genetic programming (GP) paradigm deals with evolving hierarchical computer programs that are generally represented as tree structures. Furthermore, while individual structures utilized up to this point have generally been of fixed length, programs being evolved by genetic programming generally vary in size, shape, and complexity.

One perspective is that GPs are a subset of GAs that evolve executable programs. Differences between GPs and generic GAs include the following:

■ Population members are executable structures (generally, computer programs) rather than strings of bits and/or variables.

■ The fitness of an individual population member in a GP is measured by executing it. (Generic GAs' measure of fitness depends on the problem being solved.)

The goal of a genetic programming implementation is to "discover" a computer program within the space of potential computer programs being searched that gives a desired output for a given set of inputs. In other words, a computer is figuring out how to write its own code.

Each program is represented as a parse tree, where the functions defined for the problem appear at the internal tree points, and the variables and constants are located at the external points (leaves) of the tree. The nature of the computer programs generated gives genetic programming an inherently hierarchical nature.

In preparation for running a genetic programming implementation, five steps are carried out:

1. Specify the terminal set.
2. Specify the function set.
3. Specify the fitness measure.
4. Select the system control parameters.
5. Specify termination conditions.

The terminal set comprises the variables (the system state variables) and constants associated with the problem being solved. For example, consider a "cart centering" problem, where the goal is to center a cart in the least amount of time on a one-dimensional frictionless track by imparting fixed-magnitude forces that accelerate the cart left or right. The variables would be the cart's position x and velocity v. A constant such as -1 would also be an appropriate terminal for this problem (see Koza, 1992, Chapter 6).

The functions selected for the function set are limited only by the programming language implementation that will be used to run the programs evolved by the GP implementation. They can thus include mathematical functions (cos, exp, etc.), arithmetic operations ($+$, $\#$, etc.), Boolean operators (AND, NOT, etc.), conditional operators such as if-then-else, and iterative and recursive functions. Each function in the function set requires a certain (fixed) number of arguments, known as the function's *arity*. (Terminals are functions with arity 0.) One of the tasks of specifying the function set is to select a minimal set that is capable of accomplishing the task.

This leads to two properties that are desirable in any GP application: *closure* and *sufficiency*. For the closure property to be satisfied, each function must be able to successfully operate on any function in the function set and on any value of any data type assumable by a member of the terminal set.

This occasionally requires definition of special cases for functions. For example, in arithmetic functions division by zero can be defined for the purposes of a problem as being equal to some constant value such as 1. If Boolean values returned by conditional operators are not acceptable, the conditional operator can be redefined in one of two ways: (1) Numerical values (such as 0 and 1) can be returned rather than Boolean values (such as F and T), or (2) conditional branching and conditional comparative operators can be defined to execute one of their arguments depending on the evaluation of the test involving an external state or condition, or on the comparison test outcome. Functions that are redefined so as to return acceptable values are called *protected functions*. If the closure property is not satisfied, some method must be specified for dealing with infeasible population members, and with members whose fitness is not acceptable.

For the sufficiency property to be satisfied, the set of functions and set of terminals must be sufficiently extensive to allow a solution to be evolved. In other words, some combination of functions and terminals must be capable of producing a solution. Some knowledge of the problem is generally required to be able to judge when the sufficiency property is met. In some problem domains, sufficiency is relatively easy to determine. For example, if Boolean functions are being used, it is well known that the function set comprising AND, OR, NOT is sufficient for any problem. For other problems, it can be relatively difficult to establish sufficiency.

Having more than the minimally sufficient number of functions has been found to degrade performance somewhat in some cases and to significantly improve it in others. Having too many terminals, however, usually degrades performance (Koza, 1992).

The fitness measure often is selected to be inversely proportional to the error produced by program output. Other fitness measures are also common, such as the score a program achieves in the game.

The two main control parameters are the population size and the maximum number of generations that will be run. Other parameters used include reproduction probability, crossover probability, and the maximum size allowed (as measured by the depth, or number of hierarchical levels) in the initial and final program populations.

The termination condition is usually determined by the maximum number of generations specified. The winning program is usually the best program (in terms of the fitness measure) created thus far in any generation.

Once the five preparatory steps for running a GP are completed, the GP process can be implemented as follows:

1. Initialize the population of computer programs.
2. Determine the fitness of each individual program.
3. Carry out reproduction according to fitness values and reproduction probability.
4. Perform crossover of subexpressions.
5. Go to step 2 unless termination condition is met.

The population is initialized with randomly generated computer programs comprising functions and terminals from the selected sets. In other words, each program in the initial population is created by building a rooted tree structure with randomly selected functions and terminals from the defined sets. No restrictions are placed on the size or shape (configuration) of acceptable programs, other than the maximum depth, or number of hierarchical levels, allowed. Each individual structure created is a hierarchically structured executable program. A population size of 500 has been reported to be sufficient for most problems solved with GP implementations (Koza, 1992).

The root of each program tree is a function randomly selected from the function set. The root of a randomly created program appears at the top of Figure 4.9. The number of lines, or branches, emanating from the function is equal to its arity. In the figure, the multiplication function "*" takes two arguments.

Once the root function is selected, there are a number of ways the program population can be created. Following is a description of what Koza (1992) calls the *ramped half-and-half* method. It makes use of two approaches to building program trees: the *grow* method and the *full* method.

In the *grow* approach, a random selection is made from the combined set of functions *and* terminals for placement at the end of each line emanating from the root function. If a function is selected, program creation

*Other functions and links
continue down from here*

Figure 4.9 Example of a root of a randomly created program in the initial population.

continues recursively with selections from the combined set. Whenever a terminal is selected, a leaf, or end-point, of the tree is established. Program creation along that line is thus terminated. Except for the root function, therefore, all functions are at internal tree locations. The leaves of the tree are all terminals. Anytime the maximum depth (number of hierarchical levels) is reached, the random selection is limited to the terminal set. When the grow method is used, the program tree configuration is guided by the ratio of the number of functions to the number of terminals. When the ratio is higher, the average depth of each limb is higher.

In the *full* approach, each limb of the program tree extends for the full depth. Only functions are selected for placement at the end of each line until the maximum depth is reached, at which time only terminals are selected. All programs created using the full approach thus have identical fully developed structures.

The *ramped half-and-half* approach produces a population diverse in size and shape. Koza (1992) reports using this method for almost all problems except those involving Boolean functions. The method consists of creating programs with evenly distributed depth parameters ranging from 2 to the maximum depth. For example, if the maximum depth is 5, 25 percent of the population would have depth 2, 25 percent depth 3, and so on. Within each subpopulation of a given depth, one-half of the programs are created using the grow approach, one-half using the full approach.

The fitness of each program is generally calculated for a number of cases, with the average fitness value over the cases being defined as a program's fitness. For example, if a program were being evolved to calculate y as some function of x, each individual program might be tested over 50 or 100 cases, each representing a value of x in the domain. It is important to use a sufficient number of cases to represent this domain. Although it is possible to use different cases in different generations, the same fitness cases are usually used across all generations. Fitness can be calculated in a number of ways. Koza (1992) defines four fitness metrics: raw, standardized, adjusted, and normalized fitnesses.

Steps 3 and 4 of the GP process are often carried out in parallel. A probability is assigned to reproduction, and another to crossover, so that the two sum to 1. If, for example, the probability of reproduction is 10 percent (a typical value in Koza's problems), then the probability of crossover is 90 percent. This means that once fitness calculations have been made, and it is time to build the new program population, a decision is made based on these probabilities whether to perform reproduction or crossover.

If reproduction is selected, it is often carried out in a similar fashion to roulette wheel selection used in GAs. A candidate program is selected for reproduction with a probability proportional to its fitness divided by the sum of all of the programs' fitnesses (its normalized fitness). For very large populations of 1,000 or more, highly fit individuals are sometimes given an even greater probability of selection than their normalized fitness. This is called *overselection*.

If crossover is selected, it is accomplished by first selecting two parents using a method based on normalized fitness similar to that used for reproduction. Then, one point is randomly selected in each parent as the crossover point. The point can be anywhere in each program, including the root and internal functions, or the terminals. The entire substructure consisting of the crossover point root and everything below it is exchanged between the two programs.

Note that the parent programs, as well as the exchanged substructures, are usually of different sizes and configurations. Note also that the results of some operations may not be what is usually expected of crossover. An example is when the roots of the two programs are selected as crossover points, in which case the results are identical to the two programs being selected for reproduction into the new population.

When a crossover operation would result in a program that exceeds the maximum defined depth, the program that would exceed the depth limit as a result of crossover is copied unaltered into the new population, while the crossover operation is carried out for the other program. In other words, the subtree at and below the crossover point in the unaltered program replaces the program portion at and below the crossover point in the other program.

Preprocessing and postprocessing, as typically done when working with other computational intelligence tools such as artificial neural networks and genetic algorithms, play a relatively minor role in GP implementations. The selection of the function and terminal sets significantly depends on the problem domain, however, so this selection could be thought of as preprocessing.

Formulating the approach to solving a problem with a GP implementation can be difficult. Discovering what other people have done in similar circumstances is often helpful. Chapter 26 of Koza's 1992 book presents tables to help guide a user in selection of terminal sets, function sets, population size, and so on. Genetic programming has, as have the other evolutionary algorithms we've discussed, developed a set of "advanced" operators. Included are *permutation,* in which two characters in a tree are exchanged, *editing,* in which lengthy S-expressions are compressed, and *encapsulation,* in which an entire subtree is treated as a single node, thereby insulating it from any effects of crossover or mutation.

Summary

Genetic algorithms, evolutionary programming, evolution strategies, and genetic programming have a number of qualities in common, including the fact that all evolve solutions, all utilize some kind of selection based on survival of the fittest, and all invoke some sort of evolutionary manipulation such as crossover or mutation.

The paradigms differ in their handling of infeasible points in an EC implementation. This subject has been studied and discussed extensively, but no universally applicable resolution has been developed. This lack of resolution is supported by the fairly detailed discussion of infeasible solutions that appears in Michalewicz and Michalewicz (1995). In genetic programming, however, where feasibility corresponds to closure, Koza (1992) recommends that infeasible solutions not be allowed (i.e., that closure be required).

We should emphasize (again) that the boundaries between the paradigms of evolutionary computation are becoming increasingly indistinct. Researchers in one area are adapting techniques and approaches from others. For instance, the concept of self-adaptation is now embraced by researchers in all four areas we've discussed.

In Appendix B, an implementation of an example of an evolutionary computation paradigm, a "plain vanilla" genetic algorithm, is presented. As you will see in Chapters 7 and 8, the particle swarm optimizer has some attributes similar to ES, GAs, and EP.

chapter
five

Humans—Actual, Imagined, and Implied

In painting the background for the particle swarm, we are developing a perspective of the individual mind embedded inextricably in its social context. But it is hard to establish what exactly is meant by "mind." It is a word with no definition, because all attempts to define it are necessarily circular. The existence of mind is self-evident and *only* self-evident, known only by direct experience of one's own mind and inference of other people's—or is it through inference that we know our own minds, too?

In this chapter we introduce some ideas from the social sciences, focusing as before on computer simulations and programs that instantiate relevant theoretical ideas. Our argument leads us from the individual mind up to culture, and back, and forth; it is apparent that one cannot be considered without consideration of the other. The ultimate example of the interpenetration of mind and culture is language, and so we begin by attempting to incite an appreciation for the acquisition and understanding of language in its context; in this we are developing an idea that has been seen in high-dimensional semantic spaces and also in the superficial but often-credible natterings of chatterbots. Language is not something that is held and maintained by authorities who punish grammatical violators and reward wordcraft. It is much better viewed as an emergent system, where each individual in a linguistic community participates in the creation of the language and in its maintenance. And thus with the wisdom of a culture, in all its forms.

Studying Minds

As minds cannot be observed directly, the experience of thinking and feeling can only be described in metaphorical terms, and throughout history people have used the symbols and technology of their times to describe their experience of thinking. At various times mind has been popularly conceived as operating by humors, the interventions of gods, the actions of stars and their corresponding terrestrial substances, possession by demons, and, in more recent centuries, by pneumatic and hydraulic operations of glands and other physiological systems. The metaphor used to describe mental functions comes to prescribe acceptable functions; thus the metaphor and its referent co-construct one another.

In our day the prevailing metaphor is that the cognitive system is like a computer program; brains are like hardware and minds are like the software that runs on them. While there is no precise instant when this now-common idea can be said to have begun, the rise of *information theory* in the 1950s made it possible. Claude Shannon (1948) revolutionized modern thought, not to mention technology, by proposing that information could be conceptualized mathematically as a kind of inverse function of probability (see Figure 5.1).

If information is a function of the probabilities of different events, and thinking can be somehow defined as the processing of information, then the question is how thinking is dependent on the probabilities of things. Shannon conducted an experiment with language that was interesting and prescient, and demonstrates the kinds of calculations that we make in extracting information from our environments. He was curious about the information content of the alphabet as it is used in English. As it is described in his landmark 1948 *Bell System Technical Journal* paper, "Mathematical theory of communication," Shannon took a table of the probabilities of occurrence of the letters in the English language and generated some random text by picking letters stochastically. His "first-order" approximation to English looked like this:

> OCRO HLI RGWR NMIEL WIS EU LL NBNESENYA A TH EEI
> ALHENHTTPA OOBTTVA NAH BRL

A second-order approximation was created by taking the transition probabilities of letters, for instance, the probability that *T* will be followed by *H*. In fact, Shannon used a heuristic approach here. After

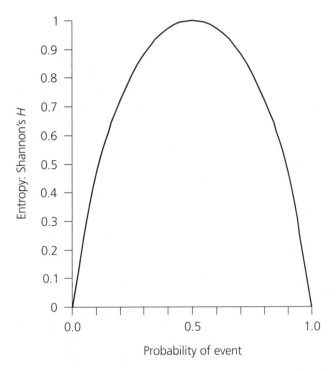

Figure 5.1 The amount of information, or entropy, in an event is a function of its probability.

picking a random starting place in a book written in English, he recorded the letter he found there, then turned to a random page in the book and ran his finger down the page until he encountered that letter. He then recorded the letter that followed it, and flipped to another random page and looked for that letter, and so on. Shannon's resulting second-order approximation was

> ON IE ANTSOUTINYS ARE T INCTORE ST BE S DEAMY ACIN D
> ILONASIVE TUCCOE AT TEASONARE FUSO TIZIN ANDY TOBE
> SEACE CTISBE

It is very interesting to see how much this random text looks like written English. Shannon's third-order approximation sampled pairs of letters, found them in a random page, took the letter that followed them, and so on. His third-order approximation was

IN NO IST LAT WHEY CRATICT FROURE BIRS GROCIDE
PONDENOME OF DEMONSTRURES OF THE REPTAGIN IS
REGOACTIONA OF CRE

Here, Shannon makes a remark whose implications he could not have
known. He says, "It would be interesting if further approximations could
be constructed, but the labor involved becomes enormous at the next
stage" (p. 8). Of course, since then, the computers that were made possible by his own theory are now able to perform the laborious calculations
almost instantaneously, with negligible human effort.

For instance, we analyzed a draft of a previous chapter of this present
volume in a similar way. The first-order approximation simply samples
the letters (including spaces and punctuation) in proportion to the frequency of their appearance in the text:

aa r ptnlomid ntm tensiidn nip nmJ irtnenflo04epnfmp tuclo sse Pt
s unhttllb rttcp opu aus t,wautihalrta.imv wflmie io ensar ih
oshotsil bss m0t r sp r.hsO lhne,.hmirrrnlriend c1f mt g ta,eoh
kmeeamsts aeekevncgsgsi ipweeeoes ariolvuhstg ertiawalnei.ii
fangmp,c aiam vrf-tlfistgcloicO Isio s eaTseoaatmtinlnnna

We agree that this passage fails to convey much meaning. The second-
order approximation samples single letters in proportion to the probability of their following the previously sampled letter:

rse me che mise by Thapan the Tulicoutoundiog ale?" Wiod hens,
atem te, wher Thoreg anof inen thuraiouskiehesithieny o alllya s
ilute Hest at oman Sthe Kichisodimborsurgly es tie ttenterofallanes
ong ia Scly trugeroajesuged Stirimant Acareiselixp as IZA
Corobisthen we, d ron theroun e ta ncotaler e ha Bomickee taces

This passage gives at least the impression of an actual language. We note
already that strings seem to have syllabic structure—you could pro-
nounce most of these words or at least make a good guess at their
pronunciation.

Because our technique is automated, and we don't have to run our
finger down a page, we can extend the paradigm to look at the third-
order approximation, that is, where pairs of characters are sampled as a
function of their following the previously selected pair (note that the
present methodology differs slightly from Shannon's, in that he sampled

the next single letter, while we take *n* letters at a time as a function of the previous *n*):

> grationgene or grepoweent a tancre pro a cal it isions or asursentin promoveme as it morgetionds, ther in hat retal embest go unds one evelowithe exed behavidualsopherends. Youtputsionmenable sual It of simizator of phen that knot ink atin the eerstices as ward peries the in es; itionarythers ife rouncts. mustife for come to

Good readers have the ability to spot misspellings on a page, as if they "jump out" at them. This effect is likely due to a powerful cognitive ability to recognize improbable character sequences in text. It is interesting then to note the failure of anything to "jump out" of the third-order random text. It "looks like" ordinary text until you try to read it.

The experiment continues to the fourth order—three-letter combinations selected probabilistically:

> 1967 paper with individual computerspect male, say for the robots of been admining comple of firstood real of animaller another ant and by simposed on seems the body, when inters frequeens when generally genes in trival, experimented you are population, fitter ches to point to waits. guage is a species' changers more able or

Suddenly, at this level of approximation, simply sampling random character strings based on the probability of their following some other strings, real English words appear. Even the strings that aren't words look like real English. It is possible to read this passage aloud with minimal difficulty.

This makes us want to sample letters in groups of four:

> is mother" in a less landscape. Hoskins uses, then there is person to us as individual's of three frameter the potential," Reynolds has program to maintaining that detect word "accept understanding individual's concept is defined with even rates thematical ant theorem of fish, 1977, p. 240). Brooks' subsumption among

This passage is *mostly* true English words. We are almost tempted to go back in our manuscript and add something about the "thematical ant theorem of fish."

Shannon left the alphabetic analysis after the third order and switched to the generation of random text based on the probabilities of the appearance of words, rather than letters. The results of this experiment are equally striking; as Attneave remarks, "the discourse approaches a sort of schizoid plausibility" (p. 17). Here is Shannon's second-order word approximation:

> THE HEAD AND IN FRONTAL ATTACK ON AN ENGLISH WRITER THAT THE CHARACTER OF THIS POINT IS THEREFORE ANOTHER METHOD FOR THE LETTERS THAT THE TIME OF WHO EVER TOLD THE PROBLEM FOR AN UNEXPECTED

A similar kind of text-generating experiment was reported by Brian Hayes in a 1983 *Scientific American* "Computer Recreations" column, deliciously titled "A progress report on the fine art of turning literature into drivel." These exercises are closely related to the contemporary generation of semantic models from sampled text by Burgess and Lund, and Landauer and Dumais. The high-dimensional semantic models extend Shannon's findings, particularly through the discovery that meaning itself emerges from the analysis of the probabilities of linguistic units' occurrence within contexts. The result is insight into the nature of language and of the use of symbols by humans; we discover that semantic processes may be driven, bottom up, by a finely tuned capacity to analyze information in its context.

Information theory gave rise to *cybernetics,* which made the leap of applying communication and control theory equally to animals—including humans—and machines, and by the 1960s the idea that human minds could be programmed like computers—they called it "psychocybernetics"—had saturated the popular belief system. During the Cold War there was great concern about "brainwashing," which was a supposed method by which Communists could take good capitalist minds and reprogram them to believe foul Red propaganda. Despite the popular belief, there was and is no scientific evidence that a person's mind can be overtaken and controlled by another without their compliance. The popular concept of mind was deteriorating from something purposive, spiritual, self-generated, and subjective to something machinelike, controllable by external forces, something that could be programmed by experts; your own mind was something that you yourself knew very little about, compared to the programmers and the psychologists ("Why did I do that, Doc?"). Mind was brain, the computer, the machine. And so we find it today.

The Fall of the Behaviorist Empire

Through most of the 20th century, mainstream academic psychologists, at least in the United States, were not allowed to talk about the mind. From the 1920s on, the influence of behaviorism was wide and profound, and mind was considered to be "the ghost in the machine." Behaviorism had emerged from the logical positivism of the early 20th century, which postulated that science can theorize only about phenomena that can be observed directly. In psychology it was clear that nothing mental could be observed directly, and therefore cognition was off the list of topics for serious research. The one thing that could be observed was the *behavior* of organisms, and so psychology (from *Psyche,* the ancient Greek goddess of the soul) came to be defined as the science of overt behavior. An undergraduate psychology education during that time consisted of lecture after lecture of well-meaning professors deriding concepts such as mind ("Show me mind!") and consciousness ("Show me consciousness!") as being unobservable and therefore unworthy of scientific study. Most research was conducted with animals, especially rats and pigeons, though generalizations to humans were offered with great optimism and confidence. Skinner's explanation of language was one such extrapolation, and the sensational thrashing it received has been partly credited with the downfall of the behaviorist dynasty.

The two branches of behavioristic doctrine perfectly well reflected the polarization of the Cold War era in which they thrived. The Russian *classical conditioning* paradigm, originating in Pavlov's laboratory, focused on the gland and smooth-muscle responses of the organism. When presented with an unconditioned stimulus such as the smell of food or presence of a threat, an animal's response is visceral, triggered by the autonomic nervous system. Salivation is of course the famous Pavlovian example of a glandular response, while increased heart rate, perspiration, and so on, are other classical responses. If a new stimulus is paired repeatedly with one that elicits a response, then that new stimulus will also come to elicit the response. Thus classical conditioning views the organism as passively reacting to events in the environment; the behavior of the organism is "pushed" by stimuli that precede it. (It is noteworthy that *all* behavior was described by the behaviorists as "responses," eliminating the possibility of curiosity, play, and other self-generated behaviors.)

American behaviorists on the other hand emphasized *operant conditioning,* in which the organism acts on its environment in order to obtain a reinforcement. The organism here is "pulled" toward a stimulus that follows the behavior. Operant behavior uses the striated muscles that move the skeleton; in other words, these responses are overt body movements or actions. American behaviorism, following the work of Watson, Hull, and Skinner, viewed all nature as a kind of capitalistic enterprise, with animals of all sorts doing whatever it takes for the almighty reinforcer. Similarities to the 20th-century American lifestyle were reflected in popular use of the term "rat race" to describe the daily life of a typical worker.

In many ways behaviorism was a step forward from the introspective psychologies of the 19th century. Though the old-time psychologists had discovered some important facts about the workings of the mind, their methods were very susceptible to error. In Wilhelm Wundt's laboratory, established in 1879, trained observers attended closely to their own sensations and reported them as carefully and objectively as possible. The memory researcher Ebbinghaus built his entire theory around experiments with himself as the only subject. Some of his findings have endured and still contribute to the literature of memory and learning, but it is clear that the introspective, single-subject, scientist-as-subject methodology was not adequate for discovering covert processes that are supposed to be universal across persons.

Behaviorism (which also relied on single-subject research designs) emphasized the importance of empirical observation in psychological research. Behaviors were defined in operational terms and were recorded carefully. From these observations the behaviorists derived elaborate formulas linking stimuli to responses. While behaviorism has all but vanished from the intellectual landscape, we can thank the paradigm for holding scientific psychology's nose to the methodological grindstone. The behaviorists themselves did not use sophisticated methods of measurement, experimental design, or statistical analysis, but they set a strict standard for empiricism that continues to influence researchers up to the present time.

The behaviorist orthodoxy continued to persist even while information science ascended in other departments of the university. Rats could be shown to strive for reinforcement consistently; once you deprived them of water for a long enough time, you could get them to do almost anything for a drink. Behaviorism fit in very comfortably with the polarized worldview of the time, and it provided professors ("Show me information!") with a self-consistent philosophy that required little justification, since everything except stimuli and responses was eliminated from the discussion, and behavior could be reduced to a set of fundamentals just large enough to provide new topics for journal publications. There did not seem to be any reason to abandon the paradigm.

The Cognitive Revolution

Different authors give different dates for the birth of the "cognitive revolution" in psychology, a precise moment when the veil was lifted and academic psychologists admitted that there was more to human behavior than simply responding to things in the environment. Howard Gardner argues that the crucial moment occurred during the Symposium on Information Theory held at MIT in September 1956. Attendees included George Miller, Allen Newell and Herbert Simon, Noam Chomsky, and others who soon became the leaders of this new field. George Miller's influential research on "chunking," cognitive compression of information by hierarchically aggregating facts, was first presented at the symposium and later published in the *Psychological Review*. Some psychologists were present at the symposium, but the field as a whole was not immediately affected by the new views.

Another event often cited as a moment when the tide turned was Noam Chomsky's 1959 review of B. F. Skinner's 1957 book, *Verbal Behavior* (Chomsky, 1959; Skinner, 1957). Skinner had tried to assimilate human language to the behaviorist paradigm, explaining it in terms of stimuli and responses. Chomsky's review tore into Skinner's views mercilessly; many readers of the day felt that he had shredded the epistemological and methodological underpinnings of Skinner's behavioristic psychology of language and of the Skinnerian explanation of human behavior generally. (In retrospect, time has been surprisingly kind to Skinner's theory of language; the effect of Chomsky's review was immediate and devastating, but new readers of Skinner often find much to agree with.)

Festinger's 1957 cognitive dissonance theory turned behaviorist assumptions on their head. In Festinger's paradigm, an experimental subject in one condition might be offered a small reward, say, 25 cents, for performing a very boring task, while subjects in another condition were offered a large amount, for instance, 10 dollars, for doing the exact same thing. When they were asked afterwards how much they had enjoyed the task, subjects in the low-reward condition reported significantly more enjoyment. Dozens, then hundreds, of experiments supported these findings. Why did a low reward have more effect than a higher one? It went totally against the premises of reinforcement theory and broke another door open for considerable research into cognitive processes.

An occasion given as the birth moment of the cognitive revolution, or at least the beginning of the end for psychological behaviorism, was the APA presidential address of Albert Bandura in 1974, when he stressed the importance of self-regulation in human behavior: "It is true that behavior is regulated by its contingencies, but the contingencies are partly a person's own making. By their actions, people play an active role in producing the reinforcing contingencies that impinge upon them." Bandura argued that cognitive processes must be considered in explaining human behavior. In that same year, Brewer published a chapter entitled, "There Is No Convincing Evidence for Operant or Classical Conditioning in Adult Humans"—the title says it all. The tide of cognitivism was pounding on the cliffs of the behaviorist island, and it was washing into the sea.

Those mainstream psychologists who had flocked around the behaviorists' roost regrouped, split, and reassembled, and by the mid-1980s were circling around another explanation for human behavior. Cognitive psychology was emerging from the ashes of behaviorism to dominate academic psychology. As the cognitive paradigm is based on the metaphor of the mind as a computer program that runs in the hardware of the brain, cognitive psychologists study individuals as information processors, with inputs and outputs, which is a relatively small step from the behaviorist view. But where behaviorism was depicted as an S-R (stimulus-response) paradigm, cognitive psychology made the transition to S-O-R; an internal mediating system was postulated (the O stands for organism), being the cerebral computations that transformed inputs into *representations,* which then were manipulated to produce the output behavior. The behaviorists had reduced psychology to a study of inputs and their effects on outputs of organisms, and the question was, "How are responses related to stimuli?" Cognitive psychology introjected another stage into the process and asked, "How are stimuli, representations, and responses related?" It was a small step really, accompanied by much debate over the acceptance of abstract, unobservable mental variables, but empirical methods were found for identifying the effects of latent variables. For instance, different kinds of problems could be shown to take different lengths of time to solve; memorizing some kinds of items resulted in predictable errors in recalling other items, and so on. The internal dynamics could not be observed directly, perhaps, but they could be *inferred* through careful measurement of observed behaviors. In the long run the transition from behaviorism to cognitivism was accomplished with a minimal number of professors losing their jobs.

Academic psychologists came around to embrace the cognitivistic view, but their field had no research tradition to base their theorizing on. There were, however, results and discussions in the field of information and cybernetic theory, and in particular in artificial intelligence, that seemed applicable to human information processing, and so in the early days of cognitive psychology there was a wholesale one-way osmosis of concepts through the membrane that separated the psychology department from the computer science department on campus. The models that had shown promise for the elicitation of intelligence from electronic machines were adapted to explain human thought.

Some of those AI models had been developed with the consultation of psychologists and with the use of psychological methods. For instance, Allen Newell and Herbert Simon's logic theorist program (Newell and Simon, 1956), which was able by 1956 to prove and even discover new mathematical theorems, was based on self-reported strategies of humans. Those researchers had asked people to explain the steps they would take to solve certain problems, and then encoded some of the reported strategies into their computer program. Thus, according to one view, these programs did represent human thinking. To tarnish the silver lining somewhat, though, it must be pointed out that Simon and Newell's programs contained *self-reported* processes, that is, processes that a person described when asked by another person to explain something about themselves. Later research has shown that people may have little or no direct access to their own cognitive processes, suggesting that perhaps the reports given to Simon and Newell represented subjects' socially acceptable rationalizations about how they "must have" solved the problem, inferences based on their preconceptions about thinking. The explanations given to Simon and Newell and captured in their AI programs were not necessarily real cognitive processes.

Bandura's Social Learning Paradigm

At Stanford University, Albert Bandura had an insight back in the early 1960s that has carried him into the current era. The main technique of behaviorist research had been to administer some schedule or system of reinforcements to an animal and to plot the changes in the animal's behavior over time as a function of the reinforcement schedule. For instance, reinforcement of *every* occurrence of the desired behavior (pulling a chain, turning around, finding the way through a maze, etc.) was found to result in a steep initial learning curve, followed by rapid extinction of the behavior when the reinforcer was removed. "Fixed interval" schedules, where the animal was reinforced for executing the behavior after some fixed amount of time, resulted in a typical scalloped curve in criterion behavior frequency, with the animal increasing the rate of responding as the time for the reward neared. Other schedules, such as variable and fixed interval and ratio schedules, resulted in characteristic patterns of learning.

In 1965 Bandura challenged the behavioristic establishment by announcing the discovery of "no-trial learning." He showed that humans can learn a task without even trying it once. Of course the trick is to let them watch somebody else do it. If the model is successful at performing the task, and in particular if they are reinforced for the behavior, then the observer can learn from them. Bandura called this *vicarious reinforcement.* Bandura's use of familiar jargon made the concept seem like an extension, rather than a rejection, of behaviorism, though it perhaps obfuscated the topic for later readers, who might think that giving a person a food pellet for performing a behavior will make it more likely that others will imitate them (some students—and yes, even their professors—today hold a view sadly close to this). In most cases, where the behavior is linguistic, for instance, the model may be demonstrating a conclusion drawn from an attitude or belief system that can be inferred by the observer, and in this case "reinforcement" comes in the form of satisfying logical validity or apparent cognitive consistency.

Bandura's most famous demonstration of vicarious learning, familiar to anyone who has taken Intro Psych in the last half-century, was his "Bobo doll" experiment (Bandura, 1962). After a child had watched another person beat up a Bobo doll, the child was much more likely himself to do it. This has been taken by ideologists of course to warn us about the importance of being good role models and about the contagiousness of aggression, but the real finding was that a person can do something they would not have otherwise thought to try by imitating the behavior of another. Other terms for the phenomenon are *modeling, observational learning,* and *social learning;* it is a very important form of learning for humans and seems hardly to exist at all among other species.

Bandura has touched almost all the psychological bases in his career. Starting as a behaviorist, he helped give birth to the cognitive revolution, has contributed very much to the practice of clinical psychotherapists, and has shifted the emphasis of theorists everywhere with his social-cognition perspective. His *Social Foundations of Thought and Action: A Social Cognitive Theory* (Bandura, 1986) reviews decades of research by psychologists on the many varieties of *imitation* in human life. The cornerstone of Bandura's theoretical perspective is the concept of *reciprocal causation.* Where previous theorists had seen behavior as a response to the presence of stimuli in the environment, Bandura saw behavior as much as a cause as an effect, and in fact he demonstrates in this landmark volume that reciprocal causation of individuals' behavior and cognition results in increased opportunities for people to exercise control over their destinies; it is, almost paradoxically, the root of free will.

Bandura has apparently never concerned himself with the larger question of what would happen if thousands, or millions, of reciprocal-causation agents interacted in a population; he was more concerned with the fate of the individual embedded in such a system. As we will see, it appears almost inevitable that the result of a complex system of interacting causal beings will be the formation of norms, and ultimately of cultures, as the individuals become more similar to one another through the effects of demonstration, persuasion, and other influence processes.

Social Psychology

Meanwhile, in another hall of the psychology department, social psychologists had been plugging along more or less ignoring the whole behavior-versus-cognition argument altogether. Social psychology, which addresses questions about the individual in a social context, never attained the dogmatic prominence of behaviorism, or even the current status of cognitive psychology, but it has persisted quietly since the 1920s and has become more visible in recent years.

Though social psychology is focused in American universities, many of its early leading proponents were German and Austrian immigrants who brought to the field a background in Gestalt theory. Gestalt psychology was primarily concerned with the ways that fragmentary perceptions become organized into wholes. This school of psychology prevailed in Germany before World War II, and the rise of Nazism led to the migration of a number of Gestalt theorists to the States—in fact, Dorwin Cartwright (1979) has written that the person who most contributed to modern social psychology was Adolph Hitler!

A fundamental principle of Gestalt psychology is *Prägnanz*, or "good form," the tendency to organize perceptions into coherent wholes. For instance, looking at Figure 5.2 it is impossible *not* to see a triangle whose corners cut into the three circles, though it doesn't actually exist. The triangle is entirely in your mind.

While every undergraduate who ever took Intro Psych has pondered textbook optical Gestalt phenomena, the reason that Prägnanz is important is often overlooked. It is important because it permits the

Figure 5.2 Gestalt processes make you see a triangle that is not there.

partitioning of the blur of the environment into *things*. Once things are identified, the organism can respond differentially to them, can generalize about things that fit the same categorical pattern, and can distinguish among categories. Cognitive scientists have expended much effort discussing distinctions between symbolic and statistical modes of thinking; something like Prägnanz, which partitions data arrays into meaningful units, including symbols, must be necessary in order to join the two types of processes. Though there has been a great amount of work done on pattern recognition, so far the kind of low-level, high-speed, data-driven cognitive operation seen in humans has not been successfully synthesized in a computer program.

Lewin's Field Theory

One of the early immigrants, whose name stands above and behind almost all the social-psychological research that followed him, was Kurt Lewin (pronounced either "Levine" or "Loo-in," both are correct) (Lewin, 1935, 1936, 1938). Lewin's mission was to produce a *field theory* of psychology that was conceptually similar to contemporary theorizing in physics, like models by Faraday, Maxwell, and Herz on the dynamics of electromagnetic fields and Einstein's new theory of relativity. Of course the psychological forces involved were themselves different from

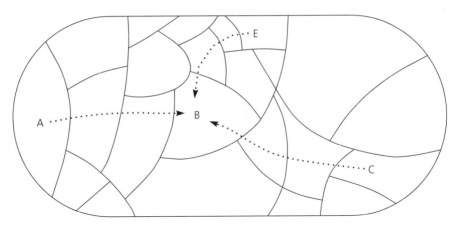

Figure 5.3 Lewin's "life space." (From Lewin, 1938.)

those in physical systems, but the method of representing them could be similar.

Lewin portrayed a "life space" wherein the individual was represented as a bounded region, acted upon and reciprocally acting through dynamical forces among elements. Using the new mathematics of topology, Lewin considered the person as being subdivided into a number of separate but interconnected regions, and the environment likewise was differentiated into numerous regions—some of which were persons. Each region represented what Lewin called a "fact," meaning anything, real or imagined, perceived or inferred, that could affect the person. *Locomotion* was Lewin's term for the person's movement through the life space. It is very important that this term does not necessarily or usually mean movement through physical space. It might better be thought of as change of state than change of position. Regions of the environment that are interconnected are able to influence one another, and locomotion from one of the regions to the other is facilitated. As seen in Figure 5.3, forces acting on the person in the life space can result in locomotion that results in equilibrium among the many regions of the person and the environment. Figure 5.3 shows paths from different beginnings to the same goal. In these cases, both the distance and the direction differ for different starting points.

Besides his high-level theoretical abstractions, Lewin was very much involved in applying psychological principles to solving social problems. He is most widely quoted for having said, "There is nothing so practical as a good theory," and he demonstrated that scientific psychology could be used to make the world a better place. For instance, during wartime he researched methods for persuading housewives to cook meats that were not popular, like brains and tripe, in order to save money for the war effort. He spearheaded group-dynamics research that resulted in the encounter groups of the 1960s as well as changing psychologists' understanding of organizations. Through his students, including Leon Festinger, Stanley Schachter, Kurt Back, Dorwin Cartwright, Hal Kelley, Morton Deutsch, John Thibaut, and others, Lewin profoundly influenced the atmosphere (one of the terms from his small-group theory) of social psychology through the latter half of the 20th century.

The dynamic theory of Lewin, for instance, as represented in 1936's *Principles of Topological Psychology* and a 1938 monograph titled *The Conceptual Representation and Measurement of Psychological Forces,* relied heavily on mathematical methods without losing touch with the humanity of its subject. The modern scholar reading Lewin's many articles and chapters can scarcely believe they were written decades ago—the Kurt Lewin bibliography spans the years from 1917 through the year of his death, 1947, and posthumously through 1960. Lewin's theorizing was visionary and prescient of the theories of complex dynamic systems that prevail today. Sadly, the only modeling tools available to him were pencil and paper, and near-hallucinatory abilities were required to imagine the dynamic implications of those static words and images. We can only guess, if Lewin had had a computer, how he could have demonstrated his vision in ways that others could really understand, how he could have explained the complex dynamics of psychological life. We think of him as a man before his time; he was in fact a man who set the stage for modern complex systems theory in psychology.

Norms, Conformity, and Social Influence

If we are to discuss something like swarming in human affairs, we should go back to 1936, when Musafer Sherif's *The Psychology of Social Norms* reported an innovative and illuminative series of experiments demonstrating the convergence of individuals' perceptions. Sherif placed subjects in a dark room with a point of light projected on the far wall. In the absence of any visual frame of reference, people tend to report that the stationary light is moving, and each person has a relatively stable characteristic rate at which it appears to them to move. This is called the *autokinetic effect,* and it results from movements of the eyes as they try to adjust to an invariant stimulus. An individual tends to report movements that are consistent across trials, though individuals' typical reports differ from one

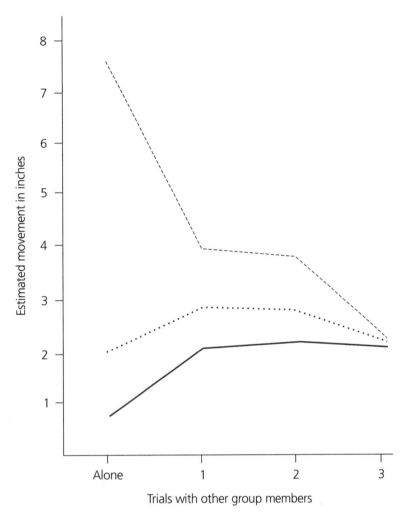

8 -

7 -

6 -

Estimated movement in inches

5 -

4 -

3 -

2 -

1 -

Alone   1   2   3

Trials with other group members

Figure 5.4 Norm formation in the autokinetic effect: When subjects announced their judgments in the presence of others, their estimates tended to converge.

another. When Sherif placed groups of individuals in a dark room together and asked them to report publicly how much the light had moved, each individual's reports drifted away from their characteristic distance toward a *norm* that was characteristic of the group (see Figure 5.4).

Sherif's research led to developments in social influence and norm formation, which really reached a peak in the 1950s. The lockstep uniformity of Nazi Germany followed by the enforced conformity of the

McCarthy era in the United States led Solomon Asch (1956), a German immigrant to America, to ask about the nature of conformity generally. In his famous experimental paradigm, a subject was led to believe that the other people in the room were also naive experimental subjects who were being asked to rate the relative lengths of some lines displayed by the experimenter (see Figure 5.5). The situation was contrived so that the real subject always answered last, after the others, who were confederates of the experimenter. On most trials, the confederates gave the correct answers, but occasionally they would unanimously give obviously incorrect ratings of the lines, saying, for instance, that a short line was the same length as one that was in fact twice as long.

Subjects in the experiment were faced with the dilemma of reporting the obvious truth versus agreeing with the group. In about a third of the trials, they chose to express the same judgments as their peers, even though the answer was plainly wrong—almost no one answered incorrectly when they were tested by themselves. Asch had expected to demonstrate that people would resist pure peer pressure to give a false report, but he found quite the opposite. People would do whatever the people around them were doing, whether it made sense or not, and even when there was no reward for conforming or punishment for deviating.

Two subsequent experiments, one by Crutchfield and another by Deutsch and Gerard, both in 1955, used automated methods to imply the presence of a group opinion without having to coordinate a large pool of confederates (and how did Asch keep them from laughing?). Subjects sat in booths that blocked their view of one another. In each booth were some lights that were said to represent the responses of the others. Each subject was informed that he or she would be responding last, giving them a chance to see how their peers had answered some questions. In this way, data could be collected from a group of subjects simultaneously, under very well controlled circumstances. Crutchfield and his colleagues asked his subjects various kinds of questions, with occasionally amusing results. Berkeley researcher Read D. Tuddenham put together a paragraph of statements that he had been able to induce people to agree with in this paradigm. Here is part:

> The United States is largely populated by old people, 60 to 70 percent being over 65 years of age. These oldsters must be almost all women, since male babies have a life expectancy of only 25 years. Though outlived by women, men tower over them, being eight or nine inches taller, on the average. The society is preoccupied with eating, averaging six meals per day, this perhaps accounting for their agreement with the assertion, "I never seem to get hungry." Americans waste

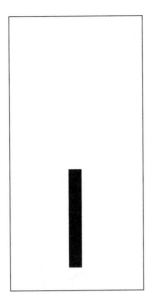
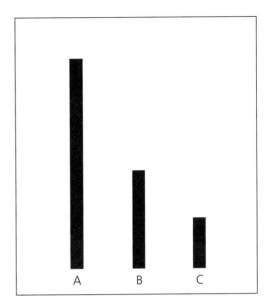

Figure 5.5 Here's a hard one: Which line on the right is the same length as the one on the left?

little time on sleep, averaging only four to five hours a night, a pattern not unrelated to the statement that the average American family includes five or six children . . . (p. 513).

Sociocognition

Research in conformity waned after the early 1960s, though group influence continued to survive as an important topic of research. In an important 1993 paper titled "Social foundations of cognition," published in the *Annual Review of Psychology,* John Levine, Lauren Resnick, and E. Tory Higgins surveyed recent research and concluded that thinking is fundamentally a social activity. For instance, findings that coordinated cognitive activities, such as some kinds of problem solving, evoke *intersubjectivity,* or shared understanding of the topic, help explain why groups are sometimes better able than individuals to perform certain kinds of tasks—and why they perform worse when group members rely too much on shared information. According to Levine and his colleagues, there are at least three well-known social aspects to memory. First, in "transactive memory," as studied by Dan Wegner at the University of Virginia, people who know each other well use one another to encode, store, and retrieve memories. For instance, a husband and wife may distribute certain kinds of memories between them, so that some details are known by one or the other, but not both (honey, isn't Georgey's birthday around this time of year?).

A second social aspect of memory is its content—a great proportion of an individual's memories refer to past social actions and experiences. Finally, research on recall in face-to-face groups has shown that the process of memory is often based on symbolic communications with other people. Collaborative recall appears to be affected by three kinds of variables: the group's consensus favoring an alternative, the correctness of the alternative, and members' apparent confidence in their recall.

Levine, Resnick, and Higgins (1993) conclude that a new field of *sociocognition* is emerging, which considers social interaction to be the occasion for the development and practice of cognition. They write:

> Although some might claim that the brain as the physical site of mental processing requires that we treat cognition as a fundamentally individual and even private activity, we are prepared to argue that all mental activity—from perceptual recognition to memory to problem solving—involves either representations of other people or the use of artifacts and cultural forms that have a social history (p. 604).

In sum, social psychologists worked more or less in parallel with, and separately from, behavioristic and cognitive psychologists through much of the 20th century, pursuing a line of inquiry that was informed by those other paradigms but relatively unaffected by them. The ongoing—and recently revitalized—tradition of social influence research examines how a person is affected by other people who seem to hold a belief or opinion, and the universal finding has been that social influence is an extremely powerful force in human affairs.

Simulating Social Influence

The particle swarm model explored in the latter half of this volume can be described as a simulation of social processes, or as a problem-solving algorithm. Computers have long been used to explore social and cognitive hypotheses; here we will describe some of the issues involved in simulating sociocognitive phenomena and some of the previous work that influenced our research.

In a 1990 *Psychological Review* paper, Bibb Latané and his colleagues (Nowak, Szamrej, and Latané, 1990) presented computer simulations to support their argument that interactions of individuals according to the principles of social impact theory will result, as an algorithmic effect, in patterns of belief and opinion that resemble those seen in real societies.

The introduction of simulation in any science is likely to raise a few eyebrows, and social psychologists had not previously shown themselves to be technological adventurers. Because by 1990 most social psychologists were aware of the disappointing inability of artificial intelligence to successfully approximate the performance of humans, there was and is today a great deal of skepticism about computational studies. This is a conservative field that would rather accumulate one more grain of good hard empirical data from a well-designed experiment with human subjects than accomplish an eye-opening leap of speculative theoretical insight. Further, few social-psychological hypotheses are specified in the kind of detail that permits mathematical description, which is necessary for simulation, and so the interpretation of parameter values and even what parameters to include are widely considered ambiguous and suspect subjects.

The field of artificial intelligence had first of all borne an interesting kind of confusion about computer simulations. It is one thing to simulate a mental process and quite another to create an artificial one. If someone were to simulate a meteorological phenomenon, for instance, by modeling the interactions of some climatic variables, no one would be likely to say there was "weather" inside the computer. It is a unique aspect of this particular topic that you could contend that the simulated mind was in fact a real mind. The question is really quite abstruse, perhaps unanswerable when you get down to it, and is a special case of the situation previously described where we are only asking whether some phenomenon belongs to a particular semantic category.

It is like someone making an edible sculpture of food. On the one hand it is a representation of food, intended to look like real food, and on the other hand, it *is* food, because you can eat it. If a mind is something that thinks, and a simulation of a mind can be shown to think, then it seems we must conclude that the simulation is not only the representation of a mind—but a mind itself. We can't really say that the difference is consciousness, because we can't prove that computer programs are not conscious, any more than we can prove that people are: consciousness is something that is self-evident to the person experiencing it, and totally unverifiable otherwise. If computer minds can perform the same kinds of cognitive operations that human minds do, and the question of consciousness is unanswerable, then the comparison between human and computer minds comes down to a simple test of skepticism.

Just to unsimplify the matter, thinking itself can be described as a covert simulation of physical or psychological reality. Simulation theorists argue that we understand events, including the thinking of other people, by running a kind of mental simulation. For instance, if I want to know how another person will respond if I say such-and-such to him, it may be that I simulate the situation, perhaps imagining myself as the recipient of the comment, and see how I would feel if it was said to me. The mental simulation view is sometimes described as one system modeling another; though it is usually discussed in terms of folk psychology, that is, the commonsense understanding of other people's minds, it can also be used to understand cognitions about physical events as well.

Mental simulation is apparent in the results of an important early experiment by cognitive psychologist Roger Shepard and his colleagues (Shepard and Metzler, 1971). Experimental participants were shown two three-dimensional images and asked whether they were the same or different (see Figure 5.6). On target trials, the second image was a rotated view of the first. The dependent variable was the time it took to give an answer for "same" pairs.

If the individual were creating a simulation of the object and rotating it in their imagination, then we would expect the response time to be a function of the amount of rotation required, the difference between the first and second views of the object. Think about this: if you were given two twisted objects and asked if they were the same shape, you would turn one of them until they lined up—and the farther you had to turn them the longer it would take you to answer the question. Shepard and his coworkers found that the amount of time required to respond was an almost perfectly linear function of the angle of rotation required. These results suggest that a mental simulation is run in order to make judgments about physical reality; a mental replica is created and manipulated as the physical object might have been.

This simulation-as-reality issue has some serious implications. Much human social behavior is imitative, yet socialized individuals come to accept the beliefs they copy from others as genuine, originating from themselves. A person may initially perform a behavior to imitate someone else; for instance, a child makes fun of someone because that is what his friend is doing. His behavior may be intended as an imitation of the friend's and is therefore "artificial" in the sense that it is not genuine, but the effects of the two children's behaviors—and the way they are interpreted by others—are identical. Other people will view the two actors as

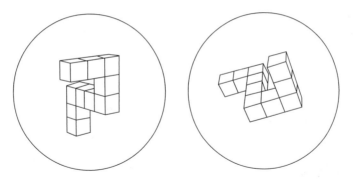

Figure 5.6 Are these two views of the same object?

equally cruel, and social-psychological theory predicts that the follower will likely actually come to dislike the child he is pretending to torment. Thus the behavioral simulation of another person's behavior is in fact "real" behavior, with real consequences.

Confusion between "real" intentions and imitation lies at the core of the *induced compliance* paradigm in social psychology, which is fundamental to cognitive dissonance theory. In the induced compliance experiment an individual is subtly persuaded to comply with the experimenter's implied request, for instance, to write an essay advocating an unpopular position. The subject is offered a choice, told they are free to choose, for instance, which side of an issue to advocate, but they may also be told that the experimenter already has enough essays expressing one view and hopes that "someone" will choose the other. In order to produce cognitive dissonance in the subject, it is necessary for them to feel that they have freely chosen to perform the behavior. Even though experimental subjects might believe at the time that they are superficially complying, it is very commonly found that their opinions actually shift in the direction of that expressed in the essay; they come to believe what they have written. The inability to distinguish between a phenomenon and a simulation of that phenomenon may be a basic pivotal feature of human nature, and the edible-sculpture-of-food analogy might well describe an important aspect of the state of knowledge and illusion of humans in the world.

The argument of "strong AI" has been that computer programs can actually think. The fact that one of the very first AI programs, Simon and Newell's logic theorist, was able to discover theorems that Russell and Whitehead had failed to prove in the *Mathematica Principia* demonstrated that a computer could have thoughts beyond those that its programmer gave it. It could make inferences and draw conclusions on its own, just like a real human mind. Thus the founders decided not to call the field "simulated intelligence," but used the word "artificial" to suggest that the intelligence was real, but man-made. In *The Sciences of the Artificial,* Herbert Simon (1969), one of the original pioneers of AI, reported that he thought the term "artificial intelligence" originated at MIT; his own research group, he says, preferred the terms "complex information processing" and "simulation of cognitive processes" to describe computer programs that emulate thinking. In interpreting the phrase "artificial intelligence" as it applies to mental simulation, Simon states, "It may be easier to cleanse the phrase than to dispense with it. In time it will become sufficiently idiomatic that it will no longer be the target of cheap rhetoric" (p. 4). In fact the phrase "artificial intelligence" never did quite rise above the assaults of cheap rhetoric and is presently fading from the vernacular, leaving behind a kind of Buck Rogers sort of nostalgic glow. There are plenty of laboratories still working on artificial intelligence projects, but the field is swinging away from the traditional symbol-processing paradigm toward other forms of information processing. (One writer has suggested that calling it artificial intelligence is like calling what airplanes do "artificial flight.")

Paradigm Shifts in Cognitive Science

Traditional AI, and cognitive science in general, has been dominated since its beginning by the symbol-processing paradigm, which is characterized [according to Eliot Smith (1996)] by

- *Separation of representation and process.* Representations are static unless acted upon by a processing unit. Representations are operated upon in two stages: retrieval from storage and use.
- *Learning.* In the symbolic paradigm, learning means constructing a new representation.
- *Retrieval and accessibility.* A representation is retrieved from storage as a function of its appropriateness given the currently available cues and its accessibility. Accessibility depends on how frequently and how recently the representation has been used.

In contrast, according to Smith, connectionist models have the following characteristics:

- *Structure.* All the representation and processing is performed by a network of interconnected processing elements.

- *Representation.* An object or concept is represented as a pattern of activations across the network of processors.

- *Learning.* Connectionist learning comprises adjustment of the weights connecting the processing elements.

- *Unity of representation and process.* Contrasted with symbol processing, representations are dynamically stored in the matrix of weights between pairs of processing elements.

- *Retrieval and accessibility.* Patterns to be retrieved are not explicitly stored anywhere. Retrieval is accomplished by activating a pattern of inputs, which is processed by the network, resulting in output of a recalled pattern. Accessibility is determined by the dynamical process of learning, such that more recently and more frequently encountered patterns have a higher probability of being recalled.

Connectionist explanations have a number of advantages for explaining psychological phenomena; further, the use of neural networks has turned out to be a rich lode for engineering and other AI-type applications. The turf battle between the two paradigms, though, is far from finished; casualties are occurring on both sides as theorists duke it out. (Of course the inevitable outcome is that it will be found that both symbolic and connectionist models explain some psychological processes better, and solve certain engineering problems, and that the matter is one of applying the right method to the particular instance.)

Newell and Simon (1977) have described the traditional AI paradigm in terms of goal-directed problem solving. For them, a problem can be broken down into an initial state, a goal state, and a set of operators for transforming one into the other via a series of intermediate steps. The operators have constraints that must be satisfied in order for the goal to be attained. The set of states, operators, goals, and constraints is called the *problem space,* and problem solving comprises the search for a path leading from the initial state to the goal state.

Some of the bad blood between the "artificial intelligentsia" and proponents of connectionist, fuzzy logic, and evolutionary methods results from the domination of funding opportunities by AI researchers through several decades in the mid-20th century. For instance, one of the first neural net models—the *perceptron* of Frank Rosenblatt—was severely criticized by Marvin Minsky and Seymour Papert, who were leading figures in the artificial intelligence movement at the time, in a 1968 book entitled *Perceptrons*. They asserted that research in perceptrons—an early form of neural network—was "without scientific value"; their critique, which has since been proven to be extraordinarily wrong, set back connectionist techniques by about two decades, while the artificial intelligentsia spent millions of dollars pursuing objectives that never came to pass.

In fact, artificial intelligence researchers never delivered what they promised. Quotes abound from the 1950s, 1960s, and 1970s, promising a Jetson future within 5, 10, or 20 years—yet we still live a lot like Flintstones; the promised future is not nearly here yet. Although they are present in every part of our lives now, almost none of the many things that computers do for us today use anything that can be called artificial intelligence. The symbol-processing paradigm may have been pretty good for such clear-cut problems as theorem proving and playing chess, but it was totally impractical when it came to such "simple" tasks—tasks that any below-average human, or dog for that matter, can do—as face or voice recognition, or drawing conclusions from noisy, ambiguous, or nonlinear data. The desire to dissociate their research from that of the past has prompted some contemporary researchers to discard the old label "artificial intelligence" and to call the newer paradigms "computational intelligence." In general, *computational intelligence* is considered to include neural networks, fuzzy logic, and evolutionary algorithms; these approaches' strengths tend to lie just where the traditional AI methods fail, that is, in dealing with complex, dynamic, poorly defined problems such as are encountered in the real world.

Doubts about traditional AI were expressed very clearly (!) in 1976 by Joseph Weizenbaum, himself a profoundly respected MIT computer science professor and contributor to artificial intelligence research (his ELIZA program was the original chatterbot):

> The various systems and programs we have been discussing share some very significant characteristics: they are all, in a certain sense, simple; they all distort and abuse language; and they all, while disclaiming normative content, advocate an authoritarianism based on expertise. Their advocacy is, of course, disguised by their use of rhetoric couched in apparently neutral, jargon-laden, factual language (that is, by what the common man calls "bullshit") (p. 248).

Parts of Weizenbaum's humanist critique of the attempted mechanization of human processes may seem somewhat Luddite in the present day, now that computers have come to play a central role in our lives, but mainly his predictions of two and a half decades ago resonate loudly; many of his nightmares have come to pass. Anyone who has dealt with a corporate or government "telephone tree" or had their records misplaced in a database will recognize that, in many aspects of our lives, we have become subservient to a system of black-and-white, in-or-out, off-or-on, this-or-that computers that are incapable of dealing with the rough edges and ambiguous cases that make life livable. Some developments in computational intelligence have the potential to soften the dehumanizing aspects of computer modeling and the imposition of computers into daily life, though perhaps Weizenbaum would not be willing to accept the omnipresence of today's computers. The earlier approach imposed strict logical formalisms, questions rephrased to fit the format, answers were crisply, distinctly right or wrong, and the human was slave to the machine. The more modern paradigms allow trial-and-error learning, and ambiguities are expected; we suspect that Weizenbaum would be more accepting of the warmer, gentler computational intelligence approach.

We suppose the scientific study of mental phenomena to be necessarily a kind of exercise in self-portraiture. What can anyone theorize about minds, unless the theory describes his or her own mind? Who would want to deliver up a theory of mind that applies to everyone but himself? Unfortunately fashions in theory of mind change, and also unfortunately the kinds of minds that ascend to positions of academic authority are, almost inevitably, unusual or unique minds—venerable professors are not like other people. Their theories of mind are likely to reflect the processes that have made them successful academicians, but may have little to say about ordinary peoples' thinking. Likewise, computer programs that model these elite professors' rational thought methodologies may not very well emulate ordinary thinking.

Social psychologists have always tended to be somewhat more humanistic than behaviorists or cognitivists. Perhaps it results from the hybridization of American pragmatism, as proposed in diverse versions by James, Dewey, and Peirce, with the holistic Gestalt principles of German psychology. While social psychologists look for laws of human behavior, they tend to hesitate to reduce people to a set of cold, deterministic principles. Assumptions of free will and personal autonomy underlie much social-psychological theorizing, and the mechanization implied by S-R and even S-O-R theory has never seemed especially palatable to social psychologists. Even Lewin's mathematical models included what he called "quasi-concepts" such as hope, expectancy, and frustration, purely humanistic causal variables. While the practice of social-psychological science requires profound statistical knowledge, that knowledge is used to analyze methodically collected experimental data and only rarely to model the characteristics of experimental subjects.

The 1968 *Handbook of Social Psychology* featured an 80-plus–page treatise by Robert Abelson (1968) called "Simulation of social behavior," which surveyed the field up to that time and summarized the notions of the day regarding the whys and hows of social-psychological simulation. It is clear from Abelson's early summary of the field that the most advanced psychological simulations of the day were cognitive; in particular Simon and Newell's general problem solver was the prime example, but a good number of social-psychological models had also been key-punched and dropped off at the campus computer center. While the *Handbook* chapter generated some interest, academic social psychologists remained fixed on laboratory experimentation with human subjects, laboriously collecting data in order to refine the details of theories, and were generally not interested in taking the large conceptual step of running computer programs to generate predictions from theoretical constructs.

The Evolution of Cooperation

By far the most important and influential computer simulation for social psychologists was the prisoners' dilemma research reported in Robert Axelrod's 1984 *The Evolution of Cooperation.* It is not overstatement to say that the publication of Axelrod's highly readable book was one of those events that turned the course of the river of science. Researchers in computer science, biology, economics, and the other social sciences read Axelrod's work with great interest; not only had he successfully applied a new research methodology to an important theoretical problem, but the results of Axelrod's investigations seemed to have implications for understanding the behavior of many kinds of systems—including systems composed of interacting humans.

The *prisoners' dilemma* is a situation where two players have opposite, symmetrical motives. Each player has the choice to cooperate or compete with the opponent: if both cooperate, their payoffs are high, and if both compete payoffs are low. If one competes (the technical term is *defecting*) while the other cooperates, the defector receives a very high reward while the cooperator's payoff is very low—the lowest in the game, called the "sucker's payoff." When the game is played just one time, the most reasonable thing to do is to defect, as there is no basis for trusting the other player and there is nothing to gain by being a sucker.

Usually, though, including in Axelrod's study, the game is iterated, a series of games is played. A player would score the highest if he always defected while his partner always cooperated—but of course no sensible player would continue to cooperate while being hammered repeatedly by a competitive opponent. Repeated trials require some consideration of strategy; for instance, a player might end up with the highest score if he lulled his opponent into cooperating, then struck with a defection, then lulled and defected, and so on. It might be that the best approach would be just to cooperate from the start—except that nothing then prevents the opponent from taking advantage. The simple game then produces opportunities for many kinds of strategies. Axelrod roughly grouped these into two kinds: "nice" strategies, which rely on coopera- tion to keep the level of payoffs high for both parties, and strategies he refers to as "mean" (specifically, that includes only the all-defect strat- egy) or "not nice." Strategies that are not nice include ones that might try to use cooperation as a way to make the opponent vulnerable, then defect for the higher payoff.

Axelrod took the interesting approach of asking a number of leading social scientists, game theorists, and others to submit computer pro- grams representing strategies that they thought would score well in iter- ated prisoners' dilemma tournaments against other strategies. A contest was then held, with programs playing against one another in pairs until a winner was found. The winner of the competition was a strategy called TIT-FOR-TAT, submitted by Anatol Rapoport. TIT-FOR-TAT is a mostly nice strategy that can be summarized as follows: start by cooperating, then do whatever your opponent does. Thus, an opponent who tries de- fecting will be defected against, but one who cooperates will be rewarded with the relatively high double-cooperation payoff.

Even though TIT-FOR-TAT won the competition, Axelrod pointed out that there were several strategies that could have beaten it. For instance, one called TIT-FOR-TWO-TATS does not retaliate until the opponent has defected on two turns in a row. This extra bit of forgiveness helps keep the game from falling into the trap of constant retaliatory defection, which produces of course no winners. A strategy called LOOK AHEAD could have won, using a method developed for an AI chess-playing pro- gram to anticipate the opponent's move—except that no one submitted the program to the contest (it was included as an example in the in- troductory materials Axelrod sent to the participants). A third strategy that was submitted, called DOWNING, would have won, Axelrod notes, with only a small modification. DOWNING was an attempt to simulate behaviors seen in human subjects in a prisoners' dilemma situation. It deliberately tried to understand the other player's approach and make the decision that seemed likely to yield the best long-term payoff. DOWNING started out by defecting on its first two moves, provoking some of the other strategies to retaliate, thus lowering its total score. When Axelrod tried a MODIFIED DOWNING strategy that began with cooperation, it easily beat all the other programs, including TIT-FOR-TAT.

Axelrod also noted that the success of TIT-FOR-TAT depended on the length of the game, and it may not do so well in the short run. If the strategy were adopted by a large number of population members, however, it could provide a safe environment for cooperation by all parties, with consequent high payoffs for all.

The results of Axelrod's tournament, and especially the superiority of TIT-FOR-TAT, were in accordance with mainstream social-psychological theories, in particular Thibaut and Kelley's (both former students of Kurt Lewin) interdependence theory, which was based on an analysis of game-theoretic payoff matrices. Partly because Axelrod's results were consistent with familiar existing theories, concepts from *The Evolution of Cooperation* were accepted by mainstream social psychologists—even though they were computer simulations.

Explanatory Coherence

Through the 1980s and 1990s, the rise of the connectionist paradigm in cognitive psychology, which is based almost entirely on simulation results, caught the attention of some social psychologists, but the field was faced with a dilemma. While it became obvious that the connectionist models were very consistent with traditional social-psychological theorizing, social-psychological methodologists were still reluctant to accept computer simulations as a legitimate research strategy. It's one thing to cite Axelrod's research and quite another to program simulations of your own model.

The reluctance diminished under the two-pronged attack of a paper and a computer program developed by a Canadian philosopher named Paul Thagard. Thagard's 1989 paper in *Behavior and Brain Sciences* described cognition in terms of a search for a coherent explanation for events, and postulated a number of aspects of coherence that needed to be accounted for. The concept of coherence does not seem to differ dramatically from Smolensky's principle of harmony (Thagard prefers to spell it "Harmany," in tribute to the Princeton philosopher Gilbert Harman), but the focus on explanation set Thagard's theory apart from other connectionist thought and placed it comfortably within the domain of social psychology. Like some previous philosophers and psychologists, Thagard theorized that people understand events by placing them in the context of a narrative or explanation—a story. This much was familiar to social psychologists, and in fact investigators had long been looking for a good, comprehensive theory of explanation. But Thagard brought in something that social psychologists were not used to. He supported his theory of explanation by encoding it in a connectionist computer program.

Some neural-net software had been distributed with the Parallel Distributed Processing volumes, but it was mostly used to demonstrate examples from the book and was not well designed for research. Thagard's ECHO program (ECHO stands for Explanatory Coherence Harmony Optimization), though, available in the LISP and C programming languages, let researchers tweak parameters, design their own models, try things. In a footnote to the *BBS* paper, Thagard made the program available on request. And as social psychologists experimented with it, they found that parallel constraint networks were almost exactly what they had been theorizing about for years, under names like cognitive dissonance theory, consistency theory, balance theory, congruence theory, and many others. Those theoretical descriptions of human thought addressed various aspects of the need for individuals to maintain consistency among their beliefs, attitudes, and behaviors—just what maximizing a harmony function does. Harmony or coherence is maximized when the nodes of a network fit together, when positively connected elements are in the same state and negatively connected ones are in opposite states—when dissonance is minimized.

Stephen Read, at the University of Southern California, and Elliot Smith, at Purdue, independently began to introduce some computer-simulated connectionist models into the social-psychological literature. An important finding of these early papers was that ECHO programs could replicate the results of many traditional laboratory experiments. Other researchers have begun to participate in this new paradigm, using simulations to investigate theoretical ideas. As of today, very few social psychologists have adopted the simulation method of research, but they are tolerating them, and simulations are accepted in most social-psychological journals.

Networks in Groups

One innovation relevant to our present discussion was reported by a cognitive anthropologist, Edwin Hutchins. Hutchins (1995) may be credited with breaking the connectionist paradigm out of its inside-the-head cognitivistic shell and opening up the possibilities of studying cognition in a social context.

A parallel constraint satisfaction network, as described in Chapter 2, depicts a number of cognitive elements and the relations between pairs of them. Some elements are consistent with others and inconsistent with yet others, and the goal is to find a pattern of activity and inactivity of elements that is most consistent, cohesive, harmonious. Some of the elements may be considered "external"— that is, they represent known facts or perceived phenomena in the environment. Of course the "internal" parts of the network should be consistent with these, as well as with one another.

Hutchins wanted to see what would happen if a group of people—in his example it was the crew of a ship—were able to communicate with one another. Each person was represented as a parallel constraint satisfaction network, and Hutchins programmed a model with four of them, where an occasional node of one was positively connected to the corresponding node of another. This was like one person communicating their belief to another person, affecting the strength with which the other person held that belief, encouraging agreement. The situation as Hutchins presented it had two globally optimal solutions, two opposite patterns of activations that satisfied the constraints equally well (see Figure 5.7).

Hutchins found that when individuals were very highly connected, what he described as a "mega-mind," the entire population gravitated toward some unsatisfactory solution, some combination of elements that could be highly inconsistent. When they operated in isolation the usual results were obtained, and each one tended to converge on one good pattern or the other. But when individuals were moderately connected, the entire population ended up in agreement, stabilized in one optimal pattern or the other.

Hutchins was more interested in observing the effects of different patterns of connections and did not remark on the fact that multiple networks could help one another optimize their activations. What we are working toward in this discussion is a model where the optimizing occurs through connections between individuals; thus Hutchins' results are especially interesting to us.

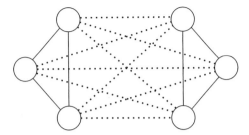

Figure 5.7 In Edwin Hutchins' parallel constraint satisfaction networks, the three nodes on each side are connected by positive links (indicated by solid lines), while connections between the two sides are negative (broken lines).

Hutchins did not explain how he created his groups of networks, but we can guess that, rather than running four ECHO programs simultaneously and adding connections between them, he designed one network and deleted most of the connections between subsets of it (see Figure 5.8). These subsets then were conceptualized as "individuals," though in actuality the group was the individual, and the so-called individuals were simply partly disconnected cells of the superorganism.

This is remindful of the American philosopher Charles Peirce's comments on the human state of ignorance. A longish, perfectly typical Peircean quote is appropriate (Peirce, 1931–35, vol. 5, p. 317):

> What anything really is, is what it may finally come to be known in the ideal state of complete information, so that reality depends on the ultimate decision of the community . . . The individual man, since his separate existence is manifested only by ignorance and error, so far as he is anything apart from his fellows, and from what he and they are to be, is only a negation . . .

Peirce is suggesting that, because reality is defined by ultimate social consensus, and because we never know everything that everyone else thinks, we are always in a negative state of knowledge, ignorance. There is a hypothetical transpersonal network out there, where all the nodes are connected, but we as individuals exist in a state of unconnectedness. We are able to transmit partial information to one another and use this incomplete information to move ourselves, and our community, into a better position to understand and deal with the world. Hutchins' analysis suggests that moderate ignorance permits not only cognitive consistency, but agreement among members of a group.

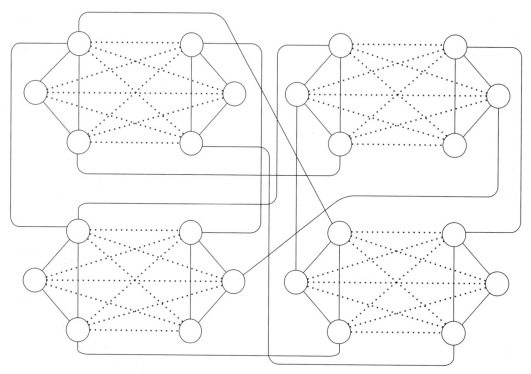

Figure 5.8 Hutchins connected subsets of nodes of networks together, simulating the effects of individuals communicating within a group. Solid lines are positive connections, dotted lines are negative connections.

Culture in Theory and Practice

We have been hinting that the mind of the individual can be considered as if it were a single cell in a larger superorganism. Because of our lack of perspective, it is most difficult for us to see clearly how the superorganism is composed, how it functions—how can the cell see the whole beast? The larger entity that we are talking about, *culture*, presents particular problems to a scientific inquirer.

The first problem is simply to determine whether culture is really something that exists or an abstraction. We have no difficulty (as we are not solipsists) stating that things exist, solid physical things, and even though we must grant some semantic latitude, we are almost unanimous that events exist. While we will readily agree that air is a real thing, we could have some trouble identifying exactly what a "wind" is, and when it does and does not exist, and whether it is a reality or an abstraction, simply moving air. But just as behavioristic psychologists accepted the reality of behavior as distinct from the physical body of the organism, we agree that wind is something that exists, and it is permissible to accept verb-things, events, as real things.

It is one little step further, having admitted motion and change as things, to admit patterns of things as things in themselves. A flock of sheep is a thing, even though you could argue that the individual sheep themselves are the *real* things—but that argument has to conclude that the cell and then the atoms of which they are composed are *really* real, and so on down to some infinitesimal quantum phantom. We do habitually accept patterns of things as things and patterns of motion and change as things, too; for instance, the falling from clouds of many drops of water can be "rain." Thus it seems entirely reasonable to view culture, the supraindividual pattern of human belief and behavior, as a thing, as something that exists.

There is one sense in which culture is undoubtedly real, and that is when it is embodied in concrete artifacts. A Navajo bracelet, a truck-stop coffee cup, ivory chopsticks—these are definitely real things, artifacts of their cultures. More interestingly, the literature of a culture is real, real ink on paper pages—but it is the abstraction embodied in the literature that lets it reflect the culture. When we talk about culture we are more likely to mean the symbols of a population of people than the hardware they share—and it is not the symbols, which might be marks or sounds or ritualistic behaviors, but their meaning that is important. The same spiral signature or cross may appear in different populations with different meanings, and the meanings are what comprise the essence of the culture. Thus even when we start with rock-solid artifacts, our search for culture leads us up and away to abstractions.

In their classic 1952 reference *Culture: A Critical Review of Concepts and Definitions*, Kroeber and Kluckhohn acknowledged and commented on these difficulties in studying culture:

As Ernst Cassirer and Kurt Lewin, among others, have pointed out, scientific progress frequently depends upon changes in what is regarded as real and amenable to objective study. The development of the social sciences has been impeded by a confusion between the "real" and the concrete. Psychologists, typically, are reluctant to concede reality in the social world to anything but individuals. The greatest advance in contemporary anthropological theory is probably the increasing recognition that there is something more to culture than artifacts, linguistic texts, and lists of atomized traits.

Structural relations are characterized by relatively fixed relations between parts rather than by the parts or elements themselves. That relations are as "real" as things is conceded by most philosophers. It is also clear from ordinary experience that an exhaustive analysis of reality cannot be made within the limitations of an atomistic or narrowly positivistic scheme (p. 122).

Anthropologists above all have had to grapple with the problem of defining and identifying culture. In trying to describe a culture, it is not enough to go into a society and record the behaviors of the people going about their business, and it is not enough to record their symbols. To collect data on real human culture, the anthropologist should understand the meanings of the behaviors and the symbols. But for a number of reasons this might be very difficult. Most importantly, because of the intimate relation—we would say identity—between the mind of the anthropologist and his or her own culture, it might be impossible to understand meanings as they are understood by members of another culture. An anthropologist interviewing a native is really a system of two people. They can only be people, bound by their own cultures to interpret their sense data the way their people do.

In her groundbreaking 1934 volume *Patterns of Culture,* anthropologist Ruth Benedict made the strong point that "society" or culture cannot be thought of as an entity separate from the individuals who comprise it. She noted that the average person conceptualizes an antagonistic relationship between the individual and society, as if something called "society" was forcing people to obey its rules. Yet, as she notes, the individual and the culture are two aspects of a single process: "No individual can arrive even at the threshold of his potentialities without a culture in which he participates. Conversely, no civilization has in it any element which in the last analysis is not the contribution of an individual" (p. 234). In order to consider human intelligence we should consider the intelligent mind as a participant in a larger cultural dynamic.

We are describing culture as if it were a kind of *norm,* and we are comfortable admitting it. As people interact they become more similar. Many of these convergences are ephemeral, disappearing overnight, it would seem; slang and trends in popular music are examples. Other norms last longer, sometimes for years before they dissipate, and others persevere for centuries, even millennia. Languages and the traditions that accompany them exemplify these long-lived norms. These latter can be considered the signs of cultures. We are not then distinguishing crisply between norms and cultures, which perhaps only differ in the breadth of their influence, the depth of their acceptance in the minds of the people, and the length of their persistence. In later sections we will look at some models of the formation of stable strategies or norms in populations of individuals as a result of their interacting, that is, the formation and maintenance of culture.

Coordination Games

A computer simulation can represent a number of individuals in a social context, with each individual implementing a cognitive strategy, based on the researcher's beliefs about how cognition works. Of course there are simulations of isolated individuals, as well; for instance, we could consider most AI programs to fit this description, but here we are mostly interested in programs that look at multiple individuals. Some of these focus on specialization and questions about how individuals choreograph their behaviors in order to minimize conflict, while others look at the spread of influence through communities and other aspects of social behavior. In this section we look at some computer paradigms for studying coordination among individuals, leading toward studies of culture in computational communities. Our version of swarm intelligence depends on an understanding of the emergence of culture and its effect on the individuals who comprise a culture.

The prisoners' dilemma game (PDG) reported previously is a very well-studied type of *coordination game,* where the outcomes of a player are determined by the combination of his or her choices and the other players' choices. In most coordination games a player could achieve high payoffs if they knew what the other person was going to do. For instance, if in a PDG you knew with certainty that the other player was going to cooperate, all you would have to do is defect and you would be assured the highest payoff. But of course the other player knows you know that. Thus for them the choice to cooperate is risky and requires trust that you will not take advantage of their vulnerability. It is common to represent coordination games in matrix form; Figure 5.9 is a PDG in the form of a *payoff matrix.*

In this kind of representation, player *A* gets the payoff above the diagonal, in the right half of the cell, and player *B* gets the bottom, left payoff. Each player chooses one of two options; that is, player *A* can choose

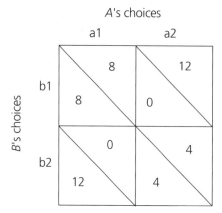

Figure 5.9 A payoff matrix for the prisoners' dilemma game.

a1 or a2, and *B* can choose b1 or b2. If *A* chooses a1 and *B* chooses b1, then they each get a payoff of 8 points; this represents cooperation by both parties. Choice 2 for either player then represents defection; they receive a high payoff when they choose 2 and the other player chooses cooperation, choice 1, and they get a low payoff if both choose to defect simultaneously.

Thus a payoff matrix shows the consequences of pairs of interdependent choices. Two players' choices affect not only their own outcomes but the other player's. In a game such as PDG there is an inherent motive, to try to get the cooperation of the other player. Whether you want to achieve the highest group outcome—the sum $8 + 8 = 16$ points is the highest joint cell in the matrix—or you want to get the other person to play sucker to your defection, you want them to choose 1. And of course the best way to do that is to play 1 yourself, to create a condition where their cooperation will not hurt them.

Payoff matrices can be used to represent an infinite number of possible interdependency situations. For instance, you might take a second to figure out why the game in Figure 5.10 is called "Chicken." Here, choice 1 is "swerving" and choice 2 is "staying on the road." The "winner" receives the highest reward by choosing 2 when the opponent chooses 1, that is, you win by staying on the road when the other driver swerves. If both swerve, that is, choose 1, then both get a moderately high payoff—you get to live, and there is not much shame, since the other person swerved, too. If you swerve and the other person doesn't, then you do get to live, but in humiliation, with a mere 6 points. If neither swerves, payoffs are zero for both, who push up daisies, but in heroic fashion.

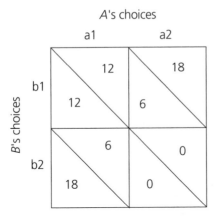

Figure 5.10 A payoff matrix for the game of Chicken.

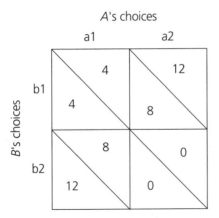

Figure 5.11 A payoff matrix for the Battle of the Sexes.

The "Battle of the Sexes" is another famous coordination game, in which a husband and wife try to work out complementary roles (see Figure 5.11). For instance, it may be that both of them strongly prefer to play the piano while the other sings, but less strongly prefer to hear the partner play the piano rather than sing. Least desirable of all is for them both to sing, though both playing the piano is pretty bad, too.

In the Battle of the Sexes, as in the other games, the objective is for the players to coordinate their choices. Here, the lowest payoffs occur where both participants do the same thing; the trick when this game is iterated is to alternate. In all these games, the meaning of what one player does is determined by the context that the other person creates for it.

The literature of coordination games is very extensive. Economists and social psychologists study the behavior of persons in coordination games for a number of reasons. By manipulating payoffs and game rules, they can study the effects of belonging to a group, the effects of communication between players, how people make decisions under uncertainty and under risk, important aspects of competition and cooperation, and so on. Importantly these studies help researchers understand exactly how and where people depart from the assumption of rationality that is traditional in economic theory as well as decision theory. A player's choice is very much bound up with his or her expectations of how the other person will play, which can reflect a great amount of information about their beliefs about human interaction in general. Many of the assumptions of game theory underlie the programs described in this section, paradigms where interacting agents affect one another's payoffs, often with conflicting constraints.

The El Farol Problem

Economist Brian Arthur (1994) noted that it is impossible for people to reason deductively in complex situations; there are just too many linkages of facts for anyone to keep them straight. In such situations people end up floundering in a pool of subjective beliefs, including subjective beliefs about subjective beliefs. The rationality then that is assumed by classical economists cannot hold. As an example, Arthur devised a diabolical situation known as the "El Farol" problem. The story is told a little differently every time, but we will accept John Casti's slightly more colorful introduction to it (Casti, 1997). According to Casti, Brian Arthur is an Irishman who likes to go to the local pub in Santa Fe, a place called the El Farol, to listen to Irish music and have a beer or two. The El Farol has Irish music only on Thursday nights. Besides Arthur, there are a lot of Irishmen in Santa Fe who enjoy a beer with Irish music in the background, and, like Arthur, they very much prefer to go on nights when the place is not crowded. In fact, they absolutely hate to go when the crowd is greater than 60 people.

But how is a thirsty Irishman supposed to predict when the bar might not be crowded? He would be glad to go to the El Farol, if only he knew those other noisy bastards would stay home. In Brian Arthur's example, the Irishmen (there are 100 of them) know that in the past weeks this many people have attended Irish Night at the bar:

44 78 56 15 23 67 84 34 45 76 40 56 22 35

A thirsty Irishman could come up with any of a number of rules, according to Arthur, for predicting this Thursday's attendance, including

- the same as last week's [35]
- a mirror image around 50 of last week's [65]
- a (rounded) average of the last four weeks [49]
- the trend in the last eight weeks, bounded by 0, 100 [29]
- the same as two weeks ago (2-period cycle detector) [22]
- the same as five weeks ago (5-period cycle detector) [76]

The paradox is that if there were in fact a good theory for predicting when a Thursday night will be quiet, all the Irishmen would figure it out, and they would all go to the El Farol on the same night, and it would end up being packed with antisocial Irishmen. Any solution to this problem in other words is self-negating.

Arthur, who is one of the founders of the Santa Fe Institute, ran a computer experiment to see how a population of computational Irishmen-agents would in fact adjust their attendance. He gave each member of the population a set of possible hypotheses, including those mentioned above, and let agents choose the ones that seemed to work for them, that is, the ones that best predicted the nights that fewer than 60 percent of the agents attended the bar. They could change hypotheses if one failed to work.

Bar attendance in Arthur's experiment fluctuated chaotically (see Figure 5.12), as the majority under- and overcorrected for previous estimation errors: there is of course no stable solution to this problem, no way all the Irishmen can go to the bar without it becoming crowded. On the other hand, Arthur noted that over time attendance was very nearly 60 percent: *average* prediction was excellent.

In his report on the experiment, Arthur suggested that perhaps a critic would argue that he had "lumbered the agents in this experiment with a fixed set of clunky predictive models" and suggested that perhaps a method such as genetic programming would be able to generate more sophisticated and flexible solutions to the problem. Recall that genetic programming is a method for evolving computer programs, using evolutionary operators. Genetically derived programs are often able to find solutions to complicated mathematical and algorithmic problems where more mundane approaches fail.

Bruce Edmonds, of the Centre for Policy Modelling at Manchester Metropolitan University in Great Britain, picked up the gauntlet and

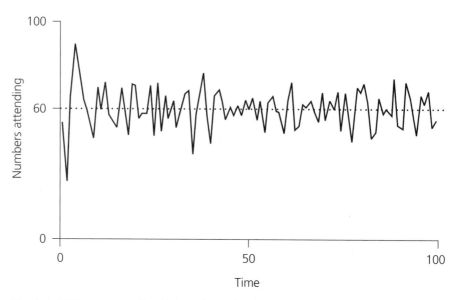

Figure 5.12 Simulated Irishmen attending Irish Night at the El Farol. (From Arthur, 1994.)

programmed a population of agents with genetic programming minds to evolve strategies for the El Farol problem. Edmonds' agents could communicate with and "recognize" other agents, and could tailor their strategies toward those others. Most interestingly, agents communicated their intentions to other agents, but were not required to report them accurately; they could evolve reporting strategies to try to persuade others to stay home. If an agent lied every time, the others could learn to ignore its reports, and the deception would be ineffective. Therefore, it seems that honesty should be the best policy—most of the time.

Edmonds' socially intelligent agents performed on average about as well as Arthur's; that is, the average attendance was very close to 60 percent. Further, no agent dominated or performed any better than any others. Upon analyzing the strategies that evolved, however, Edmonds discovered that some very sophisticated deceptive strategies had sprung up in the population. For instance, an agent could report that it was NOT-going, or that it was NOT-NOT-going, and so on, as they had been supplied with a basic alphabet including Boolean operations. Edmonds reports that one of the agents, after 100 "weeks" in the simulation, had evolved a statement of nine NOTs, so that it could sometimes trick another agent and sometimes be truthful, leaving the other in a perfect state of perplexity. What would *you* conclude if someone said they were not-not-not-not-not-not-not-not-not-going somewhere?

David Fogel and Pete Angeline, researchers for an evolutionary computation company called—what else?—Natural Selection, Inc., with UC San Diego engineer Kumar Chellapilla, attempted to evolve strategies for the El Farol problem using evolutionary programming of autoregression parameters. Populations in their experiments failed to converge, as Arthur's and Edmonds' had, on strategies that averaged near the optimum, leading them to suggest that Arthur's use of discrete problem-solving strategies, as compared to stochastic induction, had led to an artificially "ideal" result.

The El Farol problem is a beautiful paradigm for describing a kind of event that occurs often in nature, and especially often in human social life. A boy who wants to ask a girl for a date must not be too obvious, but he has to say something. She knows that he has to try not to be obvious, but cannot tell if that is his motive or if he is really not interested, so if she is interested in him her reply will have to address both interpretations. Knowing this, he will have to interpret her coolness as possibly either subdued passion or heartfelt disinterest—and so love goes. Social psychologist Ned Jones wrote a whole book on ingratiation, in which one person tries to make another person like them without appearing to be trying to ingratiate, which is of course the most unlikable thing in the world. A baseball pitcher with a count of three balls and two strikes knows he must throw a fastball right down the middle of the strike zone, and the batter knows this as well, and knows that it will be a perfect pitch to swing on—besides, he will strike out if it's in the strike zone and he doesn't swing. The pitcher, knowing that the batter knows he knows this, knows that if he throws something other than a good fastball, maybe something a little off the plate, the batter will probably swing at it, expecting a fastball, and, instead of walking him, he will strike him out. The batter knows that the pitcher knows this, though, so he will not really expect a fastball—after all, if it *is* a little off the plate, the umpire will call it a ball and he will walk. This is what makes baseball interesting. It's those not-not-not-not-not-not-not-not-not-fastballs on a full count.

Sugarscape

Josh Epstein and Rob Axtell are economists at the Brookings Institution who "grow" artificial societies in an environment they call Sugarscape (Epstein and Axtell, 1996). They compare Sugarscape to a kind of laboratory or "CompuTerrarium," which they can seed with a population of agents, an environment, and some rules for their behavior; in this way they can test whether certain phenomena of economics are necessary outcomes of dynamical principles and what parameters affect the patterns of observed behaviors.

Sugarscape is described as a cellular automaton with agents built into it. The environment is a grid containing some resources—sugar and sometimes spice—that the agents need in order to remain vital. Agents are able to "see" into the cells that are nondiagonally adjacent to theirs (sometimes farther) and are able to move into neighboring cells. Though rules can be revised by the researcher, generally agents attempt to move toward regions with higher densities of resources. Once there, they devour the resource, which (in some sets of rules) can eventually grow back.

Epstein and Axtell conducted a number of experiments demonstrating the emergence of population-level patterns from local interactions of agents with their environment and with one another. One experiment reported in *Growing Artificial Societies: Social Science from the Bottom Up* attempts to grow cultural patterns in the agent population by allowing agents to influence one another. Each agent is born with a random bitstring "label," say, 01101101. Agents are selected in turn. For each of their four neighbors (left, right, above, below), a "tag" or site on the bitstring label is selected at random; if the neighbor's tag differs from the selected agent's at that position, that is, if one has a zero and the other has a one, then the neighbor's label is changed at that position to match the selected agent's. If they match, no changes are made. As in most of the reported examples, agents move across the sugarscape toward regions where they perceive the greatest density of resources.

In this experiment, agents were colored red if more than half of their tags were ones and blue if the majority were zeroes, in order to differentiate two "cultural" subpopulations. The sugarscape contained two regions where resources were dense, in the northeast and in the southwest corners of the screen (the environment is actually a torus field); agents typically end up where the resources are. The effect of agents' taking tags from their partners is the formation of homogeneous red or blue populations within each of the resource-rich regions; one region may be occupied by the red culture and the other with the blue, or both sites are either red or blue. Sociograms depicting the patterns of connections between pairs of agents that have interacted with one another reveal dense interlinkage within clusters and almost no connections between them. In other words, agents interacting with their neighbors become more similar and gravitate together toward locations rich in resources—and cultures do not communicate with one another.

In an interesting footnote chapter to *Growing Societies,* Epstein and Axtell describe an epidemiological model with a very interesting implication. Disease agents—germs—are coded as binary strings of length five, for instance, 10110, and the "immune system" of an agent is coded as a bitstring of length 50. If any section of the agent's immune system bitstring matches an antigen's five bits, then the agent is immune to that disease. Thus an immune system that starts 10100*10110*11101010010100 . . . gives immunity from disease caused by antigen 10110 because they match in the italicized section.

Agents in the simulation were allowed to propagate genetically, evolving immune systems, with fitness being a function of how well the individual's immune system protected them from a number of diseases that existed in the population. Diseases were spread from agent to agent; when they interacted, an agent acquired one disease from its neighbor. Thus the fitness landscape was constantly evolving as diseases spread and immune systems searched for solutions through the generations.

Epstein and Axtell report that some immune systems did something that was "quite marvelous," in their words (p. 150). An immune-system bitstring could evolve that was shorter than the sum of the lengths of the antigens it guarded against. For instance, a 15-bit immune system 100101101001001 can obviously fight three 5-bit germs:

- 10010
- 11010
- 01001

as these make up the three five-bit sections of its bitstring. But further, the same immune system could prevent against germs

- 00101 (starting at the second position)
- 00100 (starting at the ninth position)

and so on. In fact, immune systems with overlapping substrings did appear on the sugarscape. The evolution of this kind of compression of information is indeed quite startling, and it is a kind of effect that deserves further investigation.

Unfortunately we do not have enough room in this volume to review the several interesting computer models of immune-system response. The immune system is a very subtle and complicated learning system found in all living things. It is able to identify antigens after one encounter, prepare a defense against them, and mount a full attack at the slightest provocation—and hardly ever when it is inappropriate. Simulations of the immune system have been developed for optimization purposes, with very good results, but are a little off the topic here.

In this short section we have seen examples of agents in the sugarscape environment interacting, pursuing resources, evolving, forming cultures, and even developing immunities. This is quite a comprehensive collection of accomplishments and indicates a good direction for the future growth of simulation software. Wide access to such a CompuTerrarium can allow testing of many interesting questions about human, ecological, and even physical systems.

Tesfatsion's ACE

Leigh Tesfatsion, at Iowa State University, conducts research in a paradigm she calls ACE, or *agent-based computational economics* (Tesfatsion, 1995, 1997). She agrees with Epstein and Axtell that simulation laboratories such as ACE might finally provide an ultimate paradigm for the study of social science and especially the study of economics.

Almost all prisoners' dilemma paradigms assign pairs of individuals to interact with one another—they almost never get to choose. Tesfatsion makes the obvious-once-you-hear-it observation that in real life, people choose whom to talk with, whom to interact with, whom to do business with. If we don't like someone's strategy, if we think they are likely to cheat us in order to beat us, we just don't interact with them. Thus she has conducted a number of simulation experiments on a model she calls "evolutionary IPD [iterated prisoners' dilemma] with choice and refusal." Choice enables individuals to increase their chances of finding a player that will cooperate with them, and refusal lets them escape defection; further, as not-nice strategies tend to result in ostracism, which leads to reduced payoffs over time, there is an incentive for mean players to adjust, to cooperate.

In one kind of choice-and-refusal ACE model, called the *trade network game* (TNG), Tesfatsion creates populations of interacting autonomous agents she calls *tradebots* with behavioral functions that enable them to trade with other tradebots and memories that allow them to identify the others that they have traded with. Thus they can have preconceptions about what to expect in dealing with others, based on experience. The tradebots are implemented in an object-oriented computer system, written by wiz programmer David McFadzean, that allows manipulation—

technically known in the field as *tweaking*—of parameters that might affect the agents' trading behaviors, their expectations, the structure of the market, the matching up of trading partners, and the way that trading behaviors are learned or evolved.

Buyers form lists of preferred sellers, and sellers maintain lists of preferred buyers. Buyers make offers to the sellers, who are privileged to select which tradebots they wish to sell to. Underlying the TNG is an assumption that the situation resembles a prisoners' dilemma. Traders receive payoffs depending on what happens in the trading process; for instance, there is a negative payoff for being refused, a "wallflower payoff" for failing to make an offer, a "sucker's payoff" for trading cooperatively with someone who defects, and so on. After each trade cycle, each tradebot updates its memory for the individual it interacted with, increasing its value if the interaction was profitable and decreasing it if the trade turned out negatively.

Evolution occurs after each cycle of trading and is conceptualized as the formation and transmission of new ideas, not biological evolution. Tradebots evolve trading strategies as a result of their experiences in the trade network. Successful trade strategies are imitated, while unsuccessful strategies are replaced with ones that more resemble successful ones.

Tesfatsion believes that the central problem for ACE researchers is to understand how order and regularity arise spontaneously in decentralized economic systems. Coordination is unplanned; buyers buy what they want, and sellers sell what they think buyers will buy. Yet an economy functions as if the famous "invisible hand" were pushing, pulling, kneading, arranging, and rearranging actors, actors' intentions, and actors' behaviors. We agree, and add another clause to the proposition: not only is it important to come to understand how local interactions can result in global effects, but it is important to understand how the global effects in turn benefit the lower-level actors.

Picker's Competing-Norms Model

Randal Picker is a law professor interested in the relationship between government and the norms that arise in social behavior. In a paper published in the *University of Chicago Law Review,* he focused particularly on the issue of competing norms (Picker, 1997). One kind of behavior might remain prevalent even while a superior behavior is available to the members of a community. Some of these kinds of norms can be very subtle; for instance, he mentions that buckling your seat belt in a Hungarian cab is considered an offense to the driver, while a French driver would take no notice. The persistence of inferior norms is of great concern to anyone interested in the welfare of society; people insist on doing dangerous things, ignoring threats, starving themselves in the name of beauty—the list could indeed be very long, and it is often not easy to change risky behaviors. Nor is it easy to enforce laws that contradict the popular way of doing things—the current American drug prohibition is an example.

Picker implemented a kind of mutable prisoners' dilemma game in a grid resembling a cellular automaton. An individual agent is represented as a cell in the CA and plays repeated games with the members of its *pay-off neighborhood*. The payoff neighborhood might be a *von Neumann neighborhood* comprising adjacent cells above, below, and to the left and right, or a *Moore neighborhood,* which also includes the diagonal adjacent cells (see Figure 5.13).

An agent in Picker's model is also conceptualized as belonging to an *information neighborhood,* which can be of the von Neumann or Moore types. The agent gathers feedback from members of the information neighborhood about what strategies other players have used, and how successful these were for them. Players then adopt the strategy that resulted in the highest payoffs in the information neighborhood. Picker manipulated the two kinds of neighborhoods, so that in some cases the information neighborhood was larger than the payoff neighborhood, in some cases they were the same size, and in other cases the payoff neighborhood was larger.

The choice of strategy for the first round, to cooperate or defect, was randomly assigned with the proportion of defecting and cooperating agents set as a parameter for the experiment. In most cases, Picker's populations converged to unanimity on one strategy or the other; the strategy selected was a function of the initial proportions and the relative goodness of the superior strategy. In general the population tended to converge on the superior strategy, in other words, they tended to optimize their decisions; but in cases where the relative benefit of the superior choice fell below a threshold, or when the initial superior population was too low, the population converged on the inferior choice. As he says, "For the inferior equilibrium to win over our players, we need either extremely bad luck or something that makes the inferior strategy especially salient" (p. 1255). Picker's data show clear phase transitions at very narrow regions near the two thresholds (payoff benefit and initial proportion), where mixed strategies are found; otherwise the population becomes uniform, all cooperators or all defectors.

An important finding in Picker's research, which almost goes unmentioned, is that by observing the strategies of some local neighbors and

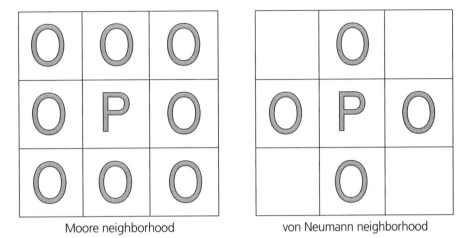

Moore neighborhood   von Neumann neighborhood

Figure 5.13 Two kinds of CA neighborhoods: Moore and von Neumann.

their consequences, his simple computational agents were able to optimize a rather complex decision function. Given a fair chance, local interactions lead to global convergence on the norm that results in the higher payoff. Knowing the meaningfulness of phase transitions, we would like to explore further the dynamics at the transition boundaries in Picker's paradigm, where competing norms coexist. It would also be interesting to experiment further with various sociometric patterns of connections among agents in the payoff and information neighborhoods; it seems that overlapping neighborhoods, as in this study, represent a special case and perhaps not an especially typical one. Picker's paradigm is rich in implications for future research.

Latané's Dynamic Social Impact Theory

In 1913 a German social psychologist named Ringelmann asked people to pull on a rope attached to a dynamometer, individually and in groups. A dynamometer measures the strength of a force applied to it; Ringelmann was measuring how hard people pulled on the rope. His interesting finding, now known as the "Ringelmann effect," was that groups of people pulled with a strength that was considerably less than the sum of their individual efforts. With each additional person pulling on the rope, it appeared that each one pulled with a smaller proportion of their full strength. The relationship between N, the number of people, and total force was *monotonic*, meaning that force increased with N, but

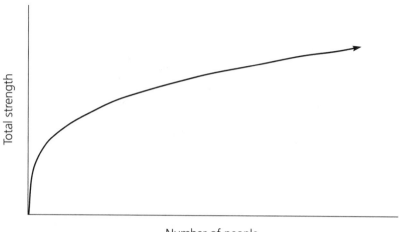

Figure 5.14 The Ringelmann effect.

increase was negatively accelerating; as the size of the group grew, the force with which they pulled grew slower (see Figure 5.14).

Through the middle of the 20th century, results similar to these were found in numerous social-psychological studies of group influence on a target individual. As the sources of influence increased, the amount of influence increased, but the increase of influence was slower than the increase of the size of the group.

In 1981 Bibb Latané summarized these results and reported on a rather large accumulation of evidence from numerous laboratory and field experiments. For instance, his research with John Darley on helping behavior and bystander intervention, exploring the conditions under which people will and will not help someone in need, had found that the probability of any individual helping someone in need decreased as the number of people present increased, similar to the Ringelmann effect (though the probability that *someone* would help continued to increase). This finding was consistent across a large number of emergency situations, both real and contrived. Latané concluded that the effect was ubiquitous and could be adopted as a parsimonious theoretical construct to explain a great amount of human conduct.

Latané's *social impact theory* was described as the "light bulb" theory of social influence. Just as the brightness on a surface is a function of the number, wattage, and nearness of light bulbs shining on it, so people are influenced by a group in proportion to the "Strength" (something like status or persuasiveness), Immediacy (the opposite of distance), and Number of the group members: $\hat{\imath} = f(SIN)$. (See Figure 5.15.)

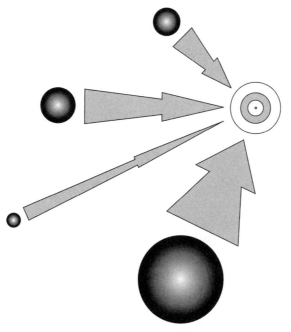

Figure 5.15 In Latané's original social impact theory formulation, impact on the hapless target was a multiplicative function of the Strength, Immediacy, and Number of sources of influence. (From Latané, 1981.)

With the emergence of complex dynamical systems research through the 1980s, Latané began to expand the predictions of social impact theory to show that the behaviors of individuals could be explained in terms of the self-organizing properties of the social system they comprised. As individuals conformed to the norms they perceived as a function of the Strength, Immediacy, and Number of people participating, patterns formed that were only apparent from the population level. Clusters of individuals began to believe similarly, and subpopulations diverged from one another.

Latané teamed up with the Polish physicist/psychologist Andrzej Nowak and some of his Polish colleagues, who had developed a computer package they called the Warsaw Simulation System (WSS) (Nowak, Szamrej, and Latané, 1990). This software allowed the user to adjust parameters and rules applied to a population of binary individuals, similar to cells in a cellular automaton, and observe their effect in terms of which states the individuals settled into. A WSS trial programmed with social impact rules progressed until change had stopped. When this occurred, it was seen that individuals resembled their neighbors, while regions of the population matrix differed from one another (see Figure 5.16). In their 1990 *Psychological Review* paper, Nowak, Szamrej, and Latané called this effect *polarization* and demonstrated that it was very similar to effects found in real societies.

This finding of polarization was important, as previous social scientists had believed that any mathematical model of interacting agents would inevitably gravitate toward homogeneity; Festinger had called it "pressure toward uniformity." But Latané had long noted that in fact people in societies, and even in small groups, do not come to agree uniformly about everything. The more common finding was just as Latané's simulations showed, that pockets of agreement formed, with differences between them.

Social impact theory evolved over the years (e.g., Latané and L'Herrou, 1996). Latané calls the current incarnation *dynamic social impact theory,* and his findings have developed beyond simple polarization. Dynamic social impact theory results, whether in simulations or studies with human subjects, are seen to possess four characteristics, as described by Latané in numerous publications. These are

- *Consolidation:* The diversity of opinions is reduced as individuals are exposed to a preponderance of majority arguments.

- *Clustering:* People become more similar to their neighbors in social space (usually correlated with physical space in Latané's view).

- *Correlation:* Attitudes that were originally independent tend to become associated.

- *Continuing diversity:* Clustering protects minority views from complete consolidation.

The dynamic social impact model differs only slightly from those described in previous chapters, depicting germs, bugs, varmints, and widgets. Now we are considering how people come to hold beliefs and attitudes, and we are not considering the process to be rational or even self-serving. Latané's model is at least approximately consistent with findings in the field of social psychology and also in sociology, economics, and anthropology. As people interact they persuade one another of things, they show one another how to do things, they impress one another, they copy one another, and the simple, obvious result is they become more similar. Their experience may be that they are making rational decisions about everything they do, but the observed effect is that they are becoming more like the people they know and interact with.

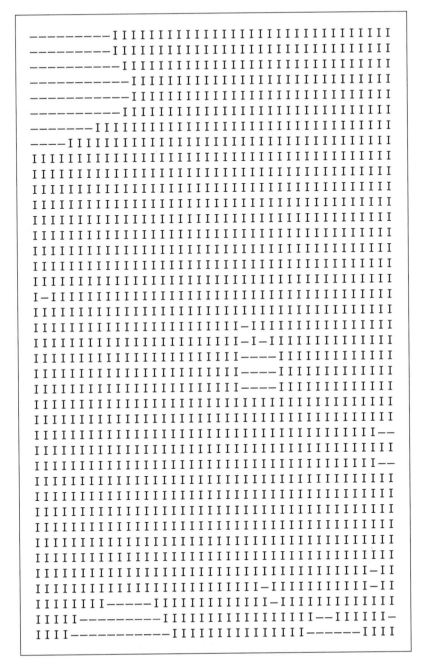

Figure 5.16 In Nowak, Szamrej, and Latané's simulations, minority attitudes (initially 30 percent of the population in this example) ended up mostly clustered within well-defined subpopulations when the system reached equilibrium. (From Nowak et al., 1990.)

Latané's simulations demonstrate some important features of collective adaptation, and they move us without much strain toward the idea of swarm intelligence as a human phenomenon. Latané's theorizing is based on a great deal of experimental evidence, both his own and others', and the effects of Strength, Immediacy, and Number are ubiquitous and easy to confirm. There is not unanimous, universal agreement among social psychologists as to the mechanisms by which polarization comes about, and neither is there complete agreement that Latané's exact formula is correct. But there is agreement at least that a nonlinear monotonic relationship exists between attitudes and beliefs on the one hand and number of sources and targets of influence on the other, and certainly there is very little disagreement about the finding that people influence one another, and in so doing, become more similar.

Though most of Latané's research reports influence on single attitudes or beliefs, the third effect reported in dynamic social impact simulations is *correlation*. By this term, social impact researchers mean that the *patterns* of beliefs held by individuals tend to correlate within regions of a population. If you agree with topic number one and disagree with topic number two, then it is likely your friends and neighbors hold those two views, too. Latané's explanation for this effect in humans and in his simulations is that individuals with high Strength, that is, relatively persuasive or influential individuals, affect their neighbors on a number of topics. Social impact theory does not predict any interaction among the various opinions or beliefs that the individual might have. Multiple attitudes are not described as a multivariate effect, but more as a set of univariate effects. The high-Strength individual affects the target's position on opinion number one, and also on opinion number two, and so on. There is no serious psychological model, however, that would assert that an individual's beliefs are independent of one another, simply blowing the same direction like a staff of flags in the wind of social influence; instead, someone's beliefs must be seen as logically interrelated, and their attitudes as affectively interrelated, as individuals strive for consistency. A truly multivariate view may seem to make social influence more complicated—we think it makes it simpler.

Boyd and Richerson's Evolutionary Culture Model

With their 1985 *Culture and the Evolutionary Process,* Robert Boyd and Peter Richerson laid out some of the implications of innumerable human interactions in a population in terms of evolution and culture and their interaction. The two ecologists were co-teaching an introductory course in environmental studies when they decided to pursue a kind of theme: the idea that humans adapt to their environments as other organisms do. Thus was born their dual-inheritance mathematical model of human behavior. On the one hand, they reasoned that some part of human behavior is determined by genetics; our biology determines our behavior or at least influences it. On the other hand, our genes predispose us to behave socially in a way that results in culture, and much of our behavior is acquired by imitation, through a process they called *cultural transmission.*

Boyd and Richerson developed a mathematical model, calculating among other things the types of environments in which social learning or individual learning might be more adaptive. An individual's learning from his or her own experience, they reasoned, should be most adaptive when the environment is relatively homogeneous and stable over time, so that lessons learned in one situation can be generalized to others. Social learning is more adaptive, they reasoned, when an individual is unlikely to sample sufficiently diverse aspects of the environment on his or her own, and can only attain a comprehensive view of the world by learning from others' experiences.

These scientists (Boyd is now an anthropology professor) presented an original and insightful view into various types and aspects of cultural transmission and how these interact with biological evolution. Recall that a genotype is the genetic coding of an organism or its traits and a phenotype is the expression of the genotype. The important fact is that the phenotype develops through the interaction of the genes with the environment. Some phenotypes are more variable than others, and some depend more on the environment for expression; for instance, phenotypic freckles do not appear unless the person spends some time in the sun. Boyd and Richerson note that in humans phenotypic expression of behavior depends on two kinds of learning—learning derived from cultural norms that the person is exposed to and learning acquired through individual experience.

As the human species evolved, individuals' adaptations—and their subsequent probability of survival and reproduction—depended jointly on their individual experiences and on what they learned socially. Boyd and Richerson hypothesize that the tendency to learn more in one way or the other was genetically evolved. Thus, genetic evolution would have produced a population that emphasized social or individual types of learning; in humans it appears that social learning was favored by evolution.

In Boyd and Richerson's view, *guided variation* occurs when evolution is informed by within-lifetime adaptation, that is, by learned behavior. Individuals who are capable of learning to adapt to their environment are more likely to survive, and natural selection is thus more likely to favor those kinds of learners. Over time the population moves toward the more adaptive learning style. When individual learning predominates over cultural transmission in the population, there is a greater amount of variation in phenotypic expression, as each individual makes his or her own errors, and the force of genetic evolution is strong. When cultural transmission is more important, then there is less variation, as all individuals behave in approximately the same way, and genetic evolution is slowed; less fit members of the population are able to survive by adapting through imitation, dulling the effect of natural selection. Thus cultural evolution is able to influence the flow of genetic evolution.

An interesting implication of the dual-inheritance model is that cultural and genetic evolution can move in opposite directions, resulting in what Boyd and Richerson call the "Freudian model." For instance, strict sexual taboos might evolve culturally even while the genetic tendency toward promiscuity or sexual impulsiveness increases. The result would be a tension between the two motives, possibly with no difference in the average prevalence of any overt behaviors. Importantly, dual inheritance implies predictions that disagree with those of sociobiologists, who argue for the importance of genetic inheritance in human behavior and downplay the ability of culture to neutralize inherited tendencies.

Boyd and Richerson note that because cultural evolution favors some variants over others, it can introduce a bias into the development of phenotypic behaviors. *Unbiased transmission* exists when individuals simply adopt the variants they are exposed to, for instance, by their parents; it does not affect the prevalence of the trait in the population. *Direct bias* occurs when the individual selects a belief or behavior based on attributes of the variant itself; that is, it may appear to be the best choice for some reason. Direct bias implies that better solutions to problems will become prevalent in a population; it suggests a tendency for improvement. *Frequency-dependent bias* exists when the individual adopts a variant because it is statistically prevalent in the population, or at least among the people that the individual interacts with. A term used by Boyd and Richerson for cultural evolution with a heavy dose of frequency-dependent bias is *conformist transmission*. Finally, *indirect bias* occurs when a variant is associated with other variants that are considered desirable. The search for cognitive consistency is highly motivating and is one of the central themes of social psychology. In Boyd and

Richerson's model it is a force that shapes human cultural evolution, and ultimately genetic evolution, through indirect bias.

In recent computer simulations testing Boyd and Richerson's hypotheses, Henrich and Boyd (1998) examined effects in a model featuring conformist transmission, that is, social influence with frequency-dependent bias. In this kind of cultural transmission, the frequency of a trait in the culture provides information about how adaptive the trait is. When the probability of adopting a trait depends on how widespread it is already, traits—beliefs, attitudes, or behaviors—that begin to become prevalent in the population will eventually take over a population or subpopulation. Therefore a culture must have other forms of transmission, in order to preserve diversity and consequent adaptability.

Henrich and Boyd (1998) hypothesized that there are two general cues for learning, social and nonsocial. Use of nonsocial environmental cues amounts to learning from your own experience, while social learning comprises the acquisition of knowledge by observing others. Thus the simulations were conducted in four phases:

- Cultural transmission, including conformist transmission of traits
- Individual learning
- Migration of individuals between cultures
- Natural selection of individuals who populate the next generation

Individuals in the simulations differed in their inherited tendency to rely on social versus individual learning, and environments differed in the stability and in the amount of information they presented to the individual, that is, how easy or hard it was to detect the state of the environment. Where an individual implemented conformist learning, their probability of acquiring a trait equaled the proportion of the population who possess it. Henrich and Boyd postulate two heritable learning parameters. One that they (as well as Boyd and Richerson) call L denotes the degree to which a person tends to use social learning. For instance, someone for whom $L = 0.0$ learns entirely from experience, where another with $L = 1.0$ learns purely by imitation. The second parameter Δ encodes the tendency to rely on biased ($\Delta = 1.0$) or unbiased ($\Delta = 0.0$) influence. Thus an individual with $L = 1.0$ and $\Delta = 0.0$ learns socially through pure unbiased imitation of observed behaviors, while if $\Delta = 1.0$, the probability of their adopting a behavior is determined by its frequency in the population. In other words, L is the social learning gene, and Δ is the conformist gene. They also manipulated the probability of individuals migrating from one population to another and the variability of the environment.

In Henrich and Boyd's simulations, individuals were given the opportunity to adapt in an environment; they searched within lifetimes for optimal behaviors and across generations for the genetic patterns that produced the best behaviors. Adaptiveness was defined in terms of a choice between two behaviors, one of which was defined as more adaptive, given a particular environment, and genetic fitness was defined by the individual's ability to select the optimal behavior. A slight reproductive advantage was given to more fit individuals. The process was run until it reached an equilibrium state where change ceased or fell below a threshold.

As suggested by Boyd and Richerson's analysis, these authors found that selection favors conformist transmission, that is, frequency-dependent social influence, over a wide range of environments. More dynamic, changing environments produced populations that relied less on social learning. Even then, frequency-dependent transmission remained advantageous; thus, though individual learning is emphasized in fluctuating environments, the social learning that is used tends to rely on imitation of popular traits. Further, a population that has discovered the value of conformist, biased learning tends to rely more on social versus individual learning in general. Regular migration of individuals between cultures reduced reliance on social learning but increased the effect of conformist transmission. In sum, when individuals take into account the frequency of behaviors in the population before deciding to imitate them, social learning is enhanced, in almost any kind of circumstance.

Henrich and Boyd note that frequency-dependent transmission is evolutionarily important because it reduces behavioral variation within groups, enabling the establishment of group boundaries and cultural differences. These differences, though they often generate conflict and strife, may also have adaptive value. As Henrich and Boyd point out, there are of course other explanations for the formation of cultures. For instance, some social scientists argue that phenotypic behavioral differences between cultures actually arise from differences in environments. It is easy, though, to point to instances of groups that, while inhabiting the same environment, maintain very different norms of behavior.

The evolution of culture may be seen in terms of movement on an environmental fitness landscape containing multiple peaks. Leaping from peak to peak is rare, and cultures tend to maintain behaviors they have found to be optimal. Some theorists have suggested that individuals conform to statistically prevalent norms in order to escape punishment. It is clear, though, that while some norms are subject to punishment, for instance, the norm of wearing clothes, others persist even though there is no sanction for violating them. The conclusion is that cultures form simply because conformist learning is adaptive.

Memetics

Since evolution was first proposed as an explanation for biological diversity, numerous social theorists have pointed out its profound similarities to human culture—in fact, it appears that Darwin may have borrowed the insight of incremental adaptation of species from the writings of Sir William Jones on the evolution of Indo-European languages, which Jones had described as "homology by descent." It is apparent that ideas, behaviors, and other changeable features of individuals propagate through populations, changing through some kind of process similar to mutation, combining with other features, evolving—there is no better word for it—over time. Cultures are a lot like species.

Contemporary thinkers who think about the evolution of culture are most likely to think of it in terms of *memetics,* a topic we briefly introduced in Chapter 1. Richard Dawkins (1987) invented the label *memes* to describe something analogous to genes but transmitted interpersonally; memes are usually treated as though they were identical to ideas, though ideas are only one kind of meme. The term was inadequately defined, but the concept has been widely—we almost said wildly—accepted by diverse observers of society. Enthusiastic proponents of the view range from Nobel winners to Internet flamers to New Age management consultants.

Paleontologist Stephen Jay Gould (1991) asserts that biological evolution is a false analogy for cultural change. In *Bully for Brontosaurus* he suggests three reasons for his skepticism. First, the speed of cultural evolution far outstrips anything that Darwinian processes can do, and as Gould notes, timing is crucial in evolution. Second, Gould asserts that cultural evolution is Lamarckian and direct, whereas biological evolution is not Lamarckian and works indirectly. The biologist Jean Baptiste Lamarck had argued, in Darwin's day, that acquired behaviors and traits could be passed from one generation to the next—the giraffe's long neck is the clichéd example, the idea being that as giraffes reached for higher and higher branches, their offspring were born with longer necks. The fact that cultural evolution involves the direct transmission of "phenotypes" is an important point in the breakdown of the metaphor; there is nothing analogous to a covert genetic medium of inheritance. Third, Gould raises the serious objection that the topologies of the two processes are different. In particular, he notes that biological evolution is a divergent process, forever branching; species never merge with other species, but can only split. Cultural evolution on the other hand seems to be forever merging, as new ideas form out of the synthesis of ones that existed previously, with a strong measure of mutation. Of course it can be argued that biological lineages do merge, at least within species, through the process of sexual recombination, so this criticism is not strictly accurate; between-species transmission, which would seem to be analogous to cross-cultural spread of ideas, does not occur, however. Gould argues that biological and cultural change, while they are not analogues, may reflect some more profound principle of organization that is common to both, and perhaps to other "historical" systems as well—a statement with which we most wholeheartedly agree.

Philosopher Daniel Dennett (1995) appreciates the richness of the idea that culture is Darwinistic. The memetic metaphor offers him a tasty soup of flavors, illuminating aspects of mind and life that might otherwise never be understood; for Dennett the comparison between cultural and genetic transmission of information is importantly insightful. At the same time, Dennett systematically reveals some serious weaknesses in the theory of memes.

For instance, Dennett notes that each mind that hosts a meme changes it greatly. We could cite evidence from social psychology, especially the literature on norms, social influence, and conformity, to argue that people do in fact tend to pick up any old idea that is floating around the social environment. Social impact theory, for instance, predicts that people will simply adopt the opinions and attitudes that are prevalent in their neighborhood. On the other hand, persuasion researchers have been somewhat able to taxonomize the conditions under which a person will accept an argument, identifying, for instance, features of the message, the source, and the target of a communication that will result in a change in belief or attitude. Obviously, in at least some cases, people pick and choose what to believe. In this way cultural transmission is unlike genetic combination, where genes are mixed and matched randomly. (Our *experience* is that we pick and choose what to accept from others— we rarely see ourselves as conformists.)

Dennett notes that memes change when they come into contact with one another in a mind. A freshly introduced idea is blended with knowledge that is already held; useful parts are kept and troublesome parts might be forgotten or rationalized away. Certainly there is nothing like this in biological evolution. As modern criminal science emphasizes, DNA is like a fingerprint; its patterns remain constant throughout an organism's lifetime, beginning at conception.

Dennett grapples with another fundamental problem of the memetic-genetic analogy, which is the question of the mapping between meme and some medium or carrier. An idea cannot be defined in terms of the language in which it is expressed, as it can be expressed in a variety of different language patterns. It almost surely cannot be defined in terms of brain structures, as it is extremely unlikely that some future neuroscientist will discover that each idea is stored in the same neurons in all people. Genes present no such problem: each gene is a distinctive pattern of amino acids on a chromosome. True, similar biological phenotypes can be generated by different genotypes, but for cultural transmission it is rarely the case that two people acquired the same belief through exposure to identical expressions of it.

Another place where the analogy breaks down, according to Dennett, is in the fact that genetic duplication is extremely precise, with very rare mutations or copying errors. On the other hand, memes, that is, ideas, are almost never copied accurately. Historians and critics may attempt to trace ideas back to their sources, through the lineage of cultural inheritance, but the quest is always ambiguous. Our culture, in fact, has come up with ways to address this problem: refereed publications and patents, one of the main purposes of which is to provide an "audit trail" for ideas.

Finally, Dennett asks about the kind of fitness that might be increased through memetic evolution. In one view, a meme might persist and spread simply because it is a good replicator—not necessarily because it helps the person who hosts it. On the other hand, Dennett argues that memes that are good replicators seem also to tend to be good for people. Here his argument is not too persuasive, as we could spend a long time listing bad ideas that have spread through populations, sometimes persisting for millennia.

We can argue from the point of view of social psychology that fitness of a "meme" or pattern of ideas for cultural evolution is measured, not in terms of survival of the organism but in terms of consistency among cognitive elements. Beliefs need to fit together logically, emotions need to fit together affectively, and emotions and beliefs need to fit together with behaviors. It is very uncomfortable to like a person with disgusting characteristics or to dislike someone who is, by all measures, good and admirable—we are cognitively obligated to find something bad about them. From this viewpoint we would say that a fitness of a meme depends on some measure of how well it fits with other memes that are already present.

Dawkins retreated somewhat from his initial formulation of memetic theory, and Dennett calls him out for it. Though Dawkins' concessions to the criticisms seem rather reasonable, at least as they are quoted in

Dennett's *Darwin's Dangerous Idea,* Dennett seems to view them as a kind of chickening out, backing down in the face of politically correct humanistic dogmatists:

> I suggest that the meme's-eye view of what happened to the meme meme is quite obvious: "humanist" minds have set up an aggressive set of filters against memes coming from "sociobiology," and once Dawkins was identified as a sociobiologist, this almost guaranteed rejection of whatever this interloper had to say about culture—not for good reasons, but just in a sort of immunological rejection (Dennett, 1995, pp. 361–362).

We are sympathetic to his argument, insofar as we have seen the march of "political correctness" in recent decades, trampling over good ideas that might have potentially disturbing implications, and we sympathize with Dennett's concerns—sociobiological theory does tend to evoke a knee-jerk response from humanist types. But antihumanism is not a meaningful support for memetic theory, and here it seems a rather weak, mildly paranoid rearguard defense. In fact there are good reasons to criticize memetic theory, and Dennett himself has listed a number of them, with no serious rebuttals.

Cultural change and biological evolution are similar in that they are dynamic, stochastic, and adaptive, and occur in populations. But where natural selection is primary in Darwinian evolution, it is a minor aspect of cultural change. Minds adapt by changing, not by survival of the fittest, not by constant replacement of weak ones. As we discussed cursorily above, it is not clear that ideas have an existence independent of minds. There seems to be little to gain by focusing on the evolution of ideas and much to gain by looking at the evolution of minds.

Memetic Algorithms

Some computational intelligence researchers are currently working in a paradigm they call *memetic algorithms,* though they bear almost no relationship to Dawkins' theoretical perspective. Memetic algorithms are defined as evolutionary algorithms with local search. That is, as a population evolves, the individuals who comprise it conduct their own individual searches, resembling the way individual organisms learn within their lifetimes even as the species evolves on a larger time scale. The idea is intuitively appealing and its implementation is very flexible—nothing says *which* evolutionary algorithm is to be used or which local search technique. Macro- and microcomponents of the paradigm can be adapted to a particular problem or situation.

Pablo Moscato is a South American researcher who has pioneered the study of memetic algorithms (e.g., Moscato, 1989). He and Michael Norman, who is now in Scotland at the University of Edinburgh, began working together in the 1980s at Caltech. In a recent paper they describe the use of a memetic algorithm for optimization of a traveling salesman problem (TSP) (Moscato and Norman, 1992). Recall that the TSP requires finding the shortest path through a number of cities, passing through each one only once. The problem has a rich history in applied mathematics, as it is very hard to solve, especially when the number of cities is large. TSP is an NP-hard problem, which suggests that if a way is found to solve it, then a large number of other problems will also have been solved. Moscato and Norman use an algorithm with both cooperation and competition among agents in the population, and implement a hybrid version of simulated annealing for local search.

A population of individuals—these researchers usually use a population size of 16—searches the problem space, which is defined by permutations of the cities, called "tours." The population is conceptualized as a ring, where each individual is attached to its two immediately adjacent neighbors, with whom it competes in the search; individuals are also connected to others on the far side of the ring, with whom they cooperate. Each individual in the population comprises a tour of the cities. Competition is seen as "challenge" and "battles" between pairs of individuals, where the tour lengths of an individual and its neighbor are compared and a probability threshold is set based on the difference. The difference between the tours' lengths affects the steepness of the *s*-shaped curve; when the difference is small or the temperature is cool, the probability distribution becomes nearly uniform, and when the difference in lengths between the two tours is great, the probability is increased that tour 1 will be deleted and replaced with a copy of tour 0.

Cooperation is used to let more successful individuals "mate" with one another, rather than with less-fit members of the population. The same rule that is used in deciding competitive interactions is used to assess the desirability of partners for crossover, which is implemented just as it is in GA. One individual "proposes" to another, and if the proposition is accepted, that is, if the stochastic decision favors their interaction, then the crossover operator is implemented. Thus the next generation is created.

Moscato and Norman implement some sophisticated twists to the coding of the chromosome, and their version of simulated annealing draws on previous research in TSP optimization. Within each time step, each individual is changed in one of several ways, for instance by swapping a pair of cities or by inserting a city between two others. Improvements are accepted depending on the "temperature," which decreases through the course of the experiment; recall that in simulated annealing the probability of taking a step that results in worse performance decreases over time. The simulated annealing cycle is repeated numerous times within each generation of the evolutionary algorithm.

The authors report excellent results on a number of different TSP examples with as many as 532 cities. They find that the most improvement in their algorithm occurs when the temperature is low, that is, in the final stages of the trial. This suggests that exploitation is the most important part of the local search strategy, which basically functions to enhance the genetic algorithm within which it is nested.

Peter Merz and Bernd Freisleben (1999a, 1999b) have reported versions of a memetic algorithm, which they have also called "genetic local search," that uses local search within a genetic algorithm framework (see Figure 5.17). They define the paradigm in terms of these four steps:

- Choose a representation and evaluation scheme for the problem.
- Choose a suitable local search algorithm.
- Choose a method for initializing the population.
- Define mutation and recombination operators.

In 1996, Merz and Freisleben's algorithm won first place in the first International Contest on Evolutionary Computation. The contest allowed researchers to submit programs to optimize a standardized set of problems. Subsequent papers have reported further successes in using the memetic algorithm for optimization.

Edmund Burke and Alistair Smith (1997, 1999) at the University of Nottingham compared various kinds of local search techniques, including simulated annealing, hill climbing, and tabu search, in a memetic algorithm framework. *Tabu search* is a kind of algorithm that maintains a list of "bad" points in the problem space, points that have been evaluated most poorly; the search algorithm then steers away from those points. They reported that hill climbing was fastest, followed by tabu search, and finally simulated annealing, for the kinds of scheduling problems they work with. Though the memetic algorithm performs about as well as other algorithms on simple, low-dimensional combinatorial problems, it is superior when problems are big.

Finally, biologist Liane Gabora (1995, 1996, 1998) has a version of memetic algorithms that she has used to optimize the design of a

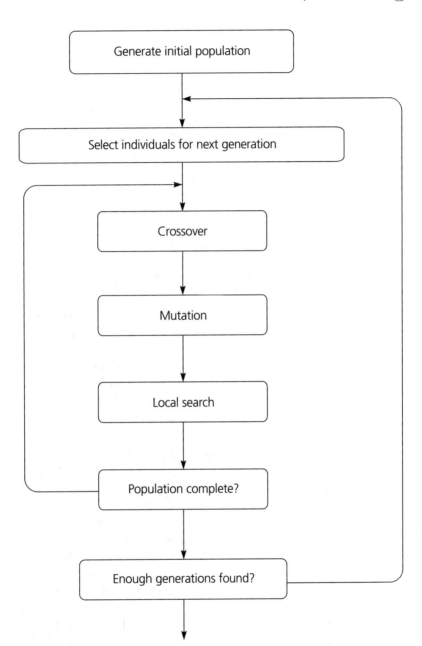

Generate initial population

Select individuals for next generation

Crossover

Mutation

Local search

Population complete?

Enough generations found?

Figure 5.17 Steps in Burke and Smith's memetic algorithm (an evolutionary algorithm with local search). (From Burke and Smith, 1999.)

hypothetical creature, which she calls a "vehicle." The vehicle has six body parts, and the problem is to evolve an internal neural network that coordinates the development and movements of the parts. Fitness is measured as a function of movement, symmetry, and head stability. There is epistasis in the fact that the movements of a limb depend on what the opposite limb is doing, and there can be numerous suboptimal combinations of features; lots of combinations of parts can be pretty good but not the best.

Vehicles in Gabora's paradigm evolve on a 10 × 10 torus grid, with one vehicle per cell. At each iteration, a vehicle is able to acquire a new meme, either by creating one or imitating a neighbor; they can also update the mutation operator or implement the new meme. To create a new meme (there is a 50 percent chance of creating versus imitating), the memetic algorithm is applied to the meme represented in the neural network; a change is accepted if it increases the fitness value. Several "rules of thumb" are implemented in determining mutational changes. In short, as the author says, "Generalizations about what seems to work and what does not are translated into guidelines that specify the behavior of the memetic algorithm." In imitation, a vehicle evaluates its eight neighbors' fitness one after the other until it finds one whose fitness exceeds its own. This neighbor's meme is copied to the vehicle's own neural network. If no neighbor has a better fitness, then nothing is done.

Gabora has studied three cultural evolution strategies:

- Mental simulation, where a new meme is not evaluated until the next time step so that the vehicle can "think about it" before expressing it

- Imitation, where memes are taken from neighbors

- Knowledge-based operators, where individuals use a mental equivalent of mutation and crossover to evolve schemas

She concluded that all three of these strategies speed up optimization, while mental simulation and imitation also increase the peak mean fitness. Further, in her experiments the ratio of creation to imitation affects peak fitness; fitness is maximized when creation occurs at about twice the rate of imitation. Increased creation also results in the population finding more of the various most-fit memes.

Gabora notes that an important difference between cultural and biological evolution is that cultural features are transmitted horizontally, that is, across individuals within generations, while genetically inherited features can only be propagated vertically, down from one generation to the next. A highly speculative theorist, she has written extensively on issues of creativity, spirituality, and other nontechnical topics, arguing that the theory of memes "plants us squarely in a garden of knowledge where spiritual truth can blossom." We agree that there is something encouraging in a point of view that optimistically theorizes that society is constantly moving toward ever more adaptive states, that sees the evolution of truth from the bottom up—rather than flowing down from all-knowing "experts" and authorities to the ignorant masses.

Cultural Algorithms

Wayne State University's Robert Reynolds is a computer scientist with a background in anthropology. He is interested in discovering how principles of the evolution of culture can be adopted to solve the kinds of problems found in computer science and applied mathematics. If culture enables human individuals to adapt to complex environments, then perhaps the processes that drive culture can be used to enable adaptation of artificial individuals to other kinds of fitness landscapes.

Reynolds' technique, which he calls *cultural algorithms,* is usually described as a kind of evolutionary algorithm (Reynolds, 1994). As in other evolutionary algorithms, a population of problem solutions is generated, usually with a random start, and operators are applied to encourage improved fitness in the population. Cultural algorithms differ from evolutionary ones, though, in an important way: they have memory. Most particularly, a population maintains a group memory or *belief space* of information about what kinds of proposed solutions have performed especially well and sometimes ones that have performed poorly and should be avoided. Individuals in the population may access this information, which provides guidance toward increasingly fitter problem solutions (see Figure 5.18).

As in human societies, culture changes over time, and its changes persist over time relative to the more rapid explorations of individuals. Reynolds has referred to this as a dual-inheritance model, not to be confused with the dual inheritance of Boyd and Richerson. The two paths of inheritance in Reynolds' view are the "micro-evolutionary" transmission of traits and behaviors between individuals and the "macro-evolutionary" formation of generalized beliefs based on individual experiences. These generalized beliefs then can operate at the group level to constrain individuals' behaviors within the population. In cultural algorithms,

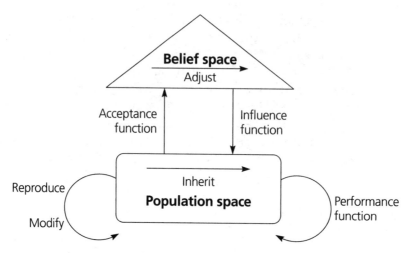

Figure 5.18 Operation of a cultural algorithm. The belief space in turn influences the population members' searches. (From Cowan and Reynolds, 1999.)

individuals interact with one another in a *population space* and are influenced by group-level generalized beliefs in the belief space.

In each step of the algorithm, individuals are evaluated using a performance function, and their fitness is determined. An acceptance function determines if the individual should influence the population's direction, that is, whether the individual will contribute to the belief space. If an individual is accepted, its state is adjusted with those of other individuals to form group beliefs, as information in the belief space is used to guide evolution in the next step.

Reynolds and his students have applied cultural algorithms to a diverse group of difficult problems in social science, computer science, and mathematics, and they have used the paradigm to improve the performance of other established algorithms. In recent years it has seemed that every conference has at least a few cultural algorithm papers, and it seems likely that this flexible and powerful approach to adaptive problem solving has a bright future.

Convergence of Basic and Applied Research

You will have noticed that the models we have just described come from two very different kinds of traditions. One kind of computer program is written by social scientists in order to see if they can understand human behavior better. A psychologist, economist, or anthropologist might have a theory about how people behave or think, and might want to study the consequences of the theory's assumptions, or might want to know how tweaking certain parameters affects the outcomes of the simulation; the scientist hopes to be able to generalize his or her findings to explain something about people.

The other tradition arises mainly from computer science, applied mathematics, and electrical engineering. This approach to modeling culture looks for better ways to solve hard mathematical problems. Parameters are tweaked not to learn anything in particular about human society, but to achieve the best results—to find the best answer to a hard problem fastest. This is similar to the situation with evolutionary algorithms, where biologists and computer scientists work together to explain and emulate the dynamical processes that comprise evolution, in order to understand biology and again to solve hard problems. It is also similar to research in neural networks, where psychologists and cognitive scientists hope to gain insights into human thinking, while computer scientists and engineers seek powerful methods for analyzing data.

This book is a psychologist/engineer joint effort and seeks a kind of middle ground that encompasses both ends. It seems most likely that nature, fine-tuning for billions of years, has found methods that really work. This suggests that it may be well worth learning about natural processes, whether you want to understand nature—in this case, human nature—or to find answers to difficult problems in applied mathematics. We expect to find convergence of basic and applied research, as application-oriented investigators find out that the best methods are those that nature herself already uses.

Culture—and Life without It

Our too-brief overview suggests that various theorists look at culture as a dynamical process similar in some ways to biological evolution and with a similar effect: adaptation. The long-term result of human culture is the survival of individuals and the propagation of future generations, with the interesting side effect noted by Boyd and Richerson that genetic evolution is slowed when natural selection is interfered with. Almost all cultures include procedures for protecting the frail and sick, the elderly and the very young, the unintelligent and the clumsy, those individuals who might not survive otherwise. Their survival suggests that their genes are passed on, along with those of the strong, the fittest, members of the population, and the evolution of the species takes on the aspect of *genetic drift,* rather than evolution toward a fitness peak.

As individuals we naturally take the perspective of individuals, and it is sometimes hard to see ourselves in the context of a culture. But the view of the individual as a cell in the cultural superorganism does not necessarily undermine the importance of individuals. In fact, though we have been considering the bottom-up effect of the individual on the culture, the *emergence* of norms and population-level effects from the local interactions of individuals, we can just as well look the other way at the top-down *immergence* of individual minds from the higher-level culture.

What would it be like to exist as a human being without a culture? Let us summarize first by saying that we would probably not be human beings, but only humans—not *Homo sapiens,* but only *Homo.*

> A man who has been alone since birth will have no verbal behavior, will not be aware of himself as a person, will possess no techniques of self-management, and with respect to the world around him will have only those meager skills which can be acquired in one short lifetime from nonsocial contingencies . . . To be for oneself is to be almost nothing (Skinner, 1971, pp. 117–118).

To consider what an individual raised without culture would be like, we can look at several known instances of feral children, that is, children raised in the wild (Candland, 1993). For instance, "Wild Peter" appeared in the countryside near the German town of Hameln in 1724. When found, he appeared to be about 12 years old, tanned and naked. He sat on his haunches like a four-footed creature, ate raw but not cooked grasses and vegetables, and liked to catch, kill, dismember, and eat (raw) birds. Though Wild Peter was passed from one protector to the other, including the royal family of England, for more than 60 years, he never learned to talk. It was reported that his best words were "ki scho," for "King George," and "qui ca," for "Queen Caroline"; other than that, his vocalizations were indecipherable.

Peter was obviously able to learn, for he had knowledge of edible foods in the wild and learned to adapt to civilized ways, even if he did not enjoy them. For instance, he would wear clothes, but preferred not to. He never was known to laugh, though he enjoyed music, and he was not interested in sex. Peter was considered a curiosity in his time and was taken care of, and though he had his favorites, he never formed any real connections to other people.

Victor, the "Wild Boy of Aveyron," appeared at the edge of the woods in France near the end of the 18th century (Lane, 1976). As a contemporary wrote: "A child of eleven or twelve, who some years before had been seen completely naked in the Caune Woods seeking acorns and roots to eat, was met in the same place toward the end of September 1799 by three sportsmen who seized him as he was climbing into a tree to escape their pursuit." After a week Victor broke free, but later voluntarily entered a house and was kept captive thereafter. Victor was largely cared for by Dr. Jean-Marc-Gaspard Itard, who by all accounts gave Victor the best attention possible and tried earnestly to civilize him. Itard described the 12- or 13-year-old Victor as "a disgustingly dirty child affected with spasmodic movements, and often convulsions, who swayed back and forth ceaselessly like certain animals in a zoo, who bit and scratched those who opposed him, who showed no affection for those who took care of him; and who was, in short, indifferent to everything and attentive to nothing."

Even as he went through puberty, Victor showed no interest in sex, though he was seen sometimes to become agitated and uncomfortable in the presence of women; he had no sense of justice, or of right and wrong, for instance, as they applied to stealing. He never learned to speak, though he could occasionally imitate the actions depicted in pictures. Victor died in 1828, at the age of 40, without ever having become socialized.

In 1920, an Indian missionary named Singh heard stories from jungle tribesmen about some "ghosts" that resided in a large ant mound in a clearing (Candland, 1993). When Singh and some helpers went to dig out the mound, they found living inside a family of wolves and two very dirty, naked little girls, who were obviously being raised by the wolves. They appeared to be about eight and one-and-a-half years old, with matted balls of hair. They were captured and raised by Reverend Singh, who was careful not to let word get out. The girls liked to sleep in a ball, walked on all fours, disliked human company, and appeared to be able to see better at night than in the daylight. Singh reported that they could smell meat from a great distance, and one of them once found the entrails of a fowl that had been thrown out 80 yards from the compound where they lived; the child was caught eating them.

The younger girl died of illness after about a year, but the older one survived until 1929. By that time, she had learned to say about 30 words—not "real" words, but sounds of her own device that apparently had meaning. Though she could name some things, she never learned to speak spontaneously, for instance, to ask for things by name.

None of these children, or any other known feral children, for there are a few other cases, ever formed social bonds with other people; they never learned to communicate or to care what others thought of them. Their intellects were so poorly formed as to be immeasurable, as they had none of the curiosity or desire to please that enables a person to perform well on a test.

In short, and in order to provoke thought and discussion, we would argue that unsocialized humans do not have what we normally think of as minds. They cannot think or communicate, and their learning is restricted to the kind of individual experience that omits the accumulation of knowledge through culture. Thus we would not conclude that culture "strengthens" the mind or "helps" the mind, but that it "creates" the mind.

Summary

Psychology traces its descent dually from philosophy and from the physical and natural sciences. The 20th century saw tremendous improvements in scientific methods for research and analysis of empirical data on psychological phenomena and a rapid expansion of scientific knowledge about human thought, feeling, and behavior. Some of the most important findings contradict our intuitive view of ourselves and are almost universally ignored by laypersons. Unfortunately, these "laypersons" include researchers in other fields who sometimes feel that introspection and common sense should be adequate for informing theories of mind and computational models of cognition. On the other hand, a tradition of computational simulation is rising in the social sciences, with demonstration of many important effects. Encouragingly, this tradition is beginning to spread to the computer sciences, where it is blossoming into varieties of useful and interesting paradigms.

Anthropologists and others have long noted that cultural change resembles biological evolution in its ability to adapt to new and diverse circumstances. Some theorists have asserted that the two processes are nearly identical, but there is plenty of reason to doubt this extreme view. Our feeling is that natural selection of memes is an imperfect analogy for the evolution of ideas through a population; it seems better to describe such phenomena in terms of changes in the states of individual minds, instead of changes in ideas themselves. We have considered some well-grounded theories of social influence—the changing of individual minds in response to perceptions about the minds around them—and have seen some powerful computer algorithms based on these social and cultural processes.

Next we begin exploring our own vision of sociocognitive computation. We start by introducing a simple algorithm called the Adaptive Culture Model, which we will use to build a base for understanding the particle swarms to be revealed in the next chapters.

chapter
six

Thinking Is Social

The linkage between social science simulations and computer programs for engineering and other applications is not always obvious. Neural networks, simulated annealing, cultural algorithms, ant colony optimization, and evolutionary algorithms are several instances where psychological, physical, and biological theories have influenced the development of computational methods for problem solving.

This chapter takes a simulation from the social sciences and shows how it can be modified slightly to perform combinatorial optimization. The adaptive culture model is extremely straightforward and easy to understand; further, it contains most of the ingredients that will be used in the more sophisticated particle swarm algorithms studied in the rest of the book. ■

Introduction

The problem with the story about the blind men and the elephant is that it assumes they were also deaf. The famous tale, set in verse in the 19th century by John Godfrey Saxe (1869/1936), describes the experiences of six blind men who, converging upon an elephant, each come to believe that the entire elephant resembles the particular part he has encountered by chance:

> The First approached the Elephant,
> And happening to fall
> Against his broad and sturdy side,
> At once began to bawl:
> "God bless me! but the Elephant
> Is very like a wall!"

The second blind man, feeling a tusk, cries that the elephant is like a spear, and so on. Of course the moral of the story is that people derive incomplete beliefs from their incomplete experiences in the world. In reality, if a group of blind men took turns announcing a description of his particular part of the elephant, it is clear that by listening to each other all the blind men would come to a rather complete and correct understanding of the heterogeneous qualities that make up an elephant. All members of the group would know that the creature has a side like a wall, tusks like spears, legs like trees, and so on. Through discussion they might even figure out how the parts are connected and how they function together. The point of this literary critique is that societies are able to benefit from the sharing of individuals' partial knowledge, resulting in a body of facts and strategies that far exceeds what any individual could have ever acquired independently.

The theme of the present chapter is that thinking is a social activity; human culture and cognition are aspects of a single process. People learn from one another not only facts but methods for processing those facts. The theme is not new: Bandura, for one, has explicitly theorized about the learning that occurs when individuals observe one another's behaviors:

> If knowledge could be acquired only through the effects of one's own actions, the process of cognitive and social development would be greatly retarded, not to mention exceedingly tedious . . . Fortunately,

most human behavior is learned through modeling . . . The capacity to learn by observation enables people to expand their knowledge and skills on the basis of information exhibited and authored by others (Bandura, 1986, p. 47).

Adaptation on Three Levels

Not only do people learn from one another, but as knowledge and skills spread from person to person, the population converges on optimal processes. The present chapter describes a system that operates simultaneously on three levels:

- Individuals learn locally from their neighbors. People are aware of interacting with their neighbors, gleaning insights from them, and sharing their own insights in turn, and local social learning is an easily measured and well-documented phenomenon.

- The spread of knowledge through social learning results in emergent group-level processes. This sociological, economic, or political level of phenomenon is seen as regularities in beliefs, attitudes, behaviors, and other attributes across individuals within a population. A society is a self-organizing system with global properties that cannot be predicted from the properties of the individuals who make it up.

- Culture optimizes cognition. Though all interactions are local, insights and innovations are transported by culture from the originator to distant individuals; further, combination of various innovations results in even more improved methods. This global effect is largely transparent to actors in the system who benefit from it.

The Adaptive Culture Model

In an earlier chapter we discussed Robert Axelrod's influential work in the 1980s with the prisoners' dilemma and the "evolution of cooperation." In 1997, this same Axelrod published a volume titled *The Complexity of Cooperation,* in which he proposed a computational model of the dissemination of culture. The present chapter extends Axelrod's computer simulation of the spread of features through a culture to demonstrate how social interaction might comprise a natural computation method that results in coherent and intelligent human thought, opinion, and action. We are considering the Axelrod model and the version we derive from it to contain the fundamental principles of the swarm algorithms that will be discussed in the following chapters. The current model has a kind of simplicity and elegance that makes it ideal for introducing the more challenging algorithms.

It has been theorized elsewhere that collaboration among individuals can result in cognitive optimization, that is, that social interaction may enable individuals to arrive at effective structures and strategies for the management of knowledge. The view has been supported in part by a paradigm called *particle swarm adaptation,* computer simulations of populations interacting in a multivariate real-number space; through imitation of successful others, individuals evolve optimal weights or activation values in connectionist and other cognitive models.

While the particle swarm algorithm has been found effective for optimization of continuous and binary phenomena, many cognitive theories include discrete variables. A major instance occurs in symbol-processing models, where discrete elements represent symbolic entities. Further, many connectionist models, especially those based on Hopfield's earlier network, are essentially combinatorial in nature; even if activations range between lower and upper limits during optimization, when a stable state is attained, activation values are usually seen to have settled in the corners of the hypercube; that is, they approach their limits of $+1$ and 0 (or -1, depending on the specification of the model), and variables are essentially discrete.

Axelrod's recent simulation of the spread of culture provides insights into the effects of social interaction and gives us a starting point for demonstrating that a small number of exceedingly simple principles can cause an artificial system to behave remarkably like a complex human society. Axelrod's culture model and the particle swarm are two branches of the same tree, with the culture model simulating societies in terms of discrete variables and the particle swarm algorithm operating on continuous or binary ones. The two approaches complement one another and point the way to a theory of culture and cognition emerging from interaction. In the present chapter, an expansion of Axelrod's algorithm called the Adaptive Culture Model (ACM) is shown to be able to optimize complex functions, and it is suggested that cognitions, attitudes, and other arrays of psychological phenomena are optimized by interaction among individuals. A series of experiments are reported that test the capabilities of the algorithm. Several cognitive theoretical models are simulated, including a new kind of multivalued combinatorial constraint satisfaction network.

Axelrod's Culture Model

Axelrod has theorized that similarity between pairs of individuals can result in the spread of culture. In his simulations, individuals are represented as strings of symbols called "features"; the number and length of the strings and the universe of symbols available to them are parameters of the system. For example, if individuals comprise five features, and these are defined as numerals in {0, 1 . . ., 9}, then one individual may be represented as 42237 and another as 99217. A two-dimensional matrix of individuals is initialized into the simulation, and they are allowed to interact.

Axelrod postulates that the probability of human interaction is a function of the similarity of two individuals: "The basic idea is that agents who are similar to each other are likely to interact and then become even more similar" (Axelrod, 1997). For instance, in the example above, the two individuals are similar in the third and fifth positions; with 40 percent similarity they have a 0.40 probability of interacting.

An interaction in ACM occurs when one individual adopts a non-matching feature of the other. An individual and one of its horizontally or vertically adjacent neighbors are selected at random. If stochastic similarity criteria are met, the selected individual will change one of its elements to match the symbol in the same position of the neighbor's string. For instance, if the two individuals above interacted, 42237 could take the 1 from its neighbor, and become 42217. The element changed is selected stochastically from the elements that are different. (This way of selecting elements to change is the same for the rest of the examples in the chapter.)

As a simulation iterates, neighbors are observed to begin resembling one another, until regions of the matrix contain identical strings. Axelrod focuses largely on the group-level effects of various parameters on the formation of regions, for instance, the numbers and sizes of regions as a function of number of features and population size. Figures 6.1 and 6.2 show "before" and "after" snapshots of a population of five-featured individuals. In this run and the others following, the beginnings and ends of rows and columns are considered adjacent, in a torus field, so that all individuals have four neighbors with whom they may interact. Note that fields do not usually wrap in Axelrod's simulations; this may result in some differences in results between the two implementations, but these do not affect the theoretical implications of the findings. Because similarity sets the threshold for instigating changes, boundaries between regions containing strings with no matching members

27217	74924	31157	53671	22660	37316	07959	57666	33206	92725
66219	08226	26707	45600	48767	39481	62784	89859	27792	35492
37262	66163	89178	60968	91098	19937	62103	07562	03500	13864
87746	66209	94122	72784	03593	16647	19776	87819	22160	48185
16880	09713	76057	30843	92125	41152	74156	98801	64760	00144
86287	66161	23271	46773	53014	44442	25424	98309	32553	16678
90624	65685	68785	32385	90770	24676	68806	25347	16640	30602
98681	11402	57304	68003	16943	01041	44693	63237	76040	61075
52249	30617	91425	92780	82342	30467	19721	84117	96595	55215
79949	70851	29089	89311	19176	67653	95954	64805	51332	74301

Figure 6.1 Initial random start for a simulation of a 10 × 10 population of individuals made up of strings of five features represented by numerals ranging from zero to nine.

22233	22233	22233	22233	22233	22233	22233	22233	22233	22233
22233	22233	22233	22233	22233	22233	22233	22233	22233	22233
22233	65955	65955	22233	22233	22233	22233	33588	33588	33588
22233	22233	22233	22233	22233	22233	22233	33588	33588	33588
22233	22233	22233	22233	22233	22233	22233	22233	22233	22233
22233	22233	22233	22233	22233	22233	22233	22233	22233	22233
22233	22233	22233	22233	22233	22233	22233	13157	22233	22233
22233	22233	22233	22233	22233	22233	22233	13157	22233	22233
22233	22233	22233	22233	22233	22233	22233	22233	22233	22233
22233	22233	22233	22233	22233	22233	22233	22233	22233	22233

Figure 6.2 Result of simulation where interaction is a stochastic function of similarity (Axelrod's paradigm).

eventually become fixed, and change in the system stops. This kind of polarization is very similar to that noted in the simulations of Nowak, Szamrej, and Latané's 1990 paper and other social impact studies.

The present chapter remains conceptually close to Axelrod's original, insofar as the simulations model populations of individuals within a society who interact by exchanging features. "Features" in both models may comprise a variety of phenomena: "Although beliefs, attitudes, and behaviors cover a wide range indeed, there are even more things over which interpersonal influence extends, such as language, art, technical standards, and social norms." Axelrod moves between levels of analogy when he simultaneously describes the symbol strings as persons and as "sites," which might be thought of as neighboring villages. The model is described in this chapter on the level of individual persons interacting with their neighbors, which does appear to be Axelrod's primary focus as well.

Experiment One: Similarity in Axelrod's Model

Axelrod theorizes that similarity is a precondition for social interaction and subsequent exchange of cultural features. In the simulations that instantiate his theory, the probability of interaction depends on similarity, and culture is seen to spread and finally stabilize through links between similar individuals. According to Axelrod and others, people become more similar as they interact; an apparent paradox, however, is that populations do not converge on unanimity. Instead, subgroups tend to become more homogeneous over time and more different from one another.

The "birds of a feather" hypothesis—that people are attracted to and influenced by others who are similar to them—has been supported by much research, especially in the attitude similarity paradigm promulgated by Byrne. On the other hand, Wetzel and Insko have shown that people are attracted to others who resemble their ideal, rather than actual, selves. Over six carefully conducted experiments, when the effects of "ideal similarity" were statistically removed from data, the effect of "self-similarity" was nonsignificant. Wetzel and Insko's conclusion was that people strive to attain their ideals; in the meantime they find others who approximate their ideals to be attractive and persuasive. At the least, there is evidence to suggest that the role of similarity is not as important as Axelrod theorizes; as he notes, few models of social influence give the factor much weight. As Axelrod purports to simulate the social influence process, that is, the propagation of beliefs, attitudes, and other features from one individual to another, it is prudent to question the causal position of similarity in the paradigm.

In the first experiment, Axelrod's model was altered slightly: the effect of similarity as a causal influence was deleted from the model. This was accomplished easily by setting the probability of interaction to 1.0

70971	70971	70971	70971	70971	70971	70971	70971	70971	70971
70971	70971	70971	70971	70971	70971	70971	70971	70971	70971
70971	70971	70971	70971	70971	70971	70971	70971	70971	70971
70971	70971	70971	70971	03593	70971	70971	70971	70971	70971
70971	70971	70971	70971	70971	70971	70971	70971	70971	70971
70971	70971	70971	70971	70971	70971	70971	70971	70971	70971
70971	70971	70971	70971	70971	70971	70971	70971	70971	70971
70971	70971	70971	70971	70971	70971	70971	70971	70971	70971
70971	70971	70971	70971	70971	70971	70971	70971	70971	70971
70971	70971	70971	70971	70971	70971	70971	70971	70971	70971

Figure 6.3 Result of a simulation of the culture model with no criterion for interaction.

for all selected pairs. Thus when an individual and its neighbor were selected they interacted, regardless of their similarity, with the individual changing one of its nonmatching elements to be the same as the neighbor's. These trials resulted in unanimity (see Figure 6.3). All 20 trials run to stability in a 10 × 10 population resulted in uniform populations of individuals with identical features.

Interestingly, *the effect of similarity as a causal influence in Axelrod's model is to introduce polarization:* dissimilarity creates boundaries between cultural regions. Interindividual similarities do not facilitate convergence, but rather, when individuals contain no matching features, the probability of interaction is defined as 0.0, and cultural differences become insurmountable. Interaction occurs, and the population converges, in the absence of any similarity criterion, but polarization was not seen; thus the effect of similarity is negative, in that its absence creates the conditions for impassable group boundaries to form.

Experiment Two: Optimization of an Arbitrary Function

It was hypothesized that the culture model might belong to a larger class of general function optimizers and that in Axelrod's implementation the function that is optimized is, in fact, similarity. Axelrod's placement of similarity as a cause in the simulations makes it essentially an objective function; if a test is passed, then a feature is adopted from a neighbor, and in the end the population maximizes the criterion. Thus, in the original versions of the model, the change rule is "if (rand < S/N) then interact," where rand is a random number between zero and one, S represents the number of similar or matching elements, and N is the number of features in a string. The following experiments substitute new terms into the parentheses on the left side of the change rule.

Experiment Two substituted a simple arbitrary function for the similarity test previously used. Rather than testing the similarity of two neighbors, the algorithm was modified so that the numerals comprising an individual's features were summed: this sum was a performance metric. The change rule became "if (the neighbor's sum is larger than the targeted individual's sum) then interact." As before, an interaction comprised the taking of a nonmatching feature from a neighbor. The question was whether the algorithm would maximize the sums of numerals comprising the individuals.

It was seen that a randomly initialized population does indeed converge on the maximum; in 20 trials of the paradigm, the population converged on the global optimum every time. Though the string "99999" was not seen in the initial population, interaction resulted in the adaptive discovery of that optimal set of features and its spread through the population. ACM is capable of optimizing a simple numerical function.

Experiment Three: A Slightly Harder and More Interesting Function

Minimizing and maximizing an entire string of digits is perhaps the simplest optimization exercise conceivable. A second task was programmed, which is at once more difficult and more interesting as a social science metaphor. The task in the next experiment was to find a set of five numbers—the individual's feature string—within which the sum of the first three numbers equaled the sum of the last two. Thus, the string 34153 would successfully accomplish the objective, since the sum of 3 + 4 + 1, the first three characters in the string, equals the sum of 5 + 3, the last two. In the program, the difference was calculated between the first and second sums, and if the neighbor's difference was smaller (the sums were more nearly equal), the target adopted a feature from the neighbor.

This task is interesting for two reasons. First, unlike the previous example, in which a string of nines or of zeroes satisfies the maximization or minimization constraints, the equal-sums task has a great number of

17769	13967	13967	13967	03764	49094	83193	83193	03434	07979
17769	18137	13967	13967	49094	49094	49094	24381	23454	07979
17567	14537	14537	12737	49094	49094	49094	13453	13453	13453
32739	34539	85599	72999	49094	49094	49094	13453	13453	13866
82789	84598	85599	72999	72081	04581	40581	40581	82459	13866
95169	85168	62127	62127	09081	04581	40141	40141	90191	92193
95187	85168	62127	62127	09081	09081	00110	92193	92193	92193
68178	19239	19239	01230	01230	21223	00110	92193	92193	92193
10955	19239	92579	02727	02727	21223	27283	92193	92193	10955
17769	13509	13509	13509	02754	23124	83193	83193	83139	07979

Figure 6.4 Result of a simulation where interaction occurred when the difference between the sums of the first three numbers and the last two was smaller for the neighbor than for the individual. Note that in all cases the sum of the first three numbers equals the sum of the last two.

perfect solutions. In the current paradigm, assuming that the algorithm would optimize the problem, it was impossible to predict whether homogeneous regions would develop or individuals would evolve idiosyncratic vectors.

Second, the task requires the complex coordination of the entire vector of elements. A "2" in the fifth position is only successful if it and the fourth element contribute together to a sum that is predicted by the first three elements. In a psychological sense, this is analogous to a model of cognitive or attitudinal consistency. Thus the task supports conceptualization of the paradigm as a model of individuals in a society, each trying to acquire and maintain a cognitive set that meets the requirements of the situation.

As seen in Figure 6.4, all individuals in the population solved the problem, and parts of solutions were distributed through contiguous regions of the population; in 20 trials, this pattern of results was seen every time. Even though the paradigm required coordination of all elements and the interaction operated on individual elements, the method successfully found solutions to this rather difficult problem. It was as if an individual picked up a hairstyle from one friend, a style of jacket from another, slang from another—and made them all fit together. It was, in other words, a depiction of a social process in which no individual embodied every single aspect of the stereotypical culture, but rather individuals adopted particular aspects of the cultural features prevalent in their area.

These results have an obvious analogue in human society. A string in the simulation may be seen as a set of features, attitudes, or beliefs held by an individual, which must be internally consistent in order to become stable. The features are also constrained to be *externally* consistent; that is, individuals strive to resemble their neighbors, at least when the neighbors are relatively successful at attaining a "good" set of features.

These experiments have taken Axelrod's model of the spread of features through a culture and modified it somewhat. In the ACM paradigm, an individual takes a feature from a randomly selected neighbor if a criterion is met. In Axelrod's writings the criterion is similarity; the present chapter substitutes other criteria and shows that the spread of culture can optimize other functions, resulting, by the way, in similarity among proximal individuals. Similarity, which was a cause in Axelrod's simulations, is now an effect.

Experiment Four: A Hard Function

A problem is considered intractable if the amount of time required to solve it increases at a faster-than-polynomial rate as the size of the problem increases. For instance, the traveling salesman problem (TSP) requires finding the shortest path through a set of nodes, or cities, without passing through any node twice, and ending up at the starting point. With each additional city the number of possible solutions grows exponentially; with N nodes there are N^N possible combinations of nodes and $N!$ "legal" tours. The TSP is a type of problem called "NP-complete" by computer scientists. The NP problems are intractable, as there is no known deterministic solution to them that reduces the search to polynomial time.

The TSP requires the ordering of discrete elements. As such it is used here to represent analogous cognitive tasks, such as determining the sequence of steps necessary to solve a complex problem, arranging conscious thoughts such that no contradictory beliefs are juxtaposed, and so on. For instance, Thagard and Verbeurgt (1995) have shown that constraint satisfaction in connectionist networks is an NP-complete problem. Rich has suggested that the essence of artificial intelligence (AI) is

CDEFGHAB	CDEFGHAB	CDEFGHAB	CDEFGHAB	CDEFGHAB	CDEFGHAB	CDEFGHAB	CDEFGHAB
CDEFGHAB	CDEFGHAB	CDEFGHAB	CDEFGHAB	CDEFGHAB	CDEFGHAB	CDEFGHAB	CDEFGHAB
CDEFGHAB	CDEFGHAB	CDEFGHAB	CDEFGHAB	CDEFGHAB	CDEFGHAB	CDEFGHAB	CDEFGHAB
CDEFGHAB	CDEFGHAB	CDEFGHAB	CDEFGHAB	CDEFGHAB	CDEFGHAB	CDEFGHAB	CDEFGHAB
CDEFGHAB	CDEFGHAB	CDEFGHAB	CDEFGHAB	CDEFGHAB	CDEFGHAB	CDEFGHAB	CDEFGHAB
CDEFGHAB	CDEFGHAB	CDEFGHAB	CDEFGHAB	CDEFGHAB	CDEFGHAB	CDEFGHAB	CDEFGHAB
CDEFGHAB	CDEFGHAB	CDEFGHAB	CDEFGHAB	CDEFGHAB	CDEFGHAB	CDEFGHAB	CDEFGHAB
CDEFGHAB	CDEFGHAB	CDEFGHAB	CDEFGHAB	CDEFGHAB	CDEFGHAB	CDEFGHAB	CDEFGHAB
CDEFGHAB	CDEFGHAB	CDEFGHAB	CDEFGHAB	CDEFGHAB	CDEFGHAB	CDEFGHAB	CDEFGHAB
HGFEDCBA	HGFEDCBA	HGFEDCBA	HGFEDCBA	CDEFGHAB	CDEFGHAB	CDEFGHAB	HGFEDCBA
HGFEDCBA	HGFEDCBA	HGFEDCBA	HGFEDCBA	HGFEDCBA	HGFEDCBA	HGFEDCBA	HGFEDCBA
HGFEDCBA	HGFEDCBA	HGFEDCBA	HGFEDCBA	HGFEDCBA	HGFEDCBA	HGFEDCBA	HGFEDCBA
HGFEDCBA	HGFEDCBA	HGFEDCBA	HGFEDCBA	HGFEDCBA	HGFEDCBA	HGFEDCBA	HGFEDCBA
HGFEDCBA	HGFEDCBA	HGFEDCBA	HGFEDCBA	HGFEDCBA	HGFEDCBA	HGFEDCBA	HGFEDCBA
HGFEDCBA	HGFEDCBA	HGFEDCBA	HGFEDCBA	HGFEDCBA	HGFEDCBA	HGFEDCBA	HGFEDCBA
HGFEDCBA	HGFEDCBA	HGFEDCBA	HGFEDCBA	HGFEDCBA	HGFEDCBA	HGFEDCBA	HGFEDCBA
HGFEDCBA	CDEFGHAB	HGFEDCBA	HGFEDCBA	HGFEDCBA	HGFEDCBA	HGFEDCBA	HGFEDCBA
CDEFGHAB	CDEFGHAB	CDEFGHAB	CDEFGHAB	CDEFGHAB	CDEFGHAB	CDEFGHAB	CDEFGHAB

Figure 6.5 Result of a simulation of TSP.

in solving these problems: " . . . one way of describing AI is that it is an attempt to solve NP-complete problems in polynomial time" (Rich, 1983, p. 104).

The present experiment implemented an eight-city tour. Thus, there are 8^8, or 16,777,216, possible combinations of eight cities, allowing for cities to be visited more than once, and there are 8!, or 40,320, legitimate tours, in which each city is visited once. Of those tours, there are 16, one starting in each of the eight cities, and going either direction around the tour, that are globally optimal, or provide the shortest possible path. Some heuristic algorithms have been devised to find relatively good tours without testing all possible permutations (for instance, terminate a tour without completing it when it exceeds the shortest distance found so far), but the problem is considered hard by all standards.

The test was set up with a set of cities defined as two-dimensional Cartesian coordinates, which were contrived so that the best tour was known to the researcher. The problem was set up so that the optimal tour was "ABCDEFGH," of course starting on any letter and going in either direction. A typical result is seen in Figure 6.5. The adaptive culture algorithm is able to optimize combinatorial functions. A penalty was added to the length of a tour if it went to a node more than once. The algorithm was run 20 times with an 18×8 population of individuals: the population converged on the globally optimal tour, which had a distance of 7.483 units, on 11 of those trials ($16/16,777,216 \approx 0.00000095$ probability of finding the optimum by chance). The mean tour was 7.909 units in length; when suboptimal tours dominated, these were seen in all cases to differ from the optimal by at most one element. Either a city was repeated (perhaps a higher penalty would have prevented this) or two neighboring cities were reversed (e.g., BDCE instead of BCDE). Note that this trial resulted in convergence on two globally optimal solutions to the problem.

In a test of the propensity for the algorithm to find multiple optima, the TSP program was run for a second set of 20 trials. Nine of these trials resulted in the globally optimal tour of length 7.483; of these, two trials resulted in convergence on two different optimal patterns, and one trial found five different series of cities that produced the shortest possible route. Other successful trials converged on a single optimum. Thus polarization was sometimes seen to occur in this paradigm.

Experiment Five: Parallel Constraint Satisfaction

ACM models the spread of features through a community. These can be beliefs and attitudes as well as tangible phenomena. A string of features then could represent a kind of cognitive system, with string symbols representing cognitive elements, using some model of cognitive goodness to optimize the vector.

Connectionist networks often implement binary coding of variables. A set of mutually inconsistent units or nodes are coded with inhibitory connections between each pair. An optimized network contains a value of one for one of the beliefs and zeroes for the others. For instance, Rumelhart, Smolensky, McClelland, and Hinton present an example of a constraint satisfaction network comprising features of rooms. Given some subset of features, such as toaster and refrigerator, the network is able to compute other features of the room that are likely to be present, such as table and sink, while deactivating features such as bed and couch that are likely to be absent. Rumelhart et al. make the point that schemata have variables, which they call "slots." As they write:

> In some cases, there are sets of units that are mutually inhibitory so that only one can be active at a time, but any of which could be combined with most other units . . . Perhaps the best example from our current data base is what might be called the size slot. In this case, the *very-large, large, medium, small,* and *very-small* units are all mutually inhibitory . . . (Rumelhart et al., 1986, pp. 33–34).

It does not seem reasonable to suppose that a person who thinks about a large room must actually inhibit thoughts about small, medium, and very large rooms. But because the network is coded in binary terms, it is necessary to code a large number of variables, and an even larger number of connections, to describe a single "slot." A primary reason for implementing binary constraints is the feasibility of optimizing the activation pattern; the Hopfield techniques work on binary nodes. ACM, however, offers a method for optimizing multiple-valued discrete nodes, as well as a sound social-psychological premise for how it could be done in reality. In the present paradigm, the "size slot" can be coded as a single node that can take on six values—the five size classes and a zero for absent or irrelevant. The model is essentially unchanged, except that it is now more comprehensible, realistic, and parsimonious.

For demonstration a parallel constraint satisfaction network was taken from a recent *Psychological Review* paper by Kunda and Thagard. The model simulates the effect of stereotypical information on a concept, in this case the descriptor "aggressive." Kunda and Thagard hypothesized that individuals are more likely to expect a stereotypical construction worker to punch someone and a lawyer to argue with someone, given that both targets are labeled "aggressive." In their paper this was demonstrated using two networks, as shown in Figure 6.6. The first network shows the constraints resulting from factors associated with a lawyer, that is, the target is stereotypically expected to be Upper middle class and Verbal, and if labeled "Aggressive," is expected to Argue rather than Punch. This effect results largely from the negative connection between Upper middle class and Punch. The second network represents the stereotype of a construction worker, i.e., Working class, Unrefined, and more likely to Punch than to Argue.

One advantage of using multivalued nodes is that comparisons can be implemented in a single network. In the example, both of Kunda and Thagard's networks have identical binary nodes representing Aggressive,

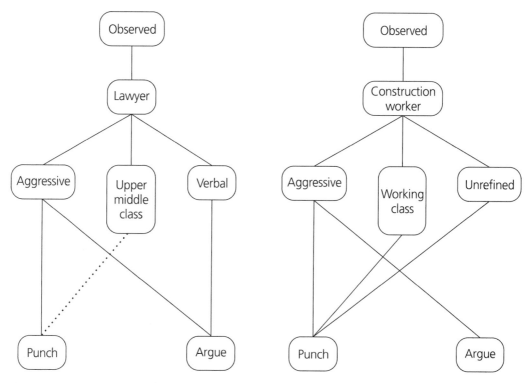

Figure 6.6 Parallel constraint satisfaction examples. Solid lines represent positive links, and the dotted line represents a negative link. (From Kunda and Thagard, 1996.)

Punch, and Argue; they differ in the other three nodes. It is clear, however, that Lawyer and Construction worker are levels of the same variable, as are Upper middle class and Working class, and Verbal and Unrefined are implicitly exclusive of one another. Further, as the question is whether Punch or Verbal satisfies the constraints better, it is possible to code them as levels of a variable called "Response." Thus, the two networks can be conceptually collapsed into one, as seen in Figure 6.7, with nodes labeled "Occupation," "Aggressive," "Socioeconomic class," "Sophistication" (for want of a better term—as usual, all labels are arbitrary), and "Expected response."

An experimental trial clamped, that is, held constant, a value of Occupation: a population was initialized with random values for the nodes and with one occupation fixed, for instance, Occupation clamped to the value "Lawyer." Individuals in the population generated patterns of activation, evaluated these, compared their own evaluations to their neighbors', and adopted a feature from the neighbor if their pattern was better.

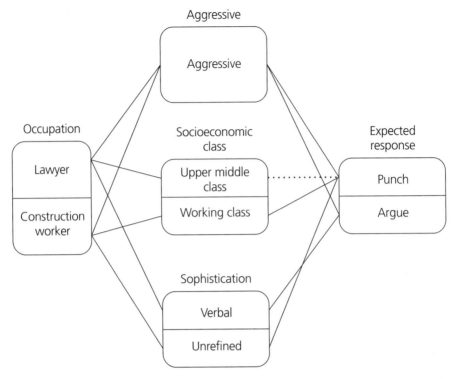

Figure 6.7 The two binary networks shown in Figure 6.6 are combined into a single multivalued network. Note that each node can also take on a zero value.

A string is composed of the variables in the following order:

1. Occupation (A = Lawyer, B = Construction worker)
2. Aggressive (value is A or 0)
3. Socioeconomic class (A = Upper middle class, B = Working class)
4. Sophistication (A = Verbal, B = Unrefined)
5. Expected response (A = Punch, B = Argue)

Thus a string such as "AABBA" represents a Lawyer who is Aggressive, Working class, Unrefined, and more likely to Punch than to Argue. In cases where a node is not connected, or the total inputs to all levels of it are zero or less, the node takes on the value of zero, indicating it is irrelevant, nonsalient, or unnecessary.

The network was coded as a series of rules of five elements. Negative connections were coded with a minus sign on the node value or on the weight, and a period represented no connection. Note that coding such a system in a two-dimensional matrix would remove all advantage and make it identical to a binary network, and thus a series of rule statements was employed.

For instance, the following was the rule set for Figure 6.7, with "Construction worker" clamped on:

```
B   .   .   .   .
A   A   A   A   .
B   A   B   B   .
.   A   .   .   #
.   .   A   .  -A
.   .   B   .   A
.   .   .   A   B
.   .   .   B   A
```

The first rule is special; it determines values to be clamped on throughout the trial. In this set the first line tells the system to assign a "B" for the first node; the dots in that row mean the other nodes are free to vary. Rules are designed so that the first column with a symbol in it represents the focal node, and other columns are nodes connected to it. The second line says that state "A" of the first node is connected to "A" for the second, third, and fourth nodes and is not connected to the fifth. The "#" symbol in the fourth rule means that the second node (the second column is the first with a symbol in it) connects to *any* value of the fifth node. A minus sign means that the element should not be present; it represents a negative or inhibitory connection.

The rule set may be accompanied by a set of weights for each element in each row, or a default weight of 1 can be used for everything. The program compares a proposed solution, for instance, "BABAB"—meaning that a Construction worker is Aggressive, Working class, Verbal, and likely to Argue—to each rule, and sums up the weights of the items that match. In the present example, the first "B" is clamped on, so all population members will have that state for their first node. The rule in the second line (A A A A .) fails to match in the first specified position, so the rest of the rule is not evaluated. The next rule (B A B B .) does match in the first specified position, so the weights for the rest of the line are summed, with a penalty for mismatches. Assuming all weights = 1.0, the sum is 1 for the matching "A" in the second position, plus 1 for the matching "B" in the third position, minus a penalty for the mismatch between the "A" in the problem solution and the "B" in the rule, minus another penalty for the final "B" in the solution, which corresponds with a "." or no connection in the rule.

The next rule says that a state of "A" for the second node corresponds with anything in the fifth node. Since the proposed solution does meet the criterion, the sum is added for a match in this row. Thus the program goes through, comparing the symbol string to the rules. A measure of goodness is defined by the sum of the weights of all constraints that are satisfied by a pattern of values:

$$G = \sum_r \sum_n W_m$$

where r is rules, n is nodes, and W_m is the weight of a relation or connection that matches, that is, is identical in the test and the rule strings.

Note that some theorists code binary node activations in $\{0, 1\}$ and some in $\{-1, +1\}$. The difference between these two implementations is seen in the effect of a node when it is turned off; a node with zero value has no effect on the units it is connected to, while a node value of -1 actively tends to inhibit them. This results in some paradoxical cases, where in order to achieve the desired result positive relationships must actually be coded using negative connections. The present model emulates connectionist networks with $\{0, 1\}$ activations and as such avoids that anomalous instance.

When the program is executed, a population of random symbol strings is generated. A string is evaluated by comparing it to each rule in the rule file. Weighted matches between the test string and the rules are summed through the rule set. A larger total indicates that more constraints were satisfied. The ACM algorithm is applied by selecting an individual and a neighbor, comparing their evaluation totals, and interacting when the neighbor's total is greater than the individual's.

This network model using a 10 × 10 population was tested 20 times with Lawyer clamped on and 20 times with Construction worker clamped on. All 40 trials resulted in the population converging on the correct stereotypical conclusion, that is, Lawyers would be thought more likely to Argue than to Punch, and Construction workers would be expected to be more likely to Punch than to Argue.

What does this approach to constraint satisfaction buy us? First, the use of multivalued nodes is a more efficient and theoretically sound way to encode slots, or sets of mutually exclusive variables. Second, the population dynamics allow important insights into the cognitive operations involved. Part of the definition, indeed, of a stereotype is that it is a belief shared by a group about members of another group. ACM shows the development of stereotyped thinking as it spreads through a population. A set of commonly held beliefs is arranged in various ways until the best explanations are found. The search is shared by the population, and the successful results spread to all members.

Experiment Six: Symbol Processing

Many kinds of problems in conventional AI are conceptualized as a search for the shortest path through some symbolic nodes. The example described here is taken from Chapter 4 of Patrick Henry Winston's classic textbook, *Artificial Intelligence* (Winston, 1992). Winston gives the example in terms of finding the shortest path from city A to city H. Symbolically this kind of problem is the same as rebuilding a carburetor, cooking an omelet, preparing for an exam—a sequence of steps must be followed, they must be in correct order, and no steps must be skipped. Many decision trees have a number of alternative successful routes, some of which are superior to others, in which case the investigator decides whether to accept a good-enough solution or to hold out for the global optimum.

Figure 6.8 depicts a network of paths, which can be transformed into a set of hierarchical paths that can be searched without looping back to a previous point. The usual discussion of this kind of problem describes strategies that can be used to reduce unproductive searches. The culture model completely ignores these rational approaches. As usual, a population of individuals is initialized, with each individual comprising a string of symbols, for instance, ADCHBEAF, in an eight-node problem (the first node will always be A).

As before, each string is evaluated, that is, the distance from the initial A to the first occurrence of the goal H is calculated. Since only a small proportion of nodes are legitimately connected to one another, a penalty is added to the length of the path for illegal connections. A neighbor is selected, the two are compared, and a feature of the neighbor is taken if the neighbor's trip is shorter.

Because trips through the graph vary in the number of nodes they pass through to reach the goal, an adjustment in the performance measure was necessary. If simple tour length were used as a performance

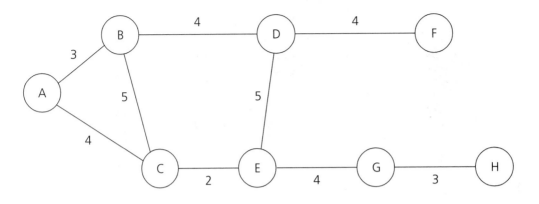

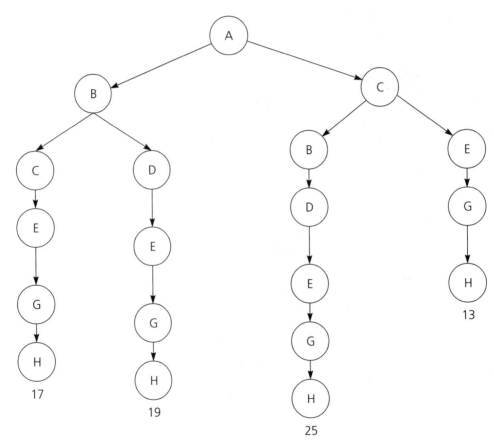

Figure 6.8 The shortest path problem. A network of nodes is transformed into a hierarchical tree structure. (Adapted from Winston, 1992.)

measure, the system would converge quickly on a shorter, more probable path, even though it is suboptimal, simply because it will be found sooner. In order to equalize the chances of finding longer, less probable, sequences of graph nodes, total distance was weighted by

$$\left(\frac{N-1}{N}\right)^{L-2} \times \left(\frac{1}{N}\right)$$

where N is the number of nodes in the graph and L is the ordinal position of the goal, for comparisons of individuals and their neighbors. The first term of this weight represents the probability that elements preceding the goal in the series will not be the goal value, and the second term is the probability that the final element will be the goal value.

For example, in the example given above, the best possible tour is ACEGH, with a distance of 13 units and the target in the fifth ordinal position of the string. The penalty for invalid sequences, that is, for going from one node to another when there is no connection between them, can be set by the user; let us assume it has been set equal to 10 units of distance added to the total for each violation. An individual then that tried the invalid tour AH would have a distance of 10 units—better than the global optimum! Thus cheating would be rewarded if it was not prevented. Weighting by the formula above, though, transforms the optimal tour into

$$13 \times \left(\frac{7}{8}\right)^{5-2} \times \frac{1}{8}$$

which evaluates to approximately 1.088, while the "cheater" solution becomes

$$10 \times \left(\frac{7}{8}\right)^{2-2} \times \frac{1}{8}$$

which is 1.25 transformed units in distance. Thus the incentive to take shortcuts is reduced. This approach enables operations on solution strings of various lengths.

In 20 trials of the Winston decision tree a 12×8 population converged on the global optimum 14 times. While it is not argued here that this trial-and-error algorithm is more efficient than traditional search procedures, it is suggested that in fact many real problems requiring the sequencing of elements are solved socially through collaboration, with each individual contributing and recombining pieces to result in improvement.

Discussion

An ACM system iterates, with individuals repeatedly interacting, until it reaches a stable point where change ceases. The simulations reported in this chapter have been seen to stop in three kinds of states: states where the population is uniform, with all feature strings identical; states with two or more regions of strings that are identical within regions and different between them; and states with a population of unique strings that satisfy the change criterion equally well.

The question of what causes groups to diverge is of course profound, important, and difficult. In Axelrod's simulations change stopped when the similarity criterion was not met anywhere in the matrix; the result was a population with one or more homogeneous regions. As was seen in the first experiment above, when the similarity criterion was deleted from the paradigm, the system evolved until no individual had a neighbor with a nonmatching feature—the entire population became homogeneous. Thus it was concluded that the similarity rule was responsible for producing boundaries between distinct groups.

In the other experiments reported here, the change rule stated that interaction occurred only when the neighbor's performance exceeded the selected individual's. In one experiment (Experiment Three), very many global optima existed, and individual feature strings tended to be correlated but unique. In others, for example, the maximization and stereotype experiments, a single global optimum existed, and the population converged on it. When a limited number of optima existed, the population would converge on as many as were discovered; when the function was hard and global optima few in number (e.g., the TSP), only one solution was found in most cases, but in some trials multiple best solutions were discovered by the population. Where two or more global optima were more easily found, regions of the population tended more frequently to converge on separate patterns. It is interesting to note the formation of "multiethnic" populations out of benign conditions, that is, individuals converged on different combinations of features; group divergence occurred even in the absence of self-categorization, as in Henri Tajfel's theory, or competition for resources, as theorized by Donald Campbell and others.

In sum, culture and cognition are seen from three simultaneous levels of phenomena. First, individuals searching for solutions learn from the experiences of others. The "problems" addressed by the present theoretical viewpoint may be epistemological, logical, ethical, aesthetic, or metaphysical; they may be emotional, political, physical, or sexual. At the social learning level of culture and cognition, individuals learn from their neighbors. This is the level that is most easily measured by social scientists, and importantly it is the level at which the system is programmed.

Second, an observer looking at a population as a whole perceives phenomena of which individual people are the parts. Opinions, beliefs, and behaviors correlate with geographical regions, as well as with ethnic, religious, socioeconomic, and other cultural dimensions. People who interact frequently become similar to one another in many ways. Nothing in the computer programs specified that regions should become homogeneous or that borders should form between groups; these global effects emerged from simple local interactions.

Third, culture affects the performance of the individuals who comprise it. Individuals gain benefit by imitating their neighbors, good patterns spread through the population, and the best patterns spread most widely through the group, so that individuals benefit from improvements that might have been discovered in quite remote regions. Further, as patterns move through a population and become integrated with existing patterns of features, the probability of even greater improvement increases. Again, nothing in the programs specified that individuals would solve problems, but only that they would imitate others who performed better than themselves.

This latter global phenomenon is largely invisible to participants in the system, for two reasons. First, an individual does not necessarily realize where an idea came from, if it originated beyond the horizon. Second, individuals are incapable of seeing the effect of culture because they are it: there is no background against which the figure can be seen, as the perceiver is an element in the perceived field.

This model appears to give individuals very little credit. Thinking, and in fact *hard* thinking, is depicted here with no assertions about, or reliance on, the intelligence of individuals. A human processing unit in these simulations functions mainly through adaptive imitation. Obviously an individual human processes a great deal of information. The present view would suggest that a relatively large proportion of cognition is concerned with evaluation and comparison of self and others. The strings of symbols processed in the current examples are highly oversimplified tokens of the multidimensional, dynamic arrays that are processed by human societies—and these experiments suggest that *societies* process information.

Summary

The ACM is able to find some combinatorial optima, but is not designed for that purpose—it was really devised only to show that a simple principle, which could be called "imitation of your betters," is able to find its way through a complex search space. The following chapters present and elaborate a paradigm that capitalizes on the ability of social interaction to result in optimization of hard problems.

The Particle
Swarm and
Collective
Intelligence

chapter
seven

The Particle Swarm

This chapter introduces the particle swarm in its binary and real-numbered forms. The book so far has been preparing a context, describing related paradigms in computer science and social science, discussing culture and norms and language and other scientific and philosophical developments that, if we have been successful, will make the particle swarm seem like an obvious thing to propose.

The Adaptive Culture Model in the previous chapter hints at what can happen as a result of the simplest imaginable interactions of the simplest imaginable agents—if these can even be called "agents." Given a large space of possibilities, the population is often able to find multivariate solutions, patterns that solve problems, through a stripped-down form of social interaction.

It is worth emphasizing that individuals in the culture model are not *trying* to solve problems. They are only following the simple rules of the algorithm, which say nothing about the existence of a problem or how to solve it. Yet through reciprocal social influence each individual betters its "fitness" (the term is less appropriate here than in discussion of evolutionary algorithms), and the performance of the population improves. We would not say that the adaptive culture algorithm is an especially powerful way to solve problems, but it is a good introduction to some social algorithms that are.

The particle swarm algorithm is introduced here in terms of social and cognitive behavior, though it is widely used as a problem-solving method in engineering and computer science. We have discussed binary encoding of problems, and the first version of the particle swarm we present here is designed to work in a binary search space. Later in the chapter we introduce the more commonly used version, which operates in a space of real numbers. ∎

Sociocognitive Underpinnings: Evaluate, Compare, and Imitate

A very simple sociocognitive theory underlies the Adaptive Culture Model and particle swarms. We theorize that the process of cultural adaptation comprises a high-level component, seen in the formation of patterns across individuals and the ability to solve problems, and a low-level component, the actual and probably universal behaviors of individuals, which can be summarized in terms of three principles (Kennedy, 1998):

- Evaluate
- Compare
- Imitate

Evaluate

The tendency to evaluate stimuli—to rate them as positive or negative, attractive or repulsive—is perhaps the most ubiquitous behavioral characteristic of living organisms. Even the bacterium becomes agitated, running and tumbling, when the environment is noxious. Learning cannot occur unless the organism can evaluate, can distinguish features of the environment that attract and features that repel, can tell good from bad. From this point of view, learning could even be defined as a change that enables the organism to improve the average evaluation of its environment.

Compare

Festinger's social comparison theory (1954) described some of the ways that people use others as a standard for measuring themselves, and how the comparisons to others may serve as a kind of motivation to learn and change. Festinger's theory in its original form was not stated in a way that was easily tested or falsified, and a few of the predictions generated by the theory have not been confirmed, but in general it has served as a backbone for subsequent social-psychological theories. In almost everything we think and do, we judge ourselves through comparison with others, whether in evaluating our looks, wealth, humor, intelligence (note that IQ scales are normed to a population average; in other words, your score tells you how you compare to others—which is really the point, isn't it?), or other aspects of opinion and ability. Individuals in the Adaptive Culture Model—and in particle swarms—compare themselves with their neighbors on the critical measure and imitate only those neighbors who are superior to themselves. The standards for social behaviors are set by comparison to others.

Imitate

You would think that imitation would be everywhere in nature; it is such an effective way to learn to do things. Yet, as Lorenz has pointed out, very few animals are capable of real imitation; in fact, he asserts that only humans and some birds are capable of it. Some slight variations of social learning are found among other species, but none compare to our ability to mimic one another. While "monkey see, monkey do," well describes the imitative behavior of our cousins, human imitation comprises taking the perspective of the other person, not only imitating a behavior but realizing its purpose, executing the behavior when it is appropriate. In *The Cultural Origins of Human Cognition,* Michael Tomasello argues that social learning of several kinds occurs in chimpanzees, but true imitation learning, if it occurs at all, is rare. For instance, an individual's use of an object as a tool may call another individual's attention to the object; this second individual may use the same object, but in a different way. True imitation is central to human sociality, and it is central to the acquisition and maintenance of mental abilities.

The three principles of evaluating, comparing, and imitating may be combined, even in simplified social beings in computer programs, enabling them to adapt to complex environmental challenges, solving extremely hard problems. Our view diverges from the cognitive viewpoint in that nothing besides evaluation, comparison, and imitation takes place *within* the individual; mind is not found in covert, private chambers hidden away inside the individual, but exists out in the open; it is a public phenomenon.

A Model of Binary Decision

Consider a bare-bones individual, a simple being with only one thing on its mind, one set of decisions to make, yes/no or true/false, binary decisions, but very subtle decisions, where it is hard to decide which choices to make. For each decision, this supersimplified individual can be in one state or the other, either in the yes state, which we will represent with a 1, or the no = 0 state. It is surrounded by other yes/no individuals, who are also trying to decide. Should I say yes? Should I say no? They all want to make the best choices.

Two important kinds of information are available to these primitive beings. The first is their own experience; that is, they have tried the choices and know which state has been better so far, and they know how good it was. But these social beings have a second consideration; they have knowledge of how the other individuals around them have performed. In fact they are so simple that all they know is which choices their neighbors have found most positive so far and how positive the best pattern of choices was. If these stripped-down beings are anything like people, they know how their neighbors have done by observing them and by talking with them about their experiences.

These two types of information correspond to Boyd and Richerson's individual learning and cultural transmission. The probability that the individual will choose "yes" for any of the decisions is a function of how successful the "yes" choice has been for them in the past relative to "no." The decision is also affected by social influence, though the exact rule in humans is admittedly not so clear. Social impact theory states that the individual's binary decisions will tend to agree with the opinion held by the majority of others, weighted by strength and proximity. But even that rule is somewhat vague, given ambiguities in the concepts of strength and proximity.

For the present introductory model we will just say that individuals tend to be influenced by the best success of anyone they are connected to, the member of their sociometric neighborhood that has had the most success so far. While we admit this is an oversimplification, it has a kernel of truth that justifies the parsimony it brings to the model.

Individuals can be connected to one another according to a great number of schemes, some of which will be mentioned in Chapter 8. Most particle swarm implementations use one of two simple sociometric principles (see Figure 7.1). The first, called *gbest*, conceptually connects all members of the population to one another. The effect of this is that each particle is influenced by the very best performance of any member of the entire population. The second, called *lbest* (g and l stand for "global" and "local"), creates a neighborhood for each individual comprising itself and its k nearest neighbors in the population. For instance, if $k = 2$, then each individual i will be influenced by the best performance among a group made up of particles $i - 1$, i, and $i + 1$. Different neighborhood topologies may result in somewhat different kinds of effects. Unless stated

The "lbest" neighborhood with $k = 2$. Each individual's neighborhood contains itself and its two adjacent neighbors. The first and last are connected.

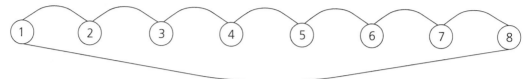

Individual #3 has found the best position so far in #4's neighborhood. Therefore, #4's velocity will be adjusted toward #3's previous best position and #4's own previous best position.

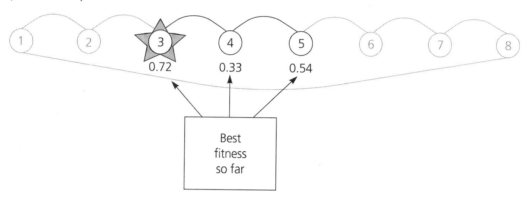

The "gbest" neighborhood. Assuming that #3 has found the best fitness so far in the entire population, all others' velocities will be attracted toward its previous best position.

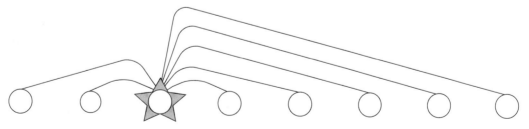

Figure 7.1 The two most common types of neighborhoods.

otherwise, the following discussions will presume lbest neighborhoods with $k = 2$ (sometimes described as "neighborhood = 3").

In a sociocognitive instance the individual must arrange an array of decisions or judgments in such a way that they all fit together, what we call "making sense" or "understanding" things. The individual must be able to evaluate, compare, and imitate a number of binary choices simultaneously.

Evaluation of binary strings can be accomplished in one step. In the psychological case, that is, if we are talking about humans, we can again use the concept of cognitive dissonance to evoke the sense of tension that exists when an array of decisions contains inconsistencies. We experience the state as discomfort and are motivated to change something to reduce the tension, to improve the evaluation. Dissonance as described by Festinger provides a single measure of cognitive evaluation, exactly as "fitness" is a single measure of genetic or phenotypic goodness.

How do we improve cognitive fitness? Of course there are plenty of theories about this. In Ajzen and Fishbein's *Reasoned Action Model,* (1980) *intent* is seen as a function of two kinds of things that should be getting familiar by now (see Figure 7.2). On the one hand, intent is affected by the person's *attitude* toward the behavior; for instance, if they believe violence is harmful or immoral, then they may intend not to act violently. This attitude is formed, in Ajzen (pronounced "eye-zen") and Fishbein's theory, by a linear combination of beliefs that the behavior will result in some outcomes (b_i) times the individual's evaluation of those outcomes (e_i):

$$A_o = \sum_{i=1}^{n} b_i e_i$$

This kind of expectancy-value model of attitude has existed in some form for many years, and we will not criticize its linearity or asymptotic issues here (never mind the decades-old debate about summing versus averaging). We are interested in the fact that intent has a second cause, which Ajzen and Fishbein call the *subjective norm*. The subjective norm regarding a behavior is also built up, in their theory, as a linear sum of products, but this time the factors entering into the formula are social. The individual's subjective norm toward a behavior is a sum of the products of their beliefs that certain others think they should or should not perform the behavior, multiplied by the motivation to comply with each of those others:

$$SN_o = \sum_{i=1}^{n} b_i m_i$$

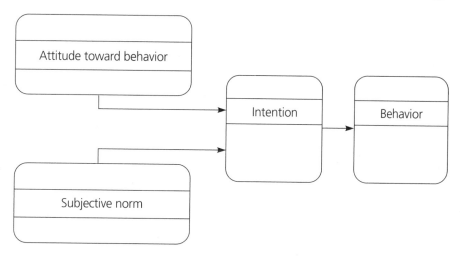

Figure 7.2 According to the Reasoned Action Model, behavior is a function of intention and only remotely of the individual's attitude about the behavior.

To point out the obvious, these two components of the theory of reasoned action map easily onto the components of Boyd and Richerson's cultural transmission model; that is, there is an individual term (individual learning or attitude toward a behavior) and a social term (cultural transmission or subjective norm). These two kinds of concepts are found in other theories as well and are represented in our decision model as the two terms that make up the change formula. We theorize that the coexistence of these two modes of knowledge, that is, knowledge acquired by the senses through experience in the world and knowledge acquired from others, gives humans the intellectual advantage; it is the source of our intelligence.

Besides their past experience and inputs from the social environment, another factor that affects the individual's decision is their current propensity or position regarding the issue. They may start with a strongly negative attitude and have subsequent positive experiences regarding the choice or attitude object—but still have a negative feeling about it. The positive experiences may make the individual *more* likely to choose the positive alternative, but in order to shift the individual's general propensity into the positive domain, the decision threshold would still have to shift upwards. If the individual's initial position is extreme, the probability is lower of its changing—for one thing, the individual is less likely to try the other alternative.

In mathematical terms, we are proposing a model wherein the probability of an individual's deciding yes or no, true or false, or making some other binary decision, is a function of personal and social factors (Kennedy and Eberhart, 1997):

$$P(x_{id}(t) = 1) = f(x_{id}(t-1), v_{id}(t-1), p_{id}, p_{gd})$$

where

- $P(x_{id}(t)=1)$ is the probability that individual i will choose 1 (of course the probability of their making the zero choice is $1 - P$) for the bit at the dth site on the bitstring

- $x_{id}(t)$ is the current state of the bitstring site d of individual i

- t means the current time step, and $t - 1$ is the previous step

- $v_{id}(t-1)$ is a measure of the individual's predisposition or current probability of deciding 1

- p_{id} is the best state found so far, for example, it is 1 if the individual's best success occurred when x_{id} was 1 and 0 if it was 0

- p_{gd} is the neighborhood best, again 1 if the best success attained by any member of the neighborhood was when it was in the 1 state and 0 otherwise

The decisions themselves will be stochastic, if for no better theoretical reason than that we never know all the forces involved—it is very unlikely that any decision is made based solely on isolated facts pertaining directly to that decision alone. A lot of randomness allows exploration of new possibilities, and a little bit allows exploitation by testing patterns similar to the best one found so far; thus we can balance between those two modes of search by adjusting the uncertainty of decisions.

The parameter $v_{id}(t)$, an individual's predisposition to make one or the other choice, will determine a probability threshold. If $v_{id}(t)$ is higher, the individual is more likely to choose 1, and lower values favor the 0 choice. Such a threshold needs to stay in the range [0.0, 1.0]. We have already seen one straightforward function for accomplishing this, when we talked about neural networks. The sigmoid function

$$s(v_{id}) = \frac{1}{1 + \exp(-v_{id})}$$

squashes its input into the requisite range and has properties that make it agreeable to being used as a probability threshold (though there is nothing magical about this particular function).

We wish to adjust the individual's disposition toward the successes of the individual and the community. To do that we construct a formula for each v_{id} in the current time step that will be some function of the difference between the individual's current state or position and the best points found so far by itself and by its neighbors. We want to favor the best position, but not so much that the individual ceases searching prematurely. If we simply added $(p_{id} - x_{id}(t-1))$ and $(p_{gd} - x_{id}(t-1))$ to $v_{id}(t)$, it would move upward when the difference between the individual's previous best and most recent states, or the difference between the neighborhood's best and the individual's most recent states, equaled 1, and would be attracted downward if either difference equaled -1. The probability threshold moves upward when the bests are ones and downward when they are zeroes.

In any situation we do not know whether the individual-learning or the social-influence terms should be stronger; if we weight them both with random numbers, then sometimes the effect of one, and sometimes the other, will be stronger. We use the symbol φ (the Greek letter phi) to represent a positive random number drawn from a uniform distribution with a predefined upper limit. In the binary version the limit is somewhat arbitrary, and it is often set so that the two φ limits sum to 4.0. Thus the formula for binary decision is

$$v_{id}(t) = v_{id}(t-1) + \varphi_1(p_{id} - x_{id}(t-1)) + \varphi_2(p_{gd} - x_{id}(t-1))$$

$$\text{if } \rho_{id} < s(v_{id}(t)) \text{ then } x_{id}(t) = 1; \text{ else } x_{id}(t) = 0$$

where ρ_{id} is a vector of random numbers, drawn from a uniform distribution between 0.0 and 1.0. These formulas are iterated repeatedly over each dimension of each individual, testing every time to see if the current value of x_{id} results in a better evaluation than p_{id}, which will be updated if it does. Boyd and Richerson varied the relative weighting of individual experience and social transmission according to some theoretical suggestions; the current model acknowledges the differential effects of the two forces without preconceptions about their relative importance. Sometimes decisions are based more on an individual's personal experience and sometimes on their perception of what other people believe, and either kind of information will dominate sometimes.

One more thing: we can limit v_{id} so that $s(v_{id})$ does not approach too closely to 0.0 or 1.0; this ensures that there is always some chance of a bit flipping (we also don't want v_i moving toward infinity and overloading the exponential function!). A constant parameter V_{max} can be set at the start of a trial to limit the range of v_{id}. In practice, V_{max} is often set at ±4.0, so that there is always at least a chance of $s(V_{max}) \approx 0.0180$ that a bit will change state. In this binary model, V_{max} functions similarly to mutation rate in genetic algorithms.

Individuals make their decision in a population, where they are influenced by the successes of their neighbors. As each individual's decision is affected by $(p_{gd} - x_{id}(t-1))$, that is, (usually) some other individual's success, they influence one another and tend to move toward a common position. As an individual begins to approximate its neighbor's best position, it may perform better and influence *its* neighbors, and on and on; good decisions spread through the population. We are comfortable calling this the formation of a culture in a computational population.

In this section we have developed an extremely parsimonious model of binary choice as a function of individual learning and social influence. Individuals tend to gravitate probabilistically toward the decisions that have resulted in successes for themselves and their colleagues. The result is optimization of each individual's decision vector and convergence of the population on an optimal pattern of choices.

The entire algorithm, maximizing goodness, is shown in pseudocode:

```
Loop
    For i = 1 to number of individuals
        if G(x̄ᵢ) > G(p̄ᵢ) then do              //G() evaluates goodness
            For d = 1 to dimensions
                pᵢd = xᵢd                        //pᵢd is best so far
            Next d
        End do


        g = i                                   //arbitrary
        For j = indexes of neighbors
            If G(p̄ⱼ) > G(p̄g) then g = j         //g is index of best performer
                                                //  in the neighborhood

        Next j
        For d = 1 to number of dimensions
            vᵢ(t) = vᵢd(t − 1) + φ₁(pᵢd − xᵢd(t − 1)) + φ₂(pgd − xᵢd(t − 1))
            vᵢd ∈ (−Vmax, + Vmax)
            if ρᵢd < s(vᵢd(t)) then xᵢd(t) = 1; else xᵢd(t) = 0;
        Next d
    Next i
Until criterion
```

Testing the Binary Algorithm with the De Jong Test Suite

It may seem confusing to jump back and forth between "cognitive models" and "test functions." We are maintaining a generous definition of cognitive models, given the lack of consensus among psychologists about the internal structure of the mechanisms of thought. Thus, to us, a cognitive model is just like any other multidimensional problem where elements interact with one another in a combination possessing some measurable goodness.

Kennedy and Eberhart (1997) tested the binary particle swarm using a binary-coded version of the classic De Jong suite of test problems. The binary versions had already been prepared for experimentation with binary genetic algorithms, so importing them into a binary particle swarm program was straightforward. A population size of 20 was used for all tests. In all cases the global optimum was at $(0.0)^n$. The algebraic forms of the functions are given in Table 7.1.

The binary particle swarm converged quickly on $f1$, also known as the sphere function, encoded as a 30-dimensional bitstring. The best

Table 7.1 Functions used by De Jong to test various aspects of optimization algorithms.

Function	Dimension
$f1(x_i) = \sum_{i=1}^{n} x_i^2; -5.12 \leq x_i \leq 5.12$	30
$f2(x_i) = 100(x_1^2 - x_2)^2 + (1 - x_1)^2; -2.048 \leq x_i \leq 2.048$	24
$f3(x_i) = \sum_{i=1}^{n} \text{int}(x_i); -5.12 \leq x_i \leq 5.12$	50
$f4(x_i) = \sum_{i=1}^{n} ix_i^4 + Gauss(0,1); -1.28 \leq x_i \leq 1.28$	240
$f5(x_i) = 0.002 + \sum_{j=1}^{25} \dfrac{1}{j + \sum_{i=1}^{2}(x_i - a_{ij})^6}; -65.536 \leq x_i \leq 65.536;$ $[a_{ij}] = \begin{bmatrix} -32 & -16 & 0 & 16 & 32 & -32 & -16 & 0 & 16 & 32 & -32 \ldots \text{etc.} \\ -32 & -32 & -32 & -32 & -32 & -16 & -16 & -16 & -16 & -16 & 16 \ldots \text{etc.} \end{bmatrix}$	34

solution the particle swarm found was 0.000002 away from the "perfect" result of 0.0, which it found on 10 of the 20 trials. It is presumed that the difference between the found optimum and the target is due to imprecision in the binary encoding rather than a failure of the algorithm to hit the target.

On the second function, De Jong's $f2$, in 24 dimensions, the particle swarm was able to attain a best value of 0.000068, compared to a target of 0.0; again, the difference is thought to derive from the precision of the encoding rather than the algorithm. This function was encoded in a 24-dimension bitstring. $f2$ was the hardest of the De Jong functions for the particle swarm; the system converged on the best-known optimum 4 times in this set of 20. The hardness of the function might be explained by the existence of very good local optima in regions that are distant in Hamming space from the best-known optimum. For instance, the local optimum

010111111101111000000111

returns a value of 0.000312, while the bitstring

110111010011101110111111

returns 0.000557, and

111000011001011001000001

returns 0.005439. The best-known optimum, returning 0.000068, was found at

110111101110110111101001.

Thus bitstrings that are very different from one another, in terms of Hamming distance, are all relatively good problem solutions. A search algorithm that relies on hill climbing is unlikely to make the leap from a locally optimal region to the global optimum. The function itself has only one optimum and is hard because of the wide flat regions where movement, whether it is toward or away from the optimum, likely results in no real change in fitness.

The third function, $f3$, is an integer function encoded in 50 dimensions whose target value was attained easily on every trial. De Jong's $f4$ function introduces Gaussian noise to the function, and performance was measured as an average over the entire population rather than a population best. Finally, on $f5$ the algorithm was able to attain a best value of 0.943665 on 20 out of 20 attempts, in 34 dimensions; we presume that to be the global optimum. The system converged rapidly on this fitness peak every time.

The five functions were implemented in a single program, where the only code changed from one test to another was the evaluation function. All other aspects of the program, including parameter values, ran identically on the various functions. Thus it appeared from this preliminary research that the binary particle swarm was flexible and robust.

No Free Lunch

There is some controversy in the field regarding the evaluation of an algorithm, and maybe our claims that particle swarm optimization is "powerful" or "effective" should be disregarded. Imagine two optimization algorithms, one that searches by following the gradient, that is, a hill-climbing algorithm, and another that searches by hopping randomly around the landscape. Now imagine two problems, one, like the sphere function, where the gradient leads inevitably to the optimum and another that has many optimal regions, some better and some worse, a landscape peppered with hills and mountain ranges.

Of course the descriptions have been contrived so that it will be obvious that each algorithm will perform better on one of the problems. It would be foolish to search this way and that when there is a clear-cut yellow brick road leading directly to Oz, and it is equally foolish to climb the nearest hill in a rugged landscape. In this particular case it is clear that the performance of the algorithm depends on the kind of problem.

In important and controversial papers in 1996 and 1997, David Wolpert and William Macready formalized and generalized this observation, and their analysis has some surprising implications. Not only are some algorithms relatively more or less appropriate for certain kinds of problems—but averaged over all possible problems or cost functions, the performance of all search algorithms is *exactly the same*. This includes such things as random search; no algorithm is better, on average, than blind guessing. Provocatively, Wolpert and Macready question whether natural selection is an effective biological search strategy and suggest that breed-the-worst might work as well as breed-the-best, except that no one has ever conducted the experiment on the massively parallel scale of natural evolution.

The No Free Lunch (NFL) theorem, as Wolpert and Macready (1997) called it, has generated considerable discussion among researchers. Where previously there had been hope that some search strategy could be found—and evolutionary computation researchers thought they had it—that would be a best first-guess approach to any class of problems, research has more recently focused on finding exactly what the strengths and limitations of various search strategies are.

Some observers take NFL to mean that no optimization algorithm can be any better than any other. Of course—that's exactly what the theorem says, isn't it? Actually, the theorem says that no algorithm can be better than any other *averaged over all cost functions*. This is a hugely important condition.

What does it mean to average over all possible cost functions? Think of it this way. We have an optimization problem: exiting a room in the dark. Our special algorithm follows these steps (see Figure 7.3(a)):

- Move in a straight line until you reach a wall.
- Move along the wall until you feel an opening.
- Go through the opening.

There could easily be other algorithms, such as stumble-around-waving-your-arms-in-the-air, ever-widening circles, and so on, and some may help us exit better, some worse, than our own proprietary algorithm.

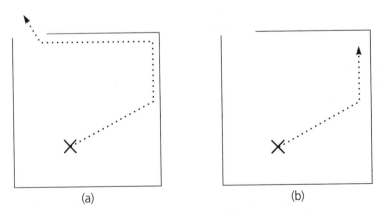

(a)   (b)

Figure 7.3 Two different algorithms, equally good at different things: the stop-in-corners algorithm is good for finding the way out of a room (a), and the find-corners algorithm is good at getting stuck in the corner of a room (b).

Now imagine that a nefarious NFL advocate has an algorithm called stop-in-corners, which he claims is just as good as our feel-for-opening algorithm. His algorithm goes like this (see Figure 7.3(b)):

- Move in a straight line until you reach a wall.
- Move along the wall until you feel a corner.
- Stop there.

But (you argue) you don't see how he will ever be able to exit a room in the dark, behaving like that! His response is, of course I won't leave the room very well, but there are four corners to the room and only one door, so I will on average be four times as successful at getting stuck in a corner as you will be at exiting. But (you insist) you don't want to get stuck in a corner, you want to exit the room—and your method is better than his for that. He admits that, but adds, my algorithm is as good as yours—averaged over all problems.

The NFL theorem says that, in order to evaluate an algorithm, you have to average it over all cost functions—and there can be very many of those. Some might be

- Exit the room (which is your problem)
- Get stuck in a corner (his problem)
- Find the center
- Find a point halfway between the center and the edge
- Find a point a third of the way between the center and the edge (et cetera, ad infinitum)
- Avoid walls altogether

and so on (another condition is that the search space must be finite, so we don't need to worry about problems from outside our domain). And averaged over all of those, his algorithm and yours are equally good. You might argue that you would never want to find the center or avoid walls altogether, and that gets to the limitation of the No Free Lunch theorem.

While it may be true that no algorithm is better than any other, when averaged over every absurd task that can possibly be imagined, it is perfectly possible that an algorithm would be better than others on the kinds of tasks that we call "problems." If anything, the NFL theorem makes us think about what it is that we try to address with an optimization algorithm. Most things, even in a finite universe, do not qualify as problems; this may be more a reflection of our way of thinking than anything inherent in mathematics or in the world. Suffice it to say, we do not feel uncomfortable saying that one algorithm, which can reliably find global optima in real problem spaces, is better than another, which can find answers to questions that no one would ever ask.

Multimodality

As part of Bill Spears' investigations of the strengths and weaknesses of genetic algorithms at the Naval Research Laboratory (NRL) and as a former graduate student of Ken De Jong's, he has assembled and posted online a collection of interesting test functions, problems that push and pull and stretch an optimization algorithm to its limit to see what it can and can't do. If there is No Free Lunch, there might be at least Some Kind of Lunch, and researchers want to know what their algorithm is good at.

In collaboration with his NRL colleague Mitch Potter, Spears designed and programmed a "multimodal random problem generator" (De Jong, Potter, and Spears, 1997). The rationale was this: obviously, if a researcher precision-tunes an optimization algorithm to work on one problem, there is a danger that it will fail on everything else. There was a need for a way to come up with different problems, but with some controllable characteristics. The random problem generator offers a way to test an algorithm on novel problems, controlling some aspects of the problems that are expected to affect performance.

Multimodality, in this context, means that a problem has more than one solution or global optimum, conceived as peaks on the fitness landscape. For instance, the problem $x^2 = 25$ is multimodal; it has two optimal solutions: $x = +5$ and $x = -5$. Since a genetic algorithm is often implemented using binary encoding, Spears wrote the program to create multimodal binary problems for the GA to solve. The concept is very straightforward. The researcher defines the dimensionality of the problem, that is, the length of the bitstring, and how many modes or peaks are desired, and the program creates that number of bitstrings, made of random sequences of zeroes and ones. For instance, imagine a researcher has specified that dimensionality $N = 10$ and multimodality or number of peaks $P = 5$. The problem generator might produce these bitstrings:

0100110111

1110010010

1101101010

0100000000

1110100101

With 10-dimensional bitstrings there are $2^{10} = 1{,}024$ possible patterns of bits. The goal for the optimizing algorithm is to find any one of the five peaks that have been defined by the program. The Hamming distance between a bitstring and the nearest optimum provides a fitness evaluation; that is, the more similar the bitstring is to one of the specified peaks, the fitter it is. For 10-dimensional bitstrings, the farthest an individual can be from a peak is 10 Hamming units, and of course a perfect match is a distance of zero from one of the peaks.

Multimodal problems can be hard for genetic algorithms. Recall that in GAs, chromosomes cross over in every generation; sections of successful ones are joined together to produce the next generation's population. In a multimodal situation it is entirely possible that the parts that are joined together come from chromosomes whose fitness derives from their proximity to different optima. For instance, the chromosome 0100110110 is only one bit different from the first optimum defined above, and 0110010010 is only one bit different from the second solution. Putting them together (we'll cut it right in the middle to be fair) could produce the child chromosome 0100110010, which is three bits different from the first optimum (Hamming distance = 3) and three bits different from the second—moving away from both of them. It is exactly the multimodality of the problem that makes crossover ineffective in this case.

GAs rely not only on recombination but on mutation (and sometimes other operators) for moving through a problem space. De Jong, Potter, and Spears tried several modifications of GAs, including one whose only operator was mutation—no crossover—in the multimodal random problem generator, calling it GA-M. In this algorithm, each site on each bitstring has a low probability of changing from a zero to a one or vice versa, usually less than 0.01. At each generation the population is evaluated, the fittest ones are selected, and mutation is applied to them.

Through this process, generations tend to improve; this amounts to a kind of stochastic hill climbing, as the more fit members of the population are more likely to be retained and mutated.

De Jong, Potter, and Spears also tested a GA implemented with crossover only—no mutation—and found that this kind tended to flounder in the early generations, but once the population started to converge on one particular peak or another, improvement came relatively fast. These crossover-only GAs, called GA-C, were very successful at finding one of the optima, if you waited long enough. The same was true of traditional GAs with both crossover and mutation. Mutation-only GAs, on the other hand, constantly improved, generation by generation, but if the dimensionality of the problem was high, the chance of mutating in a direction that led to improvement was very small and grew smaller as the population approached the optimum. When bitstrings were short, mutating chromosomes found optima quickly and efficiently, but "the curse of dimensionality" made bigger problems too difficult for them. Though they might have eventually found the global optimum, improvement

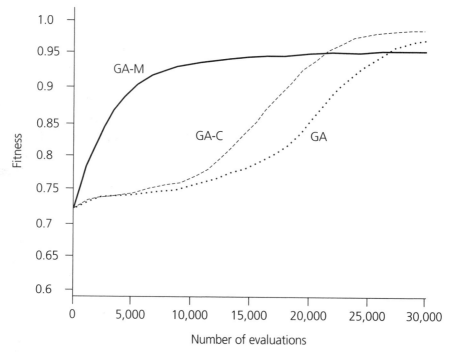

Figure 7.4 Average best-so-far performance of three types of genetic algorithms on 500-peak problems. (From De Jong, Potter, and Spears, 1997.)

decelerated over time. Figure 7.4 shows the performance of the three types of genetic algorithms.

In a follow-up to the De Jong et al. study, Kennedy and Spears (1998) compared the binary particle swarm algorithm with the three variations of GAs in the multimodal random problem generator. The study was constructed in the form of an experiment, where three independent variables were manipulated—algorithm, dimensionality, and multimodality. There were four kinds of algorithms (GA-C, GA-M, GA, and PS), two levels of dimensionality (20 and 100), and two levels of multimodality (20 and 100). In each of the 16 conditions of the experiment, there were 20 observations, and the population size was 100. (This population size, which is much bigger than a typical particle swarm, was used to make the two paradigms commensurate.)

The dependent variable in Kennedy and Spears' experiment was the shape of the best-so-far performance curves over time. This is a multivariate measure, more complicated than those found in the typical experiment, but easily computable with good statistical software. Each condition in the experiment was run 20 times for 20,000 evaluations. The mean best performance was calculated after 20 evaluations, and after 1,000, 2,000, and so on up to 20,000. It was possible to statistically compare the shapes of the performance curves for all comparisons, the question being not how well the various algorithms perform in the long run over the dimensionality and multimodality conditions, but how changes in their performance differed over time (see Figure 7.5).

GA-M performed best of all the algorithms in the early iterations of every condition, but was quickly overtaken by all the others, except in the "lite" condition, with short, 20-bit bitstrings and only 20 peaks. When either dimension or modality or both increased, however, GA-M suffered in its ability to find one of the peaks. The two GA variations with crossover, that is GA-C and GA—which implemented both crossover and mutation—started in every condition with a "dip" in performance, and then rose toward an optimum, almost always finding one of the peaks by the 20,000th evaluation.

The binary particle swarm performed the best in all conditions except the "lite" one (where it was second best); it found a global optimum on every trial in every condition and did it faster than the comparison algorithms. This is not to say that it would have performed better than *any* GA on these problems, and it may be possible to tune the parameters of a GA to optimize its performance in a particular situation. On the other hand, the significance of these results—which were statistically significant in a multivariate analysis of variance (MANOVA)—should not

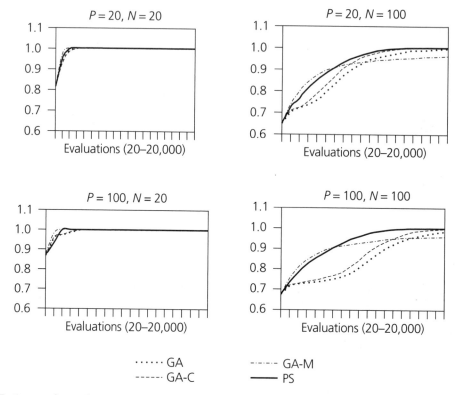

Figure 7.5 Best-so-far performance curves for four kinds of algorithms, with two levels of dimensionality (*N*) and two levels of multimodality (*P*). (From Kennedy and Spears, 1998.)

be underestimated. The binary particle swarm performed very well in the multimodal random problem generator, compared to some tough competitors.

As mentioned above, the traditional genetic algorithm with crossover often has trouble with multimodality: Crossover between parent chromosomes that are near different optima can result in a child chromosome that is not near any optimum at all. The GA with mutation simply didn't have the power to search a large space for one of the target bitstrings. For the binary particle swarm, though, multimodality just means more places to roost, more targets to hit; the effect is just the opposite of the traditional GA with crossover. Subsections of a binary particle swarm population can converge on different optima, and particles that are halfway between two optima have twice as many good directions to turn—not twice as many opportunities to fail.

Minds as Parallel Constraint Satisfaction Networks in Cultures

John J. Hopfield is a well-respected physicist. You might say that he is a "silicon-based" researcher. In 1982, Hopfield published a paper that, according to many neural network researchers, played a more important role than any other single paper in reviving the field of neural networks. One of his developments was the definition of the energy of a neural network: For a given state of the network, the energy is proportional to the overall sum of the products of each pair of node activation values (Vi, Vj) and the connection weight associated with them (T_{ij}), that is,

$$E = -0.5 \sum_{i,j;i \neq j} T_{ij} V_i V_j \quad (T_{ii} \equiv 0)$$

He showed that his algorithm for adapting the node activations, given a set of weights connecting the nodes, makes E decrease and that eventually a minimum E is obtained.

In the parallel constraint satisfaction paradigm as typically published in psychological journals, nodes in a Hopfield network represent elements, often propositions or concepts. When some subset of the nodes' states is clamped on, the network seeks a pattern of states of the other nodes that maximizes harmony or coherence or minimizes energy (E in the above equation).

Both binary- and continuous-valued Hopfield networks have been studied, but here we focus on the binary version. Nodes exist in either the active or the inactive state, represented as one and zero, and are linked by bidirectional, symmetrical connections T_{ij}. The usefulness of the network is seen when only part of an input pattern, or a noisy pattern, is introduced; the network outputs the complete pattern that best fits with the clamped pattern.

In the original Hopfield paradigm the weights between nodes are derived from data, but in most "cognitive" versions the weights are derived from the assumptions about the connections among attitudes, beliefs, or behaviors. There is usually some rationale for the choice of weight values, but these vary from one implementation to another. The common-sense fact is that some beliefs are related and others are not, they can be related positively or negatively, and some pairs of beliefs are more strongly related than others. Theorists instantiate this fact into their models with varying degrees of success and credibility.

A binary particle swarm was set up to optimize the simple network structure described by Edwin Hutchins (1995) and described in Chapter 5. A network is composed of six nodes, where the three left nodes are positively connected with one another, the three right nodes are positively connected, and there are negative connections between the right- and left-side nodes. The network is in its optimal state either when the left nodes are active and the right nodes are inactive or when the right nodes are active and the left ones are inactive; in binary terms this means that 000111 and 111000 are globally optimal patterns of node activation. These two states correspond to two interpretations of a pattern of stimuli; Hutchins introduces the model with a story of a shipwreck, where some lights in the distance were interpreted differently by various people, resulting in a collision.

As we described earlier, Hutchins had programmed a group of four networks with varying numbers of connections between them to demonstrate the effects of communication within a group. Somewhat similarly, in the particle swarm each individual is represented as a network, and their interactions allow them to find the optimal activation patterns. When the paradigm was run with 20 individuals, the globally optimal patterns were found every time by all members of the population. We admit it is not an especially difficult optimization problem, as there are only 2^6, or 64, possible network states, with two equally good global optima. The binary particle swarm has been implemented on much larger and more complicated constraint satisfaction networks, with excellent results.

In this example it is important to note the formation of cultures in the population, as seen in Figure 7.6. In the lbest particle swarm, individuals interact with their adjacent neighbors; as a result of this, neighbors become more similar to one another over time, and patterns spread from neighbor to neighbor. As commonly happens, different sections of the population settle on different optima. This seems to be a very good model of the process by which polarization forms in a society. Given the facts as they are known, both conclusions are correct: both result in patterns where the nodes with positive inputs are activated and nodes with negative inputs are inactive. Through discussing the issues among themselves, individuals not only come to agree with their neighbors but also arrive at conclusions that fit together optimally. Social interaction results in cultural convergence on patterns of beliefs, and culture results in relatively good cognitive performance.

We have described a version of particle swarm optimization that is useful when binary variables (1 or 0, true or false, etc.) are used. It is a powerful paradigm that is simple to use, easy to understand, and

000111

000111

000111

000111

000111

111000

111000

111000

111000

000111

Figure 7.6 Example of a population of binary particle swarm individuals' solutions to Hutchins' six-node parallel constraint satisfaction problem.

converges rapidly, even for high-dimensional, multimodal problems. In the real world, however, we usually must deal with continuous numbers. In the next section we show how particle swarm optimization works in continuous numbers, maintaining its simplicity, power, and ease of use.

The Particle Swarm in Continuous Numbers

The progression of ideas has been from a purely qualitative social optimization algorithm—the Adaptive Culture Model—to a model that can be interpreted as qualitative or quantitative—the binary particle swarm. In this chapter we arrive at the "real" particle swarm, which is a truly numeric optimization algorithm (Kennedy and Eberhart, 1995; Eberhart and Kennedy, 1995). The particle swarm algorithm searches for optima in the infinite search space that is often symbolized as R^n—the n-dimensional space of real numbers. (Actually, of course, it searches in computable space. And we are using PCs, so caveats related to things such as round-off errors are valid.)

The Particle Swarm in Real-Number Space

In real-number space, the parameters of a function can be conceptualized as a point. If we think of a psychological system as a kind of information-processing function, then any measure such as the psychotherapist's MMPI, a public opinion survey questionnaire, a risk-seeking inventory, a management consultant's Myers-Briggs, or the "What's Your Love-Q?" in the back of *Cosmo* will produce real numbers that can be interpreted as a point in a psychological space. In engineering applications it is customary to think of system states as points in multidimensional space. The multidimensional space is referred to by various names, depending on the situation. Names include state space, phase space, and hyperspace.

It is a small philosophical leap to suggest that multiple individuals can be plotted within a single set of coordinates, where the measures on a number of individuals will produce a population of points (see Figure 7.7). Being near one another in the space means that individuals are similar in the relevant measures; if the test is valid, there may be real similarities between the individuals. If various vectors of parameters to a mathematical function are being tested, then we would expect points in the same region to have correlated function outputs and correlated fitness.

In this view of individual minds as points in a space, change over time is represented as movement of the points, now truly *particles*. Forgetting and learning might be seen as cognitive decrease and increase on some dimensions, attitude changes are seen as movements between the negative and positive ends of an axis, and emotion and mood changes of numerous individuals can be plotted conceptually in a coordinate system. As multiple individuals exist within the same high-dimensional framework, the coordinate system contains a number of moving particles. One insight from social psychology is that these points will tend to move toward one another, to influence one another, as individuals seek agreement with their neighbors.

Another insight is that the space in which the particles move is heterogeneous with respect to evaluation: some regions are better than others. This is of course true for functions and systems as well as psychology; some points in the parameter space result in greater fitness than others. A vector of cognitive, mathematical or engineering parameters can be evaluated, and it is presumed that there exists some kind of preference or attraction for better regions of the space.

The position of a particle i is assigned the algebraic vector symbol $\vec{x}_i$. Naturally there can be any number of particles, and each vector can be of any dimension. Change of position of a particle could be called $\Delta\vec{x}_i$, but in order to simplify the notation we call it $\vec{v}_i$, for velocity. Velocity is a vector of numbers that are added to the position coordinates in order to move the particle from one time step to another:

$$\vec{x}_i(t) = \vec{x}_i(t-1) + \vec{v}_i(t)$$

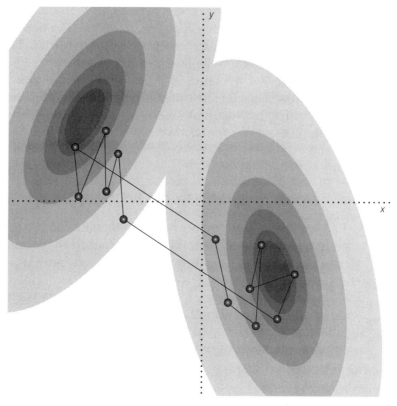

Figure 7.7 Particles in a real-number space are connected to topological neighbors, and neighbors tend to cluster in the same regions of the search space.

The question then is to define rules that move the particle in the desired way, in a way that optimally allocates trials while searching for optima. The particle swarm algorithm samples the search space by modifying the velocity term.

Social-psychological theory suggests that individuals moving through a sociocognitive space should be influenced by their own previous behavior and by the successes of their neighbors. These neighbors are not necessarily individuals who are near them in the parameter space, but rather ones that are near them in a topological space that defines the sociometric structure of the population. Who influences you is defined by your connections in the social network, not by positions in a belief space—there may be someone who agrees closely with you about everything, whom you have never met (and whom no one you know has ever met—and no one *they* know . . .), and who consequently does not influence you. Thus in the continuous-number particle swarm, as in the binary version, a neighborhood is defined for each individual, based on the position in the topological population array. The population array is commonly implemented as a ring structure, with the last member a "next-door" neighbor to the first member.

As the system is dynamic, each individual is presumed to be moving—this should really be called *changing*—at all times: this is the Lewinian concept of locomotion. The direction of movement is a function of the current position and velocity, the location of the individual's previous best success, and the best position found by any member of the neighborhood:

$$\vec{x}_i(t) = f(\vec{x}_i(t-1), \vec{v}_i(t-1), \vec{p}_i, \vec{p}_g)$$

Further, just as in the binary version, change (which is now defined in terms of velocity instead of probability) is a function of the difference between the individual's previous best and current positions and the difference between the neighborhood's best and the individual's current position. In fact the formula for changing the velocity is identical to the one used to adjust probabilities in the binary version, except that now variables are continuous, and what is adjusted is the particle's velocity and position in R^n:

$$\begin{cases} \vec{v}_i(t) = \vec{v}_i(t-1) + \varphi_1(\vec{p}_i - \vec{x}_i(t-1)) + \varphi_2(\vec{p}_g - \vec{x}_i(t-1)) \\ \vec{x}_i(t) = \vec{x}_i(t-1) + \vec{v}_i(t) \end{cases}$$

where, as before, the φ variables are random numbers defined by an upper limit. The effect of this is that the particle cycles unevenly around a point defined as a weighted average of the two "bests":

$$\frac{\varphi_i \vec{p}_i + \varphi_2 \vec{p}_g}{\varphi_1 + \varphi_2}$$

Because of the random numbers, the exact location of this point changes on every iteration.

For reasons that will be discussed in the next chapter, the system as given thus far has a tendency to explode as oscillations become wider and wider, unless some method is applied for damping the velocity. The usual method for preventing explosion is simply to define a parameter V_{max} and prevent the velocity from exceeding it on each dimension d for individual i:

$$\text{if } v_{id} > V_{max} \text{ then } v_{id} = V_{max}$$
$$\text{else if } v_{id} < -V_{max} \text{ then } v_{id} = -V_{max}$$

The effect of this is to allow particles to oscillate within bounds, although with no tendency for convergence or collapse of the swarm toward a point. Even without converging, the swarm's oscillations do find improved points in the optimal region. One approach to controlling the search is the implementation of an "inertia weight." Another method, developed by French mathematician Maurice Clerc, involves a system of "constriction coefficients" applied to various terms of the formula. These approaches are discussed in the next chapter.

Pseudocode for Particle Swarm Optimization in Continuous Numbers

In sum, the particle swarm algorithm in the space of real numbers is almost identical to the binary one, except that $\vec{v}_i$ defines increments of movement rather than a probability threshold, as illustrated by the following pseudocode:

```
Loop
    For i = 1 to number of individuals
        if G(x̄ᵢ) > G(p̄ᵢ) then do              //G() evaluates fitness
            For d = 1 to dimensions
                pᵢd = xᵢd                        //pᵢd is best so far
            Next d
        End do

        g = i                                    //arbitrary
        For j = indexes of neighbors
            If G(p̄ⱼ) > G(p̄g) then g = j          //g is index of best performer
                                                 //  in the neighborhood

        Next j
        For d = 1 to number of dimensions
            vᵢd(t) = vᵢd(t − 1) + φ₁(pᵢd − xᵢd(t − 1)) + φ₂(pgd − xᵢd(t − 1))
            vᵢd ∈ (−Vmax, + Vmax)
            xᵢd(t) = xᵢd(t − 1) + vᵢd(t)
        Next d
    Next i
Until criterion
```

Implementation Issues

One issue faced by anyone implementing the algorithm is how to initialize the population. The positions and velocities of the particles are usually initialized randomly. The initial random positions are often distributed over the dynamic range of each dimension. The initial velocities are often distributed randomly over $[-V_{max}, V_{max}]$.

Another question faced at implementation time is, "How many particles should I use?" There is no pat answer to this question, but the experience of the authors indicates that choosing somewhere between 10 and 50 usually seems to work well. One of the authors [JK] tends to run with populations at the lower end of this range, while another [RE] often prefers higher numbers of population members. If you are used to working with traditional genetic algorithms, you should probably consider starting with fewer particles than the number of GA chromosomes you would use.

The particles' flight patterns can be interesting to watch. We find that you can sometimes learn something about the problem being modeled or optimized by watching the particles fly. Pictures of the swarm that are snapshots in time, such as the figures of this book must be, cannot do justice to the system's dynamics. We therefore suggest that you watch the flight patterns for yourself. You can do this by compiling and running the source code we provide on the book's web site. But an even easier way is to go to the web site and run the Java applet. You can choose from a number of test functions and set the parameters for each function.

An Example: Particle Swarm Optimization of Neural Net Weights

The first real implementation of the particle swarm algorithm was a model that bridges psychological theory and engineering applications. The feedforward artificial neural network is a statistical model of cognition that inputs vectors of independent variables and outputs estimates of vectors of dependent variables. The network is structured as a set of weights, usually arranged in layers, and the optimization problem is to find values for the weights that make the mapping with minimal error. It is beyond the scope of this book to go into detail about neural networks; you are encouraged to refer to books that contain basic information on neural networks and their applications such as Eberhart, Simpson, and Dobbins (1996) or Reed and Marks (1999).

Normally, feedforward networks are optimized through some sort of gradient descent algorithm, traditionally using backpropagation of error. The cognitivistic interpretation is that the individual makes judgments about stimuli through adjustments based on personal experience. In contrast to this, the particle swarm view places the individual in a social-psychological context. The individual makes adjustments by integrating individual experience with the discoveries of others. Cognition in this view is a collaborative enterprise, occurring in a transpersonal milieu within which the individual is only a part.

Besides its ability to provide insights to the understanding of cognition, the feedforward neural network is also an extremely useful data-analysis tool, comprising a superset of which regression analysis is a member. These networks are often referred to as "universal function approximators" because they are able to mimic any kind of mathematical function. Networks excel in matters of estimation, for instance, when it is desirable to make a prediction from some data; their shortcoming for statistical use is related to difficulty in explaining a result, once it is obtained. This shortcoming has two levels to it: not only is there a human relations problem in trying to understand or explain how an estimate was produced by passing data through dozens or hundreds of weighted connections, but also it is so far not known how to make statistical inferences to a population from a network, that is, how to estimate the degree of confidence in a result.

A classic test problem for the feedforward network is the "XOR problem." The exclusive-or (abbreviated XOR) logical operation is true if one and only one argument is true, unlike the usual Boolean OR operator, which is true if any argument is true. This little logic puzzle accepts a pair of binary inputs, and outputs a 0 if they are the same, 1 if they are different from one another. The network is trained with a data set of inputs and outputs, shown in Table 7.2.

The feedforward neural network has been widely studied as a kind of model of human cognition, especially categorization, as it is able to simulate the human knack for lumping things together that do not seem to belong together. *Linear separability* exists when some weighted additive combination of properties can be used to classify examples, in other words, when things are graphed by their attributes and a straight line or surface can be drawn that perfectly separates the two kinds of things. Human categorization often—perhaps usually—is not linearly separable.

Table 7.2 The XOR function outputs a zero when the inputs match and a one if they are different from one another.

Inputs		Output
1	1	0
1	0	1
0	1	1
0	0	0

The XOR problem exemplifies this, as it is not linearly separable (see Figure 7.8).

Besides the two input and one output nodes required, it has been established that at least two "hidden nodes" are required for this problem, unless the inputs are to be directly connected to the outputs. Thus for purposes of illustration, we will use a network with two hidden nodes (see Figure 7.9). Weighted connections link each input node to each hidden node and each hidden node to the output node. Each hidden and output node also has a bias weight associated with it. Thus this network requires optimization of nine weights. The particles, in other words, move in nine-dimensional space, looking for patterns of weight values that effectively output the correct value for a given pair of inputs.

A population of 20 individuals is defined, where each individual starts the trial as a random vector of weights; neighborhoods are parsimoniously defined as each individual's adjacent array neighbors (borrowing from cellular automaton methodology). Inputs are entered into the network, and the mean squared difference between the obtained and observed values is calculated. The objective of course is to minimize this error measure. Error is computed for each member of the population; each particle compares its current position to its previous best and its neighbors' previous best and adjusts its velocity toward the points where those values were found. Eventually some member of the population will have found a pattern of weights that meets the criterion, for example, average sum-squared error < 0.02. One early study (Kennedy, 1997) found the network weights could be optimized to the criterion, with $V_{max} = 2$ and $\varphi_{total} = 4$, in a median of 70 iterations, using a population of 20 particles. We note that some recent adaptations of the particle swarm algorithm permit faster optimization of neural net weights.

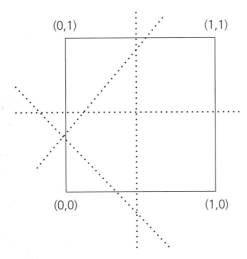

Figure 7.8 The XOR problem is not linearly separable. No line can be drawn such that the corners (0,0) and (1,1) are on one side of it and the corners (0,1) and (1,0) are on the other side.

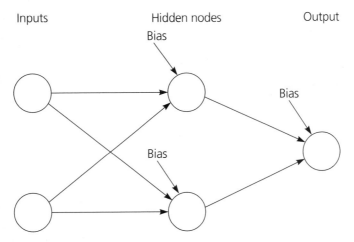

Figure 7.9 The XOR feedforward network takes a pair of inputs. It outputs 1 if the two inputs are different and 0 if they are the same.

A Real-World Application

The XOR network example has been used to ease you into the subject of neural networks. The first real-world application of particle swarm optimization to a neural network was one designed to provide an estimate of the state of charge of a pack of batteries in an electric vehicle. The network originally had eight inputs, five hidden nodes, and one output. Eventually, the network was simplified to five inputs, three hidden nodes, and one output. A team that included one of the authors [RE] was training feedforward neural networks using the backpropagation algorithm. Although the networks were fairly small, there were approximately 2,500 data sets being used to train them, and each training session using backpropagation required approximately 3.5 hours.

It was a simple matter to delete the backpropagation training algorithm and substitute particle swarm optimization. The results were dramatic. Training time required to achieve the same sum-squared error value dropped from about 3.5 *hours* to about 2.2 *minutes*. This was the first time we got a glimpse of how powerful the particle swarm algorithm is when applied to real-world optimization problems. For a more complete description of this application, see Eberhart, Simpson, and Dobbins (1996). Although the training of the network did not *depend* on particle swarm optimization to be successful, it became obvious that it was a subject worthy of further investigation.

Subsequently, many neural networks have been trained using particle swarm optimization. In these and other applications, some "rules of thumb" have evolved. The first regards the selection of V_{max}. The standard sigmoid function used in feedforward networks essentially saturates with inputs of about +/− 10. Therefore, we often consider the dynamic range of weights to be something like 20. It seems reasonable that we limit the maximum velocity to some percentage of the dynamic range. A V_{max} of 2.0 has thus been found to be successful for training feedforward networks with sigmoid activation functions. The maximum acceleration constant φ_{total} has generally been set to 4.0 for these applications. (Other ways to limit velocity, such as the damping weight and constriction factor approaches, are described in Chapter 8.)

The second rule of thumb regards the setting of the neighborhood size. Although no theoretical foundation exists, it has been found that a neighborhood size of about 10–20 percent of the population often works well. Thus, when the population comprises 40 particles, we often choose a neighborhood size of two or three adjacent neighbors on both sides of each particle, for a neighborhood size of 5 or 7, including *i*.

As will be seen in Chapter 9 on applications of particle swarms to engineering problems, the particle swarm paradigm has been used on very large networks. Further, in an exciting recent development, the particle swarm can actually be used to *design* the neural net; nodes can be simplified and deleted from the network at the same time that optimal values for weights are being found.

The training (weight adjustment) of multilayer artificial neural networks with nonlinear transfer functions was previously thought to be intractable. Though the techniques such as backpropagation of error developed for optimizing weights are known to be imperfect—they are susceptible to local optima—the methods have been considered good enough for most uses. In fact, some tricks (we are thinking here of Scott Falman's "quickprop" algorithm) have been found that propitiate the discovery of robust solutions.

As has been seen, there exists a largish library of functions for the testing of the strengths and weaknesses of optimization algorithms, including the De Jong test suite and other functions that have been collected over the years. Any new optimization algorithm, including the particle swarm, has to prove itself by triumphing over these diabolical mathematical minefields, which are contrived with multiple optima, corrugation, serration, cavities and bumps, high dimensionality, infeasible regions, epistasis, dynamic distortions, and deception (this latter is a term for problems where the gradient would move a hill climber away from the global optimum). In our studies and those of others, the particle swarm algorithms have been successful at solving these problems.

The Hybrid Particle Swarm

Now that we've shown you how binary and real-valued particle swarms work, we'd like to briefly mention one of the current subjects of our research: hybrid particle swarms. As you might guess, a hybrid swarm combines binary and real-valued parameters in one search. In this section, we give you a few ideas about *why* you might want to implement such a system and *how* to do it.

Why would we want a hybrid algorithm? Perhaps we have a problem with both continuous and binary variables; one example is an explanation facility for a multiple-symptom diagnostic system. Let's say you have designed a neural-network-based diagnostic system to diagnose various abdominal diseases such as appendicitis and nonspecific abdominal pain. Some of the symptoms such as body temperature and white blood cell count are continuous variables and are best represented as real numbers. Other inputs such as the answer to the question "Have you had your appendix removed?" are binary values. So the mix of symptoms includes real and binary numbers. This poses no trouble when training a neural network, whether you use a more traditional approach such as backpropagation or our favorite approach using particle swarm optimization.

It can be very difficult to explain how a network has made the decision it has. As the information is encoded in complex patterns of connection weights, usually linking several layers of nodes, it is rarely clear what it is about a pattern of inputs that led to the diagnosis. It is possible to construct an *explanation facility* that provides the physician with a tool to understand how the network diagnoses. We typically use an evolutionary algorithm to determine two kinds of things. First, where are the hypersurfaces that represent differential diagnoses? A differential diagnosis is that set of inputs (or a suite of such inputs) that lies on the borderline between two diagnoses. In other words, the differential diagnoses are those sets of inputs for which the chances are equal that the symptoms represent appendicitis and nonspecific abdominal pain, for example. Second, what are quintessential examples of each diagnosis? In other words, what combinations of inputs result in the maximum likelihood of a disease such as appendicitis, according to the system?

Finding these sets of inputs is an example of an "inverse problem" and is something that evolutionary algorithms are good at. So now the question is *how* we do it. You've probably already figured out the answer. We simply operate on (medical pun intended) binary inputs with the binary particle swarm algorithm and treat (oh no, another medical pun) the continuous variables with the real-valued particle swarm.

We've done some preliminary tests, and the approach seems to work well. We emphasize, however, that it is a subject of ongoing research and development.

Science as Collaborative Search

The kind of decision processes instantiated in both the binary and real-valued particle swarm algorithms exemplify a tendency that has been widely regarded as a flaw or error when seen in human cognition. Karl

Popper (1959) revolutionized scientific methodology by persuading us that it is impossible to confirm a hypothesis—it is only possible to disprove one. In his famous example, even if you have seen a million white swans, and never in your life have you seen any other color, you still have not conclusively proven that "all swans are white." On the other hand, a single black swan will *dis*prove the statement. Modern scientific methodology is based on the philosophy of *null hypothesis testing,* which takes the tack of trying to prove the hypothesis that your research hypothesis is in fact false, that is, you look for black swans. A hypothesis cannot be tested unless it is falsifiable, and scientific proof relies on identifying what would happen if the hypothesis were indeed false and then discovering if those events occur in an experimental situation.

While it is logically impossible to prove a hypothesis by accumulating support for it, this is exactly the approach people normally take. Cognitive psychologists call this tendency *confirmation bias,* the propensity to irrationally seek confirmation for our beliefs, rather than falsification. Klayman and Ha (1987) turned the issue around by pointing out that falsification is not a good strategy for determining the truth or falsehood of many hypotheses. They proposed that people tend to use a "positive test strategy," which is defined as testing cases that are expected to produce the hypothesized result, rather than testing cases that are intended to fail to produce it. They suggested that people use the positive test strategy (+testing) as a default heuristic. Further, they noted that "as an all-purpose heuristic, +testing often serves the hypothesis tester well" (p. 225).

Another way of looking at this is to compare *truth* and *certainty.* Most of the time, people solving a problem don't require knowledge that something be established as *true;* they only require that it be established to a level of *certainty.* As Karl Popper said in a recent interview with writer John Horgan (1996), "We must distinguish between truth, which is objective and absolute, and certainty, which is subjective." Adjusting your hypotheses toward the consensus position and testing cases that confirm what you already believe are methods for increasing the sense of certainty. One thing that will undermine that sense is of course contradiction by empirical facts; thus, "+testing" can only work if it is consistent with phenomena in the world. While it is possible to build up certainty in the absence of truth, the two are not independent—a fact that can be capitalized on. Strategies that increase certainty may be likely to discover truths as well.

In the model we have just described, individuals move *toward* their previous successes; confirmation bias is fundamental to this strategy. But this is an elaborated, social confirmation bias: individuals seek to confirm not only their own hypotheses but also those of their neighbors. Paradoxically, though we may be pointing out that people are not very scientific in their thinking, especially insofar as science is supposed to be mathematical and deductive, even scientists act like this. What Thomas Kuhn (1970) calls a *paradigm* is a kind of confirmatory social convergence of scientists in a theoretical decision space: "A paradigm is what the members of a scientific community share, and conversely, a scientific community consists of men who share a paradigm" (p. 176). The scientists come to agreement on the use of terminology, acceptable research methods, and other aspects of their work, and it is by intense communal focus on a narrowly defined subject domain that the scientists are able to fully exploit the learning that has preceded them. In the particle swarm analogy, a Kuhnian "revolution" occurs when an individual finds a better region of the search space and begins to attract its neighbors toward it by becoming the best in the neighborhood.

In the 1960s and 1970s some evolutionary theorists began to propose a correspondence between scientific and evolutionary processes that continues to be reiterated (Campbell, 1965, 1974; Popper, 1972; Lorenz, 1973; Atmar, 1976; Dawkins, 1987). In this view, an individual member of a species represents a hypothesis about the logical properties of the environment; the validity of the hypothesis is shown by the survival of the individual. This inductive approach to learning leads to constantly improving prediction of the important aspects of the environment. As in previous discussion of the memetic view, our objection to the too-literal acceptance of this view has to do with the difference between selection, as it occurs in evolution, and change as it appears in learning. A scientist often has a long career spanning the comings and goings of multiple paradigms. Hypotheses are ideas that are held in the minds of scientists, who are able through constant refinement, through constant adaptation—through learning—to improve the validity of their hypotheses. The evolutionary perspective looks at the mutation and selection of ideas per se, while the particle swarm view looks at the adaptive changes of individuals who hold those ideas.

In informal human social search of a problem space, little effort is typically made to carefully choose data, and both measurement and sampling error are extremely plentiful—a glance through any textbook in social or cognitive psychology will reveal dozens of "heuristics," "biases," and "errors" in human information processing. We propose that many of the biases result from the "particle swarm" tendency of individuals to move toward self- and social confirmation of hypotheses—

a tendency that, while logically invalid, in fact results in excellent information-processing capabilities. We don't agree that human thinking is faulty; we suggest on the contrary that formal logic is insufficient to solve the kinds of problems that humans typically deal with.

Emergent Culture, Immergent Intelligence

After some number of iterations the members of the particle swarm populations are found to have congregated around one or more of the optima. In cases where multiple global optima are discovered by the population, topological neighbors tend to cluster in the same regions of the search space. These clusters extend beyond hard-coded neighborhoods. When an individual finds a relatively optimal combination of elements, it draws its adjacent neighbors toward itself; if the region is superior, the neighbors' evaluations will improve as well, and they will attract *their* neighbors, and so on. If another subset of the population is attracted to a different but equally good region of the problem space, then a natural separation of groups is seen to emerge, each with its own pattern of coordinates that may easily be thought of as norms or cultures. ("We used to say 'customs' when we were talking about norms; now the norm, of course, is to say 'norm' " (Picker, 1997, p. 1233).)

When one solution is better than another, it usually ends up absorbing the lesser pattern, though in some cases mediocre "compromise" individuals on the borders of groups prevent the spreading of better solutions through the population. The polarization of these artificial populations into separate cultures appears very similar to the convergence of human subpopulations on diverse norms of attitude, behavior, and cognition. Interaction results in conformity or convergence on patterns that are similar for proximal individuals and may be different between groups.

The formation of cultures in particle swarm trials is not specified in the computer programs and is not readily predictable from the definitions of interactions in the programs. It thus would be considered an emergent effect, though we grant that "emergence" might actually be a term that represents the simplicity of our own minds; a property not of the system but of our failure to understand it.

There is another important feature of the behavior of a particle swarm, or of a human society, and that is the *immergence* of cognitive adaptation of individuals as a result of the top-down effect of *emergent*

culture. Participants in the system become intelligent, acquiring whatever qualities have been defined as "good" in the fitness function, as a result of the cultural optimization enabled by local interactions (see Campbell, 1990).

The cultural convergence of individuals in the search space allows the
intensive exploitation of optimal regions. Relatively good combinations
of elements, which in human society may be beliefs, behaviors, problem-
solving steps, opinions, and so on, receive focused attention. As a result,
the performances of individuals are improved. Culture, the emergent result of bottom-up processes, enables top-down immergent mental phenomena, optimizing the cognitive processes of individuals.

According to this perspective, minds and cultures are intimately interwoven products of the interactions of individuals. There is no need to
postulate any great distinction between "internal" and "external" information processing—it all works together. This is not to deny any individual's conscious experience, or to say that any two individual minds are
identical to one another, and certainly is not to predict the spread of in-
sipid homogeneity through a society. Rather, the diversity of minds, of
explorations and explanations and exploitations, provides the raw material for the emergence of culture and simultaneous immergence of intelligent behavior.

Emulation of superior cognitive positions allows individuals to adapt
efficiently to complex cognitive landscapes. As an optimizer, the particle
swarm algorithm has been shown to perform very well on a wide range of
hard test functions. The obvious conclusion is that mutual collaborative
emulation can result in individual adaptation: intelligence. A population of social entities evaluating, comparing, and imitating is able to zero
in on good solutions to complex problems.

Summary

The No Free Lunch theorem argues that no single algorithm can optimize better than any other, if we compare them on all possible objective
functions. But it turns out that most "possible" functions are uninteresting; they fail to qualify to be considered problems. Given the subset of
situations that researchers really do concern themselves with, it *is* possible to demonstrate that one algorithm has the advantage over another.
One way to find these differences is by trying the algorithms on sets of
problems that are known to be difficult, for different reasons; one may

have very many local optima, for instance, while another features complex interactions among variables. Several kinds of algorithms have proven themselves superior in their ability to optimize various kinds of difficult functions. Simulated annealing, various evolutionary computation methods, and now the particle swarm are among these.

The particle swarm algorithm imitates human social behavior. Individuals interact with one another while learning from their own experience, and gradually the population members move into better regions of the problem space. The algorithm is extremely simple—it can be described in one straightforward formula—but it is able to surmount many of the obstacles that optimization problems commonly present. In Chapter 8 we will see that the simple formula generates rich and complex effects. Researchers have investigated numerous ways to manipulate the search trajectories of the particles, and some of these ways have resulted in improvements and insights.

chapter eight

Variations and Comparisons

In this chapter, we explore implementations of the particle swarm paradigm, look at some variations of the algorithm, ask whether the particle swarm is an evolutionary algorithm, and compare the performance of various versions of particle swarms for optimization. This chapter is more technical than some of the others.

In the "Variations" section, we first take apart the algorithm to see what makes it tick, focusing on ways to facilitate convergence and prevent "explosion" of the swarm. Approaches include limiting the maximum allowed particle velocity, including V_{max}, a *constriction coefficient,* and an *inertia weight.* We look at the effects of varying neighborhood topologies and some other ways to modify the algorithm.

Are particle swarms really a kind of evolutionary algorithm? We consider a new way to look at evolution, focusing on the role of self-organization, and look at the processes of selection and self-organization in the context of the particle swarm. We then examine crossover and mutation processes and population topology as they relate to particle swarms.

Finally we compare versions of the particle swarm paradigm. Included are experiments in which selection is added to the particle swarm paradigm, comparisons of the inertia weight and constriction factor approaches, and a test in which particle swarms are initialized asymmetrically with respect to the global optimum. ∎

Variations of the Particle Swarm Paradigm

The particle swarm paradigm has undergone many tweaks and modifications since its discovery in 1994. Various researchers have analyzed it and experimented with it, including mathematicians, engineers, physicists, biochemists, and psychologists. In the process, a certain body of lore has emerged to provide hypotheses for research as well as guidelines for applications. The lore itself evolves as the algorithm is better understood, as theorists debate the nature of adaptive systems, and as programmers' serendipitous trials and errors result in surprising improvements. This section addresses some of the major directions of current investigations into the functioning of the algorithm.

Parameter Selection

In the particle swarm there are several explicit parameters whose values can be adjusted to produce variations in the way the algorithm searches the problem space. The most important of these are V_{max} and φ, which are set at the beginning of a trial and remain constant throughout. As will be seen, manipulation of these two parameters alone can cause surprising changes in the system's behavior.

Besides the explicit parameters, the system can be thought to contain a number of implicit ones, if we consider the terms of the formulas to be weighted, in the original version, by 1.0. We can change these implicit parameters, of course; for instance, we could eliminate the effect of a term by giving it a weight of 0.0 or increase its effect with a larger weight. In these ways we can make subtle adjustments to the system that can control important behaviors such as convergence and explosion. We can optimize the algorithm's performance by adjusting the importance of various explicit and implicit parameters.

V_{max}

The particle swarm algorithm proceeds by modifying the distance that each particle moves on each dimension per iteration. Changes in the velocity are stochastic, and an undesirable result of this is that the particle's trajectory, uncontrolled, can expand into wider and wider cycles through the problem space, eventually approaching infinity. Something needs to be done to dampen the oscillations if the particle is to search usefully. The traditional method is to implement a system constant V_{max}, with the stipulation

$$\text{if } v_{id} > V_{max} \text{ then } v_{id} = V_{max}$$
$$\text{else if } v_{id} < -V_{max} \text{ then } v_{id} = -V_{max}$$

The effect of this parameter, and other constriction methods, can be better understood by graphing the trajectory of a simplified particle (Kennedy, 1998). In order to make comprehensible graphs we will simplify the algorithm. If the particle is reduced to one dimension, and there is only one of them, then the vector notation can be thrown out, and a simple graph can display the particle's trajectory. Further, the two terms of the formula can be collapsed to one without losing any information, where p represents the weighted average of the two bests:

$$p = \frac{\varphi_1 p_i + \varphi_2 p_g}{\varphi_1 + \varphi_2}$$

and

$$\varphi = \varphi_1 + \varphi_2$$

If the weighted best point p is made constant, rather than dynamic, then the trajectory of the particle can be plotted and studied. The simplified formula is

$$v = v + \varphi \cdot (p - x)$$
$$x = x + v$$

where φ is (for now) a random number defined by its upper limit and other variables are scalars.

Keep in mind that this does not represent a realistic situation. Our population sizes are greater than one, and seldom does a particle operate on only one dimension. Furthermore, the weighted best point p is typically dynamic, often exhibiting complex behavior. So this artificial situation is contrived only to give some insight into how particles fly.

Figure 8.1 demonstrates what happens if v is undamped, that is, with no V_{max}, with $\varphi = 3.9$ for 150 iterations. The trajectory of x is started in these examples at $x = 2$ and $p = 0$, unless otherwise stated. Time moves from left to right, and the vertical axis represents the x value of the particle at each point.

Because Figure 8.1 is plotted to the scale of the most extreme values, the early cycles appear as a flat line. In fact the trajectory of x grows wider

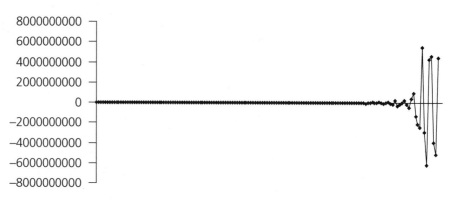

Figure 8.1 If the particle swarm algorithm is run without limiting velocity, the particle quickly explodes beyond the region of interest. (*Note:* Time runs from left to right for 150 iterations in the graphs in Figures 8.1 through 8.15.)

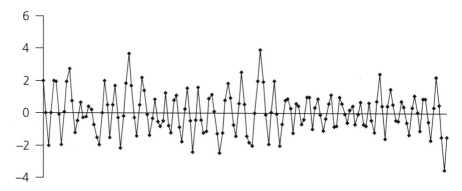

Figure 8.2 Setting a V_{max} limit of 2.0 on the particle's velocity keeps it in a useful range. Note that the particle does not converge on the optimum, which is 0.0 in this example.

and wider with time, until by 150 iterations it is in the range of $\pm 10^9$. Obviously this is not a good way to search around the area of 0.0. (We will consider causes of the explosion later in this chapter.)

Figure 8.2 shows a typical particle's behavior when the velocity is limited to $V_{max} = 2.0$. As can be seen, the explosion is prevented, and the particle cycles within a range of approximately ± 4. The search is stochastic and roughly quasi-periodic, with mean $x \approx p$.

It can also be seen that the trajectory of the particle does not converge toward p over time; instead, it cycles widely. If some cycles are seen to collapse so that amplitude is decreased, the opposite trend will just as likely be seen. All in all, this is not really a bad way to search around an

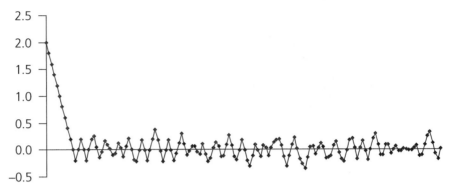

Figure 8.3 Reducing the V_{max} limit to 0.2 restrains the particle's movements to a narrower range around the optimum.

optimal region, though we would prefer the search to narrow over time, exploiting exploratory discoveries.

For comparison's sake, if we reduce the value of V_{max} to 0.2, dividing by 10, in other words, we see that after dropping from its initial value of 2.0, x cycles around $p = 0$, but in narrower cycles, within a range of approximately ±0.30 (see Figure 8.3). If we were to continue the experiment with all values of V_{max}, we would see that it simply scales the amplitude of x's oscillation around p.

Thus the system parameter V_{max} has the beneficial effect of preventing explosion and scales the exploration of the particle's search. Unfortunately the choice of a value for V_{max} depends on some knowledge of the problem. For instance, if a step larger than V_{max} is required in order to escape a local optimum, then the particle will be trapped. Further, in approaching an optimum it would be better to take smaller steps.

The Control Parameter

The control parameter φ, sometimes called the "acceleration constant," turns out to be very important in determining the type of trajectory the particle travels. If $\varphi = 0.0$, it is obvious that $v = v + 0$, and as $x = x + v$ it simply increases linearly. If φ is set to a very small value, the trajectory of x rises and falls slowly over time. Note that for all figures where V_{max} is implemented, it is set to $V_{max} = 2.0$ unless otherwise stated. Figure 8.4 shows a particle with $\varphi = 0.01$, a very small value that lets the particle wander before being pulled back by the accumulation of $(p - x)$.

Increasing φ to 0.10 increases the frequency of the waveform (see Figure 8.5). Since φ is random, the waves are uneven. The amplitude of the

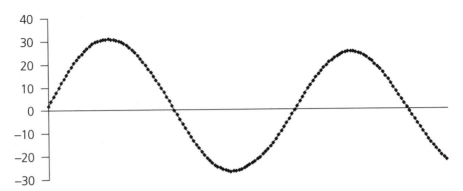

Figure 8.4 When φ is very small ($\varphi = 0.01$ in this example), the particle swings far from the optimum (which is 0.0 in this example) before the accumulated $(p - x)$ differences pull it back.

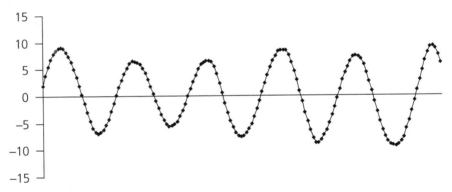

Figure 8.5 Increasing φ to 0.1 reduces the amplitude of the particle's trajectory, which is still approximately sinusoid around the optimum at 0.0.

wave is great relative to V_{max}; the distances between pairs of succeeding steps are small, though there are frequently series of V_{max}-sized steps along the sloping part of the curve.

Increasing φ to 1.0 increases the frequency of the wave even more, and overall amplitude is decreased (see Figure 8.6).

As φ increases to 10, the waveform appears to oscillate randomly (see Figure 8.7). Note that in this example there is a sequence near the beginning (and in a couple of other places) where the wave vibrates sharply back and forth, touching on the same points repeatedly. This occurs where v is repeatedly limited to $V_{max} = 2.0$, because its calculated value has exceeded the limit.

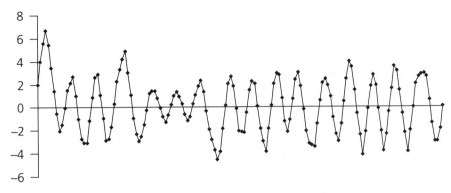

Figure 8.6 When $\varphi = 1.0$, the stochastic particle explores irregularly around the optimum. Velocity is limited in these examples to $V_{max} = 2.0$.

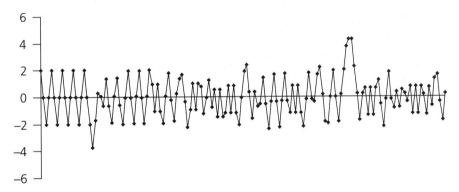

Figure 8.7 When $\varphi = 10$, the stochastic particle moves at V_{max}-sized steps much of the time.

Finally, when $\varphi = 100$, the particle's trajectory is almost entirely vibratory, dominated by steps of size V_{max} (see Figure 8.8). Though values mentioned here are specific to this particular example, in general the system behaves in this way with any V_{max} value.

In sum, in the stochastic particle swarm the system parameter φ controls the strength of the effect of $(p - x)$ on the velocity. When φ is very low, the effect is weak, and the particle's trajectory follows a wide path, only being drawn back toward p by the pull of $(p - x)$ after a large number of iterations. As φ increases, the oscillatory wavelength shortens, but individual steps lengthen until they surpass the V_{max} limit. After that point, V_{max} imposes a fixed step size on the particle trajectory. (Remember that $\varphi = \varphi_1 + \varphi_2$ and that we typically choose $\varphi_1 = \varphi_2 \sim 2.0$.)

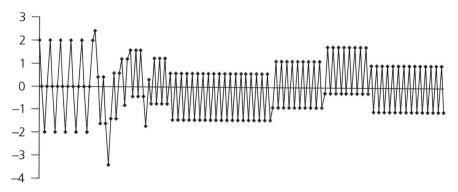

Figure 8.8 When φ is very high, for instance, 100 in this example, it hits the V_{max} limit on nearly every iteration. This ineffective trajectory searches the same points repeatedly.

The Effect of φ Varying

It is not immediately obvious why the velocity requires damping. After all, the $(p - x)$ term should become increasingly negative as x becomes positive and increasingly positive as x becomes negative—in either case pulling the particle back toward the region of p. It turns out that the system explodes because φ is varying due to being weighted with random numbers. This explosion can occur with a random φ or with φ fluctuating in almost any way. Some kind of damping is required to control it.

In order to explore this phenomenon, the examples in the preceding sections can be compared to a similar model implemented without randomness. V_{max} and random numbers temper the formal trajectory of the particle; if they are removed from the model the pure effect of φ is made evident—and that effect is interesting to see. Figures 8.9 and 8.10 demonstrate the trajectory of x, varying φ with no randomness. It can be seen that the looping sine wave of the very small values of φ gives way to complex waves of interwoven cyclicity at various values of φ and finally reaches to infinity when $\varphi \geq 4.0$. Note that explosion can also be caused when φ exceeds 4.0.

Ender Ozcan and Chilukuri Mohan (1999) analyzed the nonrandom, one-dimensional particle with constant p and concluded that "the particle does not 'fly' through the search space, but rather 'surfs' it on sine waves" (1999, p. 1943). The underlying sine waves can be clearly seen in Figures 8.9 and 8.10, crosscutting the literal trajectory of the particle with each time step; according to Ozcan and Mohan, a stochastic particle samples randomly from a sine wave and uses the underlying dynamic

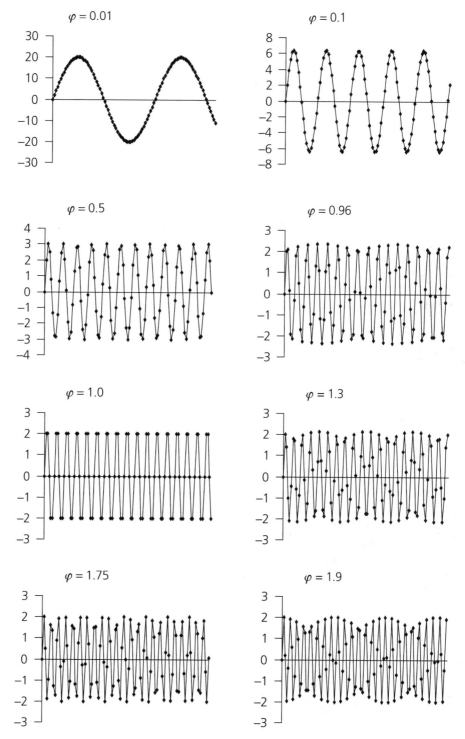

Figure 8.9 Particles "surfing the waves." The trajectories of nonrandom particles cross the underlying periodic attractor, the outline of which becomes visible with time.

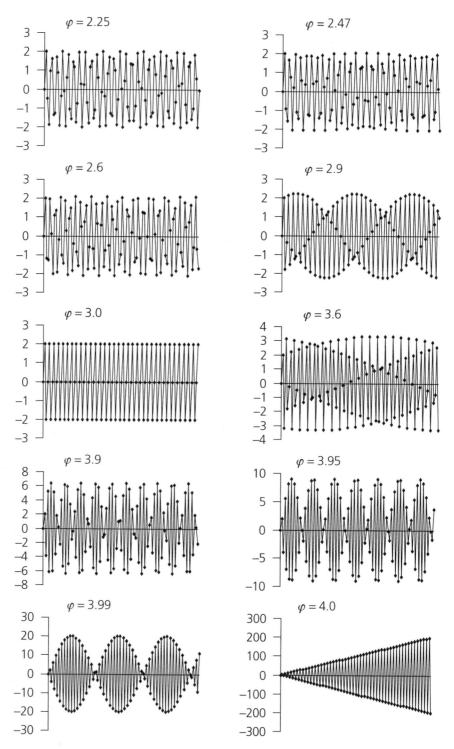

Figure 8.10 More particles surfing the waves. When $\varphi = 4.0$ and above, the amplitude expands to infinity.

pattern to search for optima. They argue that the V_{max} limit seems to help the particle jump onto a wave. Maurice Clerc showed that the particle's trajectory (whether stochastic or not) goes into complex space when time intervals are not integers, and that Ozcan and Mohan's sine waves are the "real" edges of a five-dimensional circular attractor, where two of the dimensions comprise the imaginary parts of the complex numbers and the other three dimensions are real.

When units of time are integers, as in the usual iterative computer program, the trajectory of the particle crosses the underlying sine waves at regular intervals. Clerc has determined the values of φ at which the trajectory of the nonrandom particle corresponds exactly with the sine waves; the consequence is that the particle's trajectory becomes cyclic for these values of φ.

Of course in the stochastic versions of the algorithm perfect periodicity is prevented by the damping of cycles through V_{max} as well as random weighting of φ. Doug Hoskins has compared the effect of the randomness in the system to a child's kicking in order to swing higher, and showed that the system explodes even when the "random numbers" are alternately selected from a set of two numbers. Effectively the random sampling of various patterns adds energy to the system, causing it to oscillate in an increasingly widening path.

We have seen that the deterministic system is inherently cyclical and complex—in both senses of the word. When $0 < \varphi < 4.0$, a nonrandom particle oscillates regularly around the target point *p*, with the path of the trajectory varying characteristically with φ. Randomness, though, has the effect of causing the system to explode in what has been called a "drunkard's walk." A drunk staggering randomly on the sidewalk will eventually wander into the street, especially if there is a building or hedge along the side of the sidewalk. Explosion is more probable than implosion simply because there is more area for expansion than contraction of the cycle.

Controlling the Explosion

The trajectory of the nonrandom particle is perfectly cyclic at a few special values of φ, though generally it weaves back and forth with a characteristic but nonrepeating series of values. When φ contains a random component the particle trajectory expands toward infinity: the question here is, what can be done to control the explosion?

Clerc (Clerc and Kennedy, 2000) has studied the deterministic system defined by

$$\begin{cases} v_{t+1} = v_t + \varphi y_t \\ y_{t+1} = -v_t + (1 - \varphi) y_t \end{cases}$$

where $y_t = p - x_t$. This system can be recast in terms of matrix algebra. With this kind of representation

$$P_t = \begin{bmatrix} v_t \\ y_t \end{bmatrix}$$

is the current state of the particle in R^2, and

$$M = \begin{bmatrix} 1 & \varphi \\ -1 & 1 - \varphi \end{bmatrix}$$

is the matrix of the system. Thus, $P_{t+1} = MP_t$ and, more generally, $P_t = M^t P_0$; the system is completely defined by M.

Clerc has further produced a generalized particle swarm model, comprising a set of coefficients that can be adjusted to affect swarm behavior:

$$\begin{cases} v_{t+1} = \alpha v_t + \beta \varphi y_t \\ y_{t+1} = -\gamma v_t + (\delta - \eta \varphi) y_t \end{cases}$$

$$\varphi \in R_+^*$$

$$\forall t \in N, \{y_t, v_t\} \in R^2$$

where $\alpha, \beta, \gamma, \delta,$ and η are coefficients that can be manipulated in order to influence the particle trajectory.

Based on this system, he showed that a generalized particle swarm system can be created in which explosion—and convergence as well—can be controlled. There are, Clerc says, an infinite number of ways to accomplish this, and he has worked out several methods.

Simplest Constriction

Clerc's simplest constriction coefficient, called Type 1″, requires application of coefficients to both terms of the velocity formula. This constriction method is described by the simplified system

$$\begin{cases} v(t) = \chi(v(t-1) + \varphi(p - x(t-1))) \\ x(t) = x(t-1) + v(t) \end{cases}$$

where φ must be greater than 4.0.

A simple formula to compute the constriction coefficient is

$$\chi = \frac{2\kappa}{\left|2 - \varphi - \sqrt{\varphi^2 - 4\varphi}\right|}$$

The variable κ can range in $[0,1]$; a value of 1.0 works fine, as does a value of $\varphi = 4.1$.

Thus, if $\varphi = 4.1$ and $\kappa = 1$, then $\chi \approx 0.73$, simultaneously damping the previous velocity term and the random φ. The Type 1" constriction coefficient is not defined for $\varphi \leq 4.0$. As φ increases above 4.0, χ gets smaller; for instance, if $\varphi = 5$, then $\chi \approx 0.38$, and the damping effect is even more pronounced.

This constriction method results in particle convergence over time (see Figure 8.11); that is, the amplitude of the individual particle's oscillations decreases as it focuses on a previous best point. When $\kappa = 1$, convergence is slow enough to allow thorough exploration before the search converges.

Though this kind of particle converges to a point over time, another factor in the paradigm prevents collapse of the trajectory—that is the fact that the target "best" point is actually a stochastically weighted average of two points, p_i and p_g. If those two points are near one another, then the particle will cycle around a singular center, eventually converging on the region of the two points (see Figure 8.12). On the other hand, if the swarm is still exploring various optima, and a particle's own previous best is in a different region from the neighborhood's previous best, that is, p_i is distant from p_g, then the particle's cycles will remain wide; it cannot converge on a target that keeps moving around the search space (see Figure 8.13). Thus a particle with a built-in tendency to converge will continue to explore when the "social" conditions are not conformist.

As the members of a neighborhood begin to cluster in the same optimal region, the particle trajectories will become narrower, intensely exploiting a focused region of the search space. If it should happen that another member of the neighborhood discovers a new optimum, the particle's trajectory is free to expand again to search the region between p_i and the new p_g. As can be seen in Figure 8.14, the Type 1" particle has no difficulty switching from exploratory mode to exploitative and back again.

The Inertia Weight

Eberhart and Shi have published several papers describing research with a version of the particle swarm algorithm that incorporates what is called

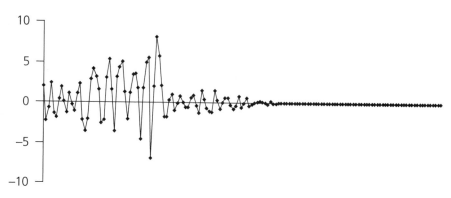

Figure 8.11 Clerc's Type 1" constriction coefficient causes the particle trajectory to converge over time.

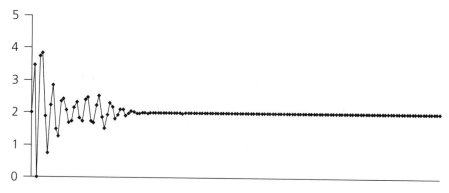

Figure 8.12 If there are two bests and they are near one another, as in this case where both equal 2.0, the Type 1" particle's trajectory converges. The amplitude of the particle's path, in other words, is scaled to the distance between the two best points.

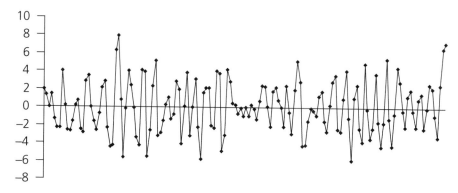

Figure 8.13 When there are two bests, for instance, the individual's best and the neighborhood's best, the spread between them causes the Type 1" particle to continue searching in a wide area, without converging.

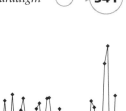

Figure 8.14 If a new best point is found after the particle has converged, the trajectory will be rescaled to the new distance between the two bests. Here a new best is introduced after 100 iterations.

an inertia weight (Shi and Eberhart, 1998). The inertia weight can be conceptualized within Clerc's generalized model as $\alpha = \gamma \,; \beta = \delta = \eta = 1$, or in simplified form:

$$\begin{cases} v(t) = \alpha v(t-1) + \varphi(p - x(t-1)) \\ x(t) = x(t-1) + v(t) \end{cases}$$

As a thought experiment—this is never done in practical applications—to help us understand the inertia weight, we can consider what the system does when $\alpha = 0$. In this extreme case, the particle trajectory is defined as

$$\begin{cases} v = \varphi \cdot (p - x(t-1)) \\ x(t) = x(t-1) + v \end{cases}$$

or, simplified:

$$x(t) = x(t-1) + \varphi(p - x(t-1))$$

It is apparent in this case that if $\varphi < 1.0$, the particle simply approaches p asymptotically without passing it. If $\varphi > 1.0$, the stochastic particle can fly past p, but the absolute value of $(p - x)$ decreases over time if $\varphi < 2.0$, so the particle converges on p. If $\varphi = 2.0$, then on average the particle moves to the other side of p, the same distance from it as when it began, and cycles back without converging or expanding. $\varphi > 2.0$ can cause the system to expand over time.

A nonzero inertia coefficient introduces the preference for the particle to continue moving in the same direction it was going on the previous iteration. If the absolute value of v increases on average over time, then the particle will oscillate out of bounds. As both α and φ affect the change in v, it is possible to prevent explosion by selecting values of these parameters that will control the convergence or de-convergence of the particle system.

Clerc's generalized constriction model suggests that the inertia-weighted particle swarm will eventually converge when $0 < \varphi \le 2\alpha + 2 - \varepsilon$ and $0 \le \alpha < 1$. Using these convergence criteria it is possible to parameterize the inertia-weighted particle swarm so that no V_{max} is necessary for convergence, though it may still turn out to be useful as a problem-solving heuristic.

Eberhart and Shi usually implement the inertia weight so that it decreases over time, typically from approximately 0.9 to 0.4, with $\varphi = 4.0$. V_{max} is set as a function of the dynamic range of the variables. The effect of the time-decreasing coefficient is to narrow the search, to induce a shift from an exploratory to an exploitative mode.

Neither the inertia coefficient with appropriate weight values nor the constricted models require any V_{max} parameter; convergence is an innate consequence of the mathematical properties of the algorithm. The constricted model has the advantage that it can recover from the shift to exploitation; the time-decreasing inertia weight is not as able to recover, but with a well-chosen "cooling schedule" the contraction of the particle search can perform effectively. Experience with the particle swarm algorithm indicates that inclusion of a V_{max} when using a constriction factor may be a good idea and costs very little computationally.

Particle Interactions

The effectiveness of the particle swarm algorithm comes from the interactions of particles with their neighbors. As one particle discovers a local optimum, it becomes the "best" in its neighbors' neighborhoods, and they too are attracted to the optimal region. As they move toward the new optimum, their search may uncover new regions that are even better, and they may end up attracting the first particle toward their best positions, and so on.

Figure 8.15 shows two particles searching for the minimum of the one-dimensional sphere function:

$$f(x) = \sqrt{x^2}$$

Since this is only one dimension, it can be plotted like previous examples. A simple constriction coefficient of approximately 0.7298 was

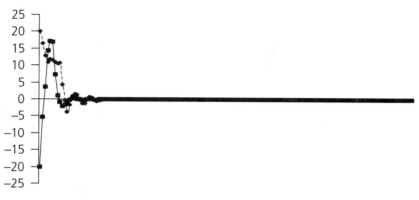

Figure 8.15 Two particles interact to optimize the sphere function.

implemented, with $\varphi = 4.1$. One particle starts out at $+20$ and the other at -20, with initial velocities of 0.0. As can be seen, in this trial the particles' trajectories weave around one another and quickly converge on the optimum $x = 0.0$. Naturally, every trial follows a different path; in this example, the solid-lined particle is initially attracted upward toward the dotted-lined one's better value, and then both race down toward the optimum.

In a typical particle swarm, neighborhood size can vary from three (the particle and its two neighbors) to the population size (in which case there is just one neighborhood).

Neighborhood Topology

Human social interaction occurs in the context of a group or social structure, often depicted by social scientists as a network of connections between pairs of individuals. Research since the 1940s has shown that communication within a group, and ultimately the group's performance, is affected by the structure of the social network. Particle swarm research has relied on several simple social structures, in particular the interaction of individuals with their immediate adjacent neighbors and interaction of all individuals with the best-performing individual in the population, though other social structures are possible. It has been shown that *isolated* particle swarm individuals perform very poorly: the interactions between particles make the algorithm work. Is there a best social structure for particles? If the analogy between particle swarms and human populations holds, then the answer to this question also has relevance for human organizations.

Sociometrics of the Particle Swarm

Particles have historically been studied in two general types of neighborhoods, called gbest and lbest. In the gbest neighborhood every individual is attracted to the best solution found by any member of the population. This structure then is equivalent to a fully connected social network; every individual is able to compare the performances of every other member of the population, imitating the very best. In the lbest network each individual is affected by the best performance of its k immediate neighbors in the topological population—a regular ring lattice. In one common lbest case, $k = 2$, the individual is affected by only its immediately adjacent neighbors.

The choice of social structures used has been frequently a matter of individual artistry, with some lore and little data to help the researcher choose a strategy. The lore suggests that gbest populations tend to converge more rapidly on optima than lbest populations, when they converge, but are also more susceptible to convergence on local optima.

An experiment was conducted where populations of 20 individuals were configured into Circles (lbest), where each individual is connected to its k immediate neighbors only, and Wheels, where one individual is connected to all others and they are connected to only that one (see Figure 8.16) (Kennedy, 1999). A gbest condition was also run.

In the Circle topology, parts of the population that are distant from one another are also independent of one another, but neighbors are closely connected. Thus one segment of the population might converge on a local optimum, while another segment converges on a different optimum or keeps searching. Influence spreads from neighbor to neighbor in this topology, until, if an optimum really is the best found by any part of the population, it will eventually pull all the particles in. Circles were defined with $k = 2$.

The Wheel topology, on the other hand, effectively isolates individuals from one another, as all information has to be communicated through the focal individual. This focal individual compares performances of all individuals in the population and adjusts its trajectory toward the very best of them. If adjustments result in improvement in the focal individual's performance, then that performance is eventually communicated to the rest of the population. Thus the focal individual serves as a kind of buffer or filter, slowing the speed of transmission of good solutions through the population. (It should be noted that the highly centralized Wheel is a common configuration for many business and government organizations.) The buffering effect of the focal particle

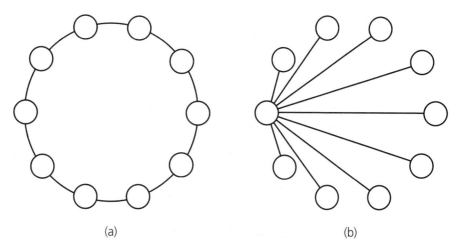

(a)   (b)

Figure 8.16 In the Circle topology (a), individuals interact with their *k* nearest neighbors (here *k* = 2). The Wheel (b) is like a lot of business and government organizations, where all information is filtered through one central individual.

should prevent overly rapid convergence on local optima; it is a way to preserve diversity of potential problem solutions, though it was expected that it might entirely destroy the ability of the population to collaborate.

Standard test functions were taken from the literature of evolutionary computation, including De Jong's *f*1 sphere, Griewank, Rastrigin, and Rosenbrock functions (see Table 8.1). All functions were implemented in

Table 8.1 Functions used in the experiment.

Function	Formula
Sphere	$f_0(x) + \sum_{i=1}^{n} x_i^2$
Rosenbrock	$f_1(x) = \sum_{i=1}^{n} (100(x_{i+1} - x_i^2)^2 + (x_i - 1)^2)$
Rastrigin	$f_2(x) = \sum_{i=1}^{n} (x_i^2 - 10 \cos(2\pi x_i) + 10)$
Griewank	$f_3(x) = \frac{1}{4000} \sum_{i-1}^{n} x_i^2 - \prod_{i-1}^{n} \cos(\frac{x_i}{\sqrt{i}}) + 1$

30 dimensions, and the dependent variable used was best performance on the test function after 1,000 iterations. All trials used populations of 20 individuals, with $\varphi = 4.1$. Other factors were manipulated in the experiment, though this chapter will report only on effects of sociometric structure on functions.

Analysis of variance showed that the interaction of neighborhood type with function was very strong, as it was seen that populations performed better on three of the functions when they were in the Circle rather than the Wheel configuration. Performance on the Rastrigin function, however, was just the opposite; Rastrigin populations performed better in the Wheel topology.

Discussion of Sociometry Experiment

The sociometry of the particle population interacts significantly with function; in other words, the optimal pattern of connectivity among individuals depends on the problem being solved. The study did not systematically manipulate aspects of test functions, but there are grounds for speculation as to an explanation for the interaction. The sphere and Rosenbrock functions are unimodal, with relatively smooth surfaces. The Griewank function is depicted in two dimensions as a bumpy bowl-like surface, a gradual overall slope toward the global optimum textured with many slight local optima. The Rastrigin function, though, features very many steep gradients toward local optima, with the depth of minima gradually increasing toward the global optimum. Failure occurs when the population clusters on a local optimum and is unable to leap from that hill to a better one. One explanation for the current results—Wheels performed better than Circles on Rastrigin only—might be that the buffering effect of communicating through a "hub" slows the population's attraction toward the population best, preserving diversity and preventing premature convergence on local optima. Thus a hypothesis can be proposed for future testing: centralized Wheel topologies may perform better by maintaining a diverse population on landscapes with many local optima.

It had been thought that the Star or gbest structure would be better for easier functions but that perhaps populations would "follow the leader" into locally optimal solutions. In the present data, with 30-dimensional mean-and-nasty test functions, Star performed very well. In fact, on all functions the Star was either best or nearly so. (The main disadvantage of the gbest topology is that it is unable to explore multiple optimal regions simultaneously.)

Substituting Cluster Centers for Previous Bests

After a few iterations, particles in the particle swarm are seen to cluster in one or several regions of the search space. These clusters indicate the presence of optima, where individuals' relatively good performances have caused them to attract their neighbors, who in moving toward the optimal regions improved their own performances, attracting *their* neighbors, and so on. It seemed reasonable to investigate whether information about the distribution of particles in the search space could be exploited to improve particle trajectories.

If clusters in regions of the search space indicate the location of optima, then perhaps the mean position of a clustered subpopulation will be closer to the optimum than any of their individual positions. One of the authors [JK] programmed a variation on the particle swarm that identified clusters of particles in the search space, and used the centers of clusters as substitutes for the bests (Kennedy, 2000).

In a population of 20 particles, it was arbitrarily determined to use five clusters. Clusters were found by the following steps:

1. Select C individuals as proposed cluster centers (evenly distributed topologically in the population).
2. Calculate the distance of all N individuals from the centers.
3. Assign all N individuals to the nearest cluster center.
4. Calculate the mean point in vector space for each cluster.
5. Loop to step 2 until centers stabilize.

Steps 2–5 were iterated three times based on the observation that little change typically occurred after three iterations. The cluster centers, identified on each iteration of the algorithm, could substitute for either the individual best or neighborhood best, using i's cluster center or g's.

Experimental Design

Four variations were tried:

1. Both terms used i's and g's individual previous bests.
2. The individual best term was replaced with i's cluster center.
3. The neighborhood best term was replaced with g's cluster center.
4. Both best terms were replaced with cluster centers.

The first of these versions is the traditional particle swarm. An lbest sociometry was defined, with each neighborhood comprising the individual and its two adjacent neighbors. For version 1, then, $\vec{p}_i$ and $\vec{p}_g$ were simply the best points found by the individual and by its best-performing neighbor (including itself).

In version 2, the individual's previous best performance was stereotyped, that is, particle i was attracted toward its cluster's average previous best performance as well as the performance of the best neighbor.

Version 3, replacing $\vec{p}_g$ with g's (the best neighbor's) cluster center, attracted particle i toward the center of g's cluster, while i still gravitated toward its own previous best performance. Social-psychologically this was like conforming to a group norm rather than the actual behaviors of group members.

The fourth version used cluster centers for both previous best terms. Here the individual stereotyped both self and other, and information from aggregate performance was used wholly.

First Experiment

Particle swarms were run in each of the four conditions on the four test functions given above, plus Shaffer's $f6$:

$$f_6(x) = 0.5 - \frac{\left(\sin\sqrt{x^2 + y^2}\right)^2 - 0.5}{\left(1.0 + 0.001\left(x^2 + y^2\right)\right)^2}$$

with Clerc's constriction factor and $\varphi = 4.1$. These versions implemented the $V_{max} = X_{max}$ parameterization described later in this chapter. Rosenbrock, Rastrigin, Griewank, and sphere functions were implemented in 30 dimensions, and $f6$, as usual, was run in 2 dimensions. Each trial consisted of 1,000 iterations, and the dependent measure was best fitness at that time. The experimental design was viewed in terms of manipulation of two independent variables in two levels each, crossed with a five-level function factor (the factor is called FUNC). The individual term (CLUSI) of the particle swarm formula could be the individual particle's best performance or a cluster center, as could the social influence term (CLUSG) (see Table 8.2).

The design is conceptually a $2 \times 2 \times 5$ factorial experiment. Because the test functions were not comparable, for instance, their ranges are quite different, it was not reasonable to compare performance on them using their unstandardized output. Instead, scores within each function were standardized by subtracting the mean and dividing by the standard

Table 8.2 Experimental design of the first experiment.

	individual-g	*cluster*-g
individual-*i*	1	3
cluster-*i*	2	4

Table 8.3 The mean best performance for all functions, for all experimental conditions, after 1,000 iterations (each entry represents 20 trials).

	individual-g	*cluster*-g
Sphere		
individual-*i*	0.00000	0.3178
cluster-*i*	0.00000	0.0002
Rosenbrock		
individual-*i*	39.6499	51.2138
cluster-*i*	25.2236	26.8999
Rastrigin		
individual-*i*	58.8001	9.2720
cluster-*i*	8.6626	10.3473
Griewank		
individual-*i*	0.0012	2.0669
cluster-*i*	0.0151	0.1580
Shaffer's *f6*		
individual-*i*	0.0000	0.0020
cluster-*i*	0.0017	0.0038

deviation. Using this method it is not possible to test for a main effect of function; that is, we cannot identify if one function or another was harder or easier overall. We can, however, compare the effects of the experimental manipulations across the various functions to see if using cluster centers in either term works better with some problems than with others.

The results are shown in Table 8.3. Analysis of variance on the results found that several effects were significant. The largest effect, by far, was the main effect for CLUSI: results were significantly better when $\vec{p}_i$ was a cluster center than when it was the individual particle's previous best position. The second-largest effect, in terms of the amount of variance explained, was CLUSG's main effect; performance was significantly worse when $\vec{p}_g$ was a cluster center. CLUSG interacted with FUNC, with the effect being that cluster-g conditions did worse than individual-g on all functions except Rastrigin, where they performed better than the average; this effect seems to be due largely to the poor performance of the standard version 1 particle swarm on that function. CLUSI also interacted significantly with FUNC; while it was better overall, it was not better than the traditional version on either Griewank or $f6$ functions. CLUSI and CLUSG interacted significantly; while performance was better when $\vec{p}_i$ was a cluster center, algorithms with individual-i performed especially badly when $\vec{p}_g$ was a cluster center. Finally, the significant three-way interaction suggests that, in some sense, everything depends on everything else. While $\vec{p}_i$ cluster centers may result in better problem solutions in general, the improvement to be expected is moderated by whether $\vec{p}_g$ is an individual's best or a cluster center, depending on the function.

Second Experiment

In human cognition, clustering or categorization of persons and things happens very fast. In fact, recognition seems to be largely a matter of sensing that information is processed faster for familiar stimuli than for unfamiliar ones; the term "perceptual fluency" is used in cognitive psychology to describe this effect (e.g., Jacoby and Dallas, 1981). In a computer, though, clustering requires some work, which takes some time. The substitution of cluster centers resulted in better average results over a fixed number of iterations; another question is, how does the clustering affect length of time required to reach a criterion?

To test this question, the program was run with a timer until a criterion was met. Each experimental condition was run 20 times, as before; trials were terminated after 3,000 iterations if the criterion had not been reached or after 500 iterations without improvement. As before, the constricted version with $V_{max} = X_{max}$ was used, with populations of 20 particles (see Table 8.4).

Table 8.5 gives the median time to meet the criterion in seconds. Numbers in parentheses give the percentage out of 20 trials that met the criterion. The right column gives the ratio of the median times for the two individual-g conditions for each function. As can be seen, the

Table 8.4 Functions used, their initialization ranges, and criteria.

Function	X_{max}	Criterion
Sphere	100	0.01
Rosenbrock	10	100
Rastrigin	5.12	100
Griewank	300	0.05
Shaffer's *f6*	100	0.00001

Table 8.5 Results of the second experiment.

	Median time (seconds)		Ratio
	individual-g	*cluster-g*	
Sphere			1.301
individual-*i*	1.479 (100)	∞ (0)	
cluster-*i*	1.924 (100)	2.852 (100)	
Rosenbrock			0.739
individual-*i*	1.334 (100)	∞ (30)	
cluster-*i*	0.9902 (100)	1.260 (100)	
Rastrigin			1.994
individual-*i*	0.631 (100)	1.283 (100)	
cluster-i	1.258 (100)	0.990 (95)	
Griewank			1.432
individual-*i*	1.895 (95)	∞ (0)	
cluster-*i*	2.713 (90)	∞ (50)	
Shaffer's *f6*			1.670
individual-*i*	0.574 (60)	∞ (30)	
cluster-*i*	0.959 (70)	∞ (15)	

cluster-*g* condition struggled to meet the criteria, and the individual-*g* versions failed to meet them sometimes for two of the functions. The median amount of time required for version 2, cluster-*i* and individual-*g*, relative to the standard version 1 particle swarm ranged from 74 percent to nearly 200 percent. Thus, as expected, adding the clustering steps to the program generally made it take longer.

Discussion

The empirical evidence from these experiments suggests that average performance per a fixed number of iterations can be improved by substituting cluster centers for the individual's previous best positions. The clustering algorithm implemented in this experiment generally added some time.

These preliminary results should not be taken as a recommendation to use the cluster-analyzed version instead of the standard incarnations, but they do point to a potentially useful research direction. The clustering algorithm used here was not intended to produce perfectly calculated centers, nor was it chosen for its speed and efficiency; it is expected that improved clustering methods will cut down the extra time added and may improve performance as well by discovering points in the search space that are nearer the optima.

It was seen that particle swarm search is relatively effective when individuals are attracted toward the centers of their own clusters and is not good when they are attracted to neighbors' cluster centers. It is interesting to think about *how* substituting a cluster center for the individual's previous best can result in improvement. The explanation is to be found in consideration of the probability that any individual's performance will be better than its cluster's center. As seen in Figure 8.17, if the individuals comprising a cluster are distributed around a local optimum, as in Cluster *A*, then it is entirely possible that the evaluation of the cluster center will be better than that of any of the particles that make it up. If, on the other hand, the particles are approaching an optimum from one side, as exemplified by Cluster *B*, then the cluster center's fitness will be closer to the average of the particles' evaluations. It is not likely, if the clusters do in fact indicate local optima, that the cluster center will be worse than the average of the particles that make it up. Thus the average cluster center's fitness will be greater than the average individual's.

Even though the average cluster center might be better than the average individual, it seems that the average cluster is not better than the best of a number of individuals. Note that in Cluster *B*, some individuals perform better than their cluster center; this is surely a common occurrence. In the lbest version with $k = 2$, the neighborhood best is determined through comparison of individuals $i - 1$, i, and $i + 1$. Taking the best of the three obviously increases the chances that $\vec{p}_g$ will be better than $\vec{p}_i$. From these results it appears that the neighborhood members' bests will have been better than their clusters' centers often enough to facilitate search. It appears in the present data that the performance of the cluster

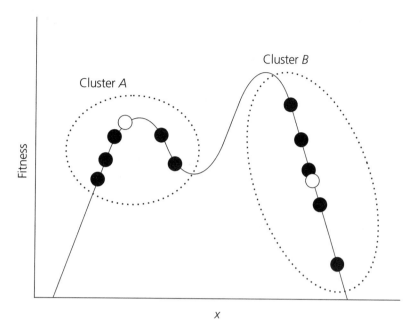

Figure 8.17 Cluster *A*'s center (white circle) performs better than any of the members of the cluster, while Cluster *B*'s center performs better than some and worse than others.

center is usually better than an average individual, but worse than the best of a group of three.

Adding Selection to Particle Swarms

Peter J. Angeline (1998a) investigated the relationship between particle swarms and evolutionary methods by introducing selection to particle swarms. At every iteration the worst half of the population, as determined by tournament selection, was replaced by clones of the better half. Social-psychologically, this simulates the effect of changing group membership, for instance, in an organization when poor workers are fired or quit, to be replaced by new and better individuals, or in informal groups where deviants are rejected by the group, to be replaced by better-fitting members.

Results of testing of the hybridized algorithm are intriguing. Angeline compared a version of the original V_{max} particle swarm to the hybridized version on four test functions: the sphere, Rosenbrock, Rastrigin, and Griewank functions. Angeline found that the hybridized algorithm performed significantly better on the first three functions, though the standard particle swarm performed better on the Griewank function—significantly better, according to *t*-tests, when functions comprised 20 and 30 dimensions. At 10 dimensions, no statistically significant difference was found in performance.

Though no analysis of variance was performed, Angeline's results appear to demonstrate the causal interaction of algorithm with function; that is, the effectiveness of the algorithms depended on the function to be optimized. Adding selection to the particle swarm algorithm resulted in improvement on some test functions, but performance was worse on at least one other. Hybridization had an effect—but the effect depended on the test function. This kind of hybridization of methods certainly suggests some interesting new directions for research.

Comparing Inertia Weights and Constriction Factors

In optimization, as in other areas of computer science and other endeavors, the investigator is often forced to make a choice between speed and power. For instance, a brute-force search of every point in the problem space is certain to provide the best possible answer, but it is not used on "real" problems because it would take too long. The two most common ways to run the particle swarm as of this writing have been the "inertia weight method," multiplying $v(t-1)$ by a coefficient, and Clerc's constriction method, where the entire right side of the formula is weighted by a coefficient. Both approaches have been described above. In this section we report an empirical comparison between the two approaches and a variation that appears to significantly improve performance (Eberhart and Shi, 2000).

Method

The two versions were tested on a standard set of five test functions used in the cluster experiment above (see Table 8.6). All functions except *f6* were run in 30 dimensions; *f6* is a two-dimensional function.

Procedure

Particle swarms were run until some individual in the population met the criterion for the function and the number of required iterations was

Table 8.6 Functions used in the comparison of inertia and constriction versions.

Function	Xmax	Criterion
Sphere	100	0.01
Rosenbrock	30	100
Rastrigin	5.12	100
Griewank	600	0.05
Shaffer's f6	100	0.00001

recorded. In all trials the population size was set to 30, and the maximum number of iterations was set to 10,000. Each of these values is somewhat arbitrary. The maximum number of iterations was set higher than ever found necessary in previous applications and at the upper limit of the author's [RE's] patience given the speed of his home computer. The neighborhood size was set to include the particle and its two neighbors on each side; the total neighborhood size was thus five, or about 17 percent of the population size.

In all cases for which the inertia weight was used, it was set to 0.9 at the beginning of the run and made to decrease linearly to 0.4 at the maximum number of iterations. Inertia weight cases used a V_{max} set to the maximum range X_{max}. Each of the two $(p - x)$ terms was multiplied by an acceleration constant of 2.0 (times a random number between 0 and 1).

In all cases for which Clerc's constriction method was used, φ was set to 4.1 and the constant coefficient to $\chi = 0.729$, as calculated in the formula given earlier in this chapter. In the initial comparisons, V_{max} was set to 100,000, since it was believed that V_{max} isn't even needed when Clerc's constriction approach is used, and it was expected that no step would approach this unreasonably large value.

Results

The results are shown in Table 8.7. Several trials of the constriction method failed to achieve the criterion. On the Rastrigin function the specified error value was not achieved in one of the 20 runs; that run was terminated after 10,000 iterations with an error of about 125. Also, on the Griewank function the constriction method failed to meet the criterion after 10,000 iterations on three trials.

Table 8.7 Mean number of iterations required to meet the criterion for the two versions on each of the five functions, and the minimum and maximum number of iterations required.

	Inertia version			Constriction version		
	Mean iterations	**Min**	**Max**	**Mean iterations**	**Min**	**Max**
Sphere	1537.8	1485	1615	552.05	503	599
Rosenbrock	3517.35	2866	4506	1424.1	475	4793
Rastrigin	1320.9	743	1704	943*	233	7056
Griewank	2900.5	2556	3891	437*	384	663
f6	512.35	339	748	430.55	105	899

*These means include only trials that did meet the criterion, including 19/20 for the Rastrigin function and 17/20 for the Griewank.

Observations and Improvements

One of the authors [RE] observed that the variance using the constriction method and an essentially infinite V_{max} of 100,000 was much greater than when using the inertia weight method. In fact, the scale of the area used to observe the particles had to be increased by 10 times on both the x and y scales (100 times in area) in order to keep constriction method particles on the screen. It was like watching spacecraft explore the Milky Way galaxy in order to find a target known to be in the solar system.

If you know that the target is in the solar system, it makes sense to limit the distance that can be covered in one time step to the largest dimension of (distance across) the system. We call this maximum distance, that is, the upper limit of allowable parameter values, X_{max}. Note that if we limit our maximum velocity to X_{max}, we are not limiting our exploration to the solar system; our spacecraft particles can still overshoot the system, sometimes by a wide range. But we are limiting our search to at least some reasonable vicinity of the system. It is, of course, assumed that the optimum we are seeking is somewhere within the dynamic range defined by X_{max}.

It was therefore decided to try the constriction method on all of the test functions configured as before except to set $V_{max} = X_{max}$. The results are presented in Table 8.8. The swarm's performance was surprisingly better. All trials met the criterion in this condition.

Table 8.8 Mean number of iterations required to meet the criterion for the constriction version with $V_{max} = X_{max}$ on each of the five functions, and the minimum and maximum number of iterations required.

	Mean iterations	Min	Max
Sphere	529.65	495	573
Rosenbrock	668.75	402	1394
Rastrigin	213.45	161	336
Griewank	312.6	282	366
f6	532.4	94	2046

Conclusion

The constricted particle swarm met the criterion faster on average than the inertia-weighted versions on all functions, but tended to get stuck in local optima. The results of these experiments suggest that an implementation that uses Clerc's constriction coefficient may perform better if it also has a V_{max} limit to keep particles in bounds. The dynamic range of variables X_{max} suggests a reasonable limit that seems to improve performance significantly, and it is usually an available value; that is, it does not require extra calculations or parameterization.

Asymmetric Initialization

Up to this point, we have been initializing the particle swarm symmetrically with respect to the origin, and in most of the test functions the optimal point has also been at the origin. Real life doesn't work that way. Very few optima are at 0,0 and seldom do we know where the optimum is when we initialize our system.

So how does particle swarm function when we initialize it "off-center," that is to say, asymmetrically? To answer this question, one of the authors [YS] ran four of the benchmark functions described earlier (sphere, Rosenbrock, Rastrigin, and Griewank) but initialized them off-center, as indicated in Table 8.9 (Shi and Eberhart, 1999).

Table 8.9 Asymmetric initialization ranges.

Function	Asymmetric initialization range
Sphere	$(50, 100)^n$
Rosenbrock	$(15, 30)^n$
Rastrigin	$(2.56, 5.12)^n$
Griewank	$(300, 600)^n$

Each function was tested in 10, 20, and 30 dimensions. The maximum number of iterations was set to 1000, 1500, and 2000, corresponding to the dimensions 10, 20, and 30, respectively. Different population sizes were used for each function with different dimensions: 20, 40, 80, and 160. A linearly decreasing inertia weight was used that started at 0.9 and ended at 0.4, with acceleration constants φ of 2.0 for each $(p - x)$ term. V_{max} and X_{max} were set to be equal, with values identical to those in the previous section. A total of 50 runs for each experimental setting were conducted.

Results

Table 8.10 lists the mean fitness values of the best particle found in 50 runs for each condition.

Discussion

From these results it appears that asymmetrically initialized particle swarms are indeed able to find the optima of these difficult functions. Increasing population size not surprisingly improves performance in terms of the overall best solution, though of course it has a cost. From these data it cannot be determined whether a better strategy is to distribute the computational cost across a larger population for a smaller number of iterations or to use a smaller population size and a greater number of iterations.

We note that the two experiments reported above took different approaches to benchmarking. In the previous experiment, the number of iterations to a given error value was investigated. In the second one, the error values found after a fixed number of iterations were tabulated. The preferred approach will depend somewhat on knowledge of the problem being investigated. For example, if you don't know much about the

Table 8.10 Mean fitness values for the various functions, with asymmetric initialization.

Population size	Dimensions	Iterations	Mean best fitness			
			Sphere	Rosenbrock	Rastrigin	Griewank
20	10	1000	0.0000	96.1715	5.5572	0.0919
	20	1500	0.0000	214.6764	22.8892	0.0303
	30	2000	0.0000	316.4468	47.2941	0.0182
40	10	1000	0.0000	70.2139	3.5623	0.0862
	20	1500	0.0000	180.9671	16.3504	0.0286
	30	2000	0.0000	299.7061	38.5250	0.0127
80	10	1000	0.0000	36.2945	2.5379	0.0760
	20	1500	0.0000	87.2802	13.4263	0.0288
	30	2000	0.0000	205.5596	29.3063	0.0128
160	10	1000	0.0000	24.4477	1.4943	0.0628
	20	1500	0.0000	72.8190	10.3696	0.0300
	30	2000	0.0000	131.5866	24.0864	0.0127

optimum, running to a constant (large) number of generations can provide information about the performance of the algorithm over time. If the optimum value is known, you may want to see how long it takes (how many iterations) to get to some small error value. Training to an error value may also be preferred if you are designing a system to a specification. For example, if you must meet certain performance criteria for a diagnostic system, you will need to train until these criteria are met, which is analogous to training to a given error. Finally, if your goal is to compare your results with others, you should use the same approach they used.

Some Thoughts on Variations

Several features distinguish the particle swarm from related adaptive dynamical systems. First, in especial contrast to Darwinistic evolutionary computation paradigms, the particle swarm is a cooperative approach to population problem solving. Evolutionary methods are typified by a

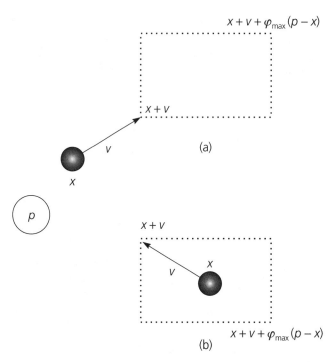

Figure 8.18 When a particle is moving away from *p*, its next step takes it into an area that does not include its current position (a). If it is returning toward *p*, though, it may end up at, or very near to, its current position (b). Note that it may be in different parts of its cycle on different dimensions.

fundamentally competitive philosophy, evoked by the phrase "survival of the fittest." In particle swarms, on the other hand, all individuals survive, all persist from one time to the next. Rather than adapting through differential reproduction, individuals (who are guaranteed survival) change over time, and one particle's successful adaptation is quickly shared and reflected in the performance of its neighbors. Axelrod might call the particle swarm algorithm "nice."

A particle with coordinates $\vec{x}_i$ will, on the next step, move into a rectilinear box that is bounded by corners $\vec{x}_i(t) + \vec{v}_i(t)$ and $\vec{x}_i + \vec{v}_i(t) + \varphi_{1\max}(\vec{p}_i - \vec{x}_i) + \varphi_{2\max}(\vec{p}_g - \vec{x}_i)$; the target region is bounded by corners where the random numbers are 0.0 and those where they are 1.0. As seen in Figure 8.18, this box may contain the current position of the particle, or exclude it, in which case the particle has to "think outside the box." (Attention future readers: please forgive this quaint turn-of-the-century colloquialism.) Or perhaps the box has to think outside the particle.

Some "swarm" and social-theoretic methods are said to be derived from the cellular automaton concept, where the locations of agents are defined in terms of cells in a grid. There is no theoretical gap between the CA and particle swarm views; in fact it seems that thinking of the swarm as a cellular automaton illuminates some aspects of the system's behavior. But first—it is not clear to us that some models where agents are defined as objects moving across a lattice are really cellular automata at all. In a traditional CA, rules determine what happens next for a fixed cell depending on its neighborhood configuration, and the miracle is the emergence of phenomena such as gliders—that is, individual entities—from the rules. Some applications, though, define an agent as a cell that is turned on, and operate by moving the "glider" or agent across the lattice directly. In other words, a cell on the grid is a location, and an agent moves from one location to another—but it seems to us that in a traditional CA the cells in the grid are the individuals themselves, and the dynamics of the system comprise their changes of state.

In the particle swarm each individual is conceived to be fixed in a physical or topological location. It interacts with the same neighbors throughout the trial and does not move in either topological or physical space. All its movements take place in the parameter space that defines its state and that is essentially independent of the topological space in which the particle resides. Thus in some paradigms the movements of individuals on the lattice determine the sequence of problem solutions to be tested. In the particle swarm the individuals stay where they are, as do cells in a CA, but they change states—as do cells in a CA—depending on their previous states and the states of their neighbors.

Are Particle Swarms Really a Kind of Evolutionary Algorithm?

The particle swarm algorithm has been influenced by concepts from evolutionary computation (EC) since its inception, and continues to be. The use of populations of simple individual problem-solution vectors derives directly from the traditions of EC, and features of the particle swarm algorithm can be seen to resemble evolutionary operators in important ways. There are also important differences. The EC community has welcomed particle swarming as an alternative way of doing things, and particle swarm researchers consider themselves to be coparticipants in an important joint enterprise—to the point of considering particle swarms to be a kind of evolutionary algorithm.

Evolution beyond Darwin

We have commented that the distinguishing feature of traditional Darwinistic evolution is selection, the survival of the fittest members of each generation, with the subsequent propagation of characteristics in proportion to their fitness. This is not exactly an uncontested viewpoint. Some theorists, of whom Stuart Kauffman is probably the most visible, contend that self-organization and emergence are as important as selection in the coming together of life and its subsequent evolutionary development.

The Neo-Darwinian View of Evolution and Its Shortcomings

What is usually described as the Darwinian view of evolution is perhaps better described as the neo-Darwinian view. For example, chromosomes weren't even known in Darwin's time, so the presently prevailing view is a sort of an amalgam of Darwinian and Mendelian ideas. (In 1865 Gregor Johann Mendel, an Augustinian priest in Brno Monastery in the Czech Republic, described to the Brno Natural Science Society the transfer of genetic material in pea plants. Unfortunately, the fundamental importance of Mendel's finding was not understood by the Society. Until about 1900 it was not recognized that Mendel had discovered the "law of heredity.")

The neo-Darwinian view of evolution reflects three main observations. First is that chromosome composition is determined by the parents (at least in animals and humans). Second is that random mutation expands the search space of the species, providing the desirable attribute of diversity. Third, fitter individuals have a higher probability of surviving to the next generation.

According to modern researchers including Kauffman, there are two fundamental shortcomings of the existing theory. The first is that the origin of life by "chance" or mutation is highly improbable in the time frame of earth's history. The second is that evolution of complex life forms solely through mutation is also highly improbable. A detailed discussion of these points is beyond the scope of this book. For compelling arguments, see Kauffman (1993, 1995).

Self-Organization and the New View of Evolution

Enter self-organization. There are almost as many ways to define self-organization as there are writers on the subject, but a summary of attributes and descriptions of self-organization usually include the following points:

- Self-organizing systems usually exhibit what appears to be spontaneous order.

- Self-organization can be viewed as a system's incessant attempts to organize itself into ever more complex structures, even in the face of the incessant forces of dissolution described by the second law of thermodynamics.

- The overall system state of a self-organizing system is an emergent property of the system.

- Interconnected system components become organized in a productive or meaningful way based on local information.

- Complex systems can self-organize.

- The self-organization process works near the "edge of chaos."

For example, anthropologist Jeffrey McKee (2000) has described the evolution of the human brain as a self-organizing process. He uses the term *autocatalysis* to describe how the design of an organism's features at one point in time affects or even determines the kinds of designs it can change into later. For example, when early hominids began walking upright, the angle of the skull on the top of the spine—instead of level with it—left some extra room for the brain to expand. Thus the evolution of the organism is determined not only by selection pressures but by the constraints and opportunities offered by the structures that have evolved so far.

This leads to a new view of evolution, in which, due primarily to self-organization, complex systems can "appear" over a relatively short time frame compared with Darwinian evolution. In this new perception of evolution, it appears that natural selection and self-organization work hand in hand, that is,

$$evolution = natural\ selection + self\text{-}organization$$

Selection and Self-Organization

Selection is not generally considered to play a part in the particle swarm algorithm. Whereas in evolution, individuals' differential survival determines the nature of the next generation, all particle swarm individuals survive. The word "learning" is sometimes applied to evolutionary populations, as if the population itself were learning to solve a problem, but learning is more usually a kind of change in an individual over time. It is the acquisition and retention of knowledge. The word "adaptation" seems more appropriate to evolutionary populations. A particle swarm models changes in—not replacement of—individuals in time; in this important way the particle swarm approach differs from all the EC approaches that have been discussed in this book (Angeline, 1998b).

If you were to try to fit particle swarms into the evolutionary mold, it *could* be argued that all of the particles from one iteration are replaced by those of the next. That is, all of the particles could be said to "die" at each point in time and be replaced by their "children" in the next moment—as Freud commented, "The child is the father of the man." In this view, particle swarms are somewhat similar to evolutionary programming, in which each population member is mutated to produce a (candidate) population member for the next generation. In EP, as in particle swarm optimization, there is a clear "ancestral" path for each population member across time, except that in EP some parents may leave no descendants and some may survive to the next generation alongside their offspring. In particle swarms there is a one-to-one correspondence between individuals at one iteration and those in the next. Thus, in EP lineages die out, which cannot happen in a particle swarm. In the (1,1)-ES strategy (which is seldom used in evolution strategies but is nonetheless a valid variation), each parent produces one child that replaces it in the next generation. This replacement strategy can be seen as very similar to that used in particle swarms, as can the (1+1)-ES, where the survivor is either the child or the parent (now considered to be a clone of the parent).

One major difference between particle swarms and traditional evolutionary computation methods is that particles' *velocities* are adjusted, while evolutionary individuals' *positions* are acted upon; it is as if we were altering the "fate" rather than the "state" of particle swarm individuals. In EAs that self-adapt step sizes, the variance of the step and not its direction is optimized—it is not meaningful to speak of an evolutionary individual's "trajectory," even if individuals were considered to persist across generations. If the model above where the inertia weight was set to 0.0 were considered, that is, where position alone was acted upon by the particle swarm formula, the similarity between particle swarms and evolutionary methods, especially EP and (1,1)-ES, would be even greater. The main differences would be the directional component of the "muta-tion," repeated interaction with the same neighbors, and the fact that EP and ES step size is evolved in response to the fitness of the current param-eters, while particle swarm velocities are adjusted in relation to a previ-ously discovered relatively optimal point. Explicit velocity cannot exist when selection is practiced, as it presumes the conservation of an object's identity over time.

The distinctions and similarities between evolutionary and socio-cognitive processes provide the central theme of this book. If we are to seek analogues and similarities, we note that particle swarms implement a kind of operator that mixes the effects of crossover and mutation, and adds directionality as well as a feature that is similar to momentum—a tendency for the particle to continue in the direction it was going on the previous time step. In the particle swarm algorithm the mutation and crossover operations are performed simultaneously. Not only is the effect of the "borrowed" information $\vec{p}_g$ randomly weighted, but the individ-ual's previous best $\vec{p}_i$ is too. We can consider the first of these to be a kind of crossover and the second to be a kind of mutation, if we wish to stretch the evolutionary metaphor to cover this other thing.

Particle swarm step size is controlled by three factors. First, the step size of the current iteration is derived from the step size in the previous one. Whichever direction and distance per iteration the particle was moving, it has a tendency to continue. Second, the distances from the current position to the previous bests $(p - x)$ scale the change in velocity for the time step. Often this means that the particle moves faster when it is near the optimum, then slows as it moves farther from it, until the par-ticle changes direction and approaches the weighted average best from the other side. This is because, when x is near p, the term $(p - x)$ is small, and as it is added to v, the smaller term produces a greater tendency for v to remain as it was. Third, all particle swarm implementations have some method for limiting velocity, whether it is the inertia weight, a constric-tion coefficient, or simply V_{max}. The common usage of the first two of these results in a convergence of the individual particle's trajectory to-ward the optimum, while V_{max} limiting only prevents it from escaping its orbit, without inducing convergence; V_{max} versions are better at explora-tion than exploitation.

Particle swarms have a unique way of using gradient information to guide their search. In fact the $(p - x)$ terms in the algorithm define a gra-dient; they show the direction and distance from the point that is cur-rently being tested and the point that has previously been found to be best. As the effect of these terms is to adjust the velocity of the particle, the effect on its trajectory is not to move it with the gradient, but to cause the particle to oscillate around the previous best point—sometimes with the gradient and sometimes against it. The gradient then is used simply to keep the particle oriented in the region of previous successes, similar to methods in EC, and is not followed in the traditional way of hill climbers.

A particle moves in a stochastic oscillatory trajectory through the problem space, sampling around relatively optimal local points, while an evolutionary individual searches by changing position through mutation and crossover. This perpetuated directional movement through the search space gives particles their characteristic behavior; their interaction results in effective search for optima.

A particle swarm is a self-organizing system whose global dynamics emerge from local rules. As each individual trajectory is adjusted toward the successes of neighbors, the population converges or clusters in optimal regions of the search space. The search would fail if individuals did not influence one another; because a number of them are sharing information locally, it is possible to discover optima in the landscape. Because of this reliance on self-organization, we are comfortable comparing particle swarms to evolutionary algorithms and usually do not see any reason to draw a line between the two approaches, as long as the differences, especially regarding selection, are understood. The two approaches are just not linearly separable.

This is the last example in this book of an argument about whether a phenomenon should be correctly categorized as a such-and-such. We have seen throughout the discussion that sociocognitive and Darwinian processes are similar enough to lead a good number of very knowledgeable people to wonder if they are identical. We quibble with the memetic theorists, but the issues are probably academic; culture and mind are a lot like the evolution of species, even if they are a lot different.

Ergodicity: Where Can It Get from Here?

It is theoretically possible for a GA chromosome to reach any point in the problem space via mutation. It is, however, unlikely, particularly near the end of a run. This is because a number of mutations will likely be needed to reach a distant point. Since mutation rates are typically quite low (0.1–1.0 percent is a common range), several generations of favorable mutations may be needed. Near the end of a run, however, when the population has converged and the average fitness value is high, mutation will quite likely result in a low-fitness chromosome that does not survive the selection process. In fact, the probability of survival decreases geometrically with generations. So even though a number of mutations could bring the chromosome into a high-fitness region, the chromosome never gets there because it doesn't survive selection.

So, even though a GA is *theoretically* ergodic (there exists a nonzero probability that a chromosome can occupy any state, or, stated another way, there is a zero probability that any given state cannot recur), it is not ergodic in a practical sense because of the multiple steps required. An evolutionary programming (EP) system is truly ergodic, since there is a finite probability that an individual can reach any point in problem space with one jump (in one generation).

The behavior of particle swarm systems seems to fall somewhere between GA and EP systems in this regard. It may be that a particle cannot reach any point in problem space in one iteration, although this might be possible at some point of the run, given sufficiently large velocity. But any particle can eventually go anywhere, given enough iterations and an appropriate set of parameters. A stronger case can thus be made for the ergodicity of particle swarms than for GAs.

Convergence of Evolutionary Computation and Particle Swarms

In reviewing the literature of recent EC research, it occurs to us that many of the changes being attempted with evolutionary algorithms have a sociocognitive or "persisting-individuals" aspect to them. Various researchers have implemented versions of evolutionary algorithms, for instance, "group memories," in which some subset of superior individuals is allowed to persist across generations, to interact with population members. In GA, "elitism" is a term for the retention of the most-fit individual from a generation to the next; we tend to think of these collective-memory methods as elaborations on elitism. Parasite and coevolutionary methods, such as the "virus" algorithm, provide another way that successful problem solutions and schemata can be retained for future interaction with population members. Numerous independent researchers have come up with methods for keeping individuals across generations.

In sum, evolutionary and social algorithms have much in common, and their fusion seems inevitable. Some evolutionary operations are markedly analogous to particle swarm methods, but the differences between approaches are considerable.

Summary

In this book we are looking at a paradigm in its youth, full of potential and fertile with new ideas and new perspectives. Researchers in many countries are experimenting with particle swarms, applying their imaginations to this simple system. The algorithm is really just a pair of formulas, one for velocity and one for position, a two-line computer program with some housekeeping (in fact these two lines can be collapsed into one). Even so, the trajectories of particles are rich and interesting, even when they are stripped down to the bare minimum. Who knows what will be tried next?

Many of the questions that have been asked have not yet been satisfactorily answered. The interaction of sociometry with function, for instance, obviously depends on some aspect of the problem—but what? After more investigation perhaps researchers will be able to parameterize the swarm appropriately depending on particular problem characteristics—or perhaps the parameterization can be built into the algorithm itself, so that a program can assess aspects of the problem and structure the swarm topology and relevant parameters to optimize performance. Researchers who want to use the algorithm for solving problems might want to know how best to set up the swarm so that explosion and convergence are held in optimal balance. For instance, it may turn out to be best to keep the system at the edge of chaos, where it's gnarly (Rucker, 1999).

chapter
nine

Applications

In the last chapter, we saw how particle swarm optimization works. But we were still in the world of mathematics. In this chapter, we move into the practical world of engineering and take a brief look at some applications of particle swarm optimization.

There are various kinds of applications. Some are more general; others are more specific. By "general" we mean approaches that can be used across a wide range of specific applications. Specific applications are those that are tailored to a specific requirement. Even though they are focused on solving a particular problem, however, we can often see how to apply the concepts to other problems.

We begin with a section on a general application: evolving neural networks. Evolving artificial neural networks, including not only the network weights but also the network structure, is one of the most exciting early applications of particle swarm optimization. The method is so simple and efficient that we have almost completely ceased using traditional neural network training paradigms such as backpropagation. Instead, we evolve our networks using particle swarms. The approach works for any network topology (e.g., feedforward, recurrent, etc.) and any training algorithm (e.g., radial basis function, backpropagation, etc.).

In the next section, we look at a specific application in the area of biomedical engineering: diagnosis of human tremor. Here, we use particle swarm optimization to evolve the weights of a feedforward neural network that can differentiate between normal (physiologic) and abnormal (pathologic) tremor.

Finally, in the third section, we briefly review four other engineering applications in widely diverse areas: ingredient mix optimization, battery state-of-charge estimation, electric power and voltage management, and computer numerical control machine optimization.

It's time to put on your hard hats! We tried to minimize the "engineer-ese," but this chapter is decidedly applications oriented, and the writing unashamedly reflects that.

■

Evolving Neural Networks with Particle Swarms

Neural network and evolutionary computation methodologies have each been proven effective in solving certain classes of problems. For example, neural networks are very good at mapping input vectors (patterns) to outputs in such applications as diagnostic systems; evolutionary algorithms are very good at optimization in applications such as scheduling systems. It was therefore natural for engineers and scientists to combine the methodologies to develop hybrid computational tools that are even more effective than either methodology by itself.

Since the popularization of the backpropagation algorithm in the mid-1980s (Rumelhart and McClelland, 1986; Werbos, 1974), there has been a significant increase in research and development in the area of applying evolutionary computation (EC) techniques for the purposes of evolving one or more aspects of artificial neural networks. Publications you might want to look at that review these efforts include Schaffer, Whitley, and Eshelman (1992), Yao (1995, 1997), and Fogel (1998).

These EC techniques have usually been used to evolve neural network weights, but sometimes have been used to evolve network structure or the network learning algorithm. In the next section we provide a brief review of previously published work. In the following section, we discuss advantages and disadvantages of the approaches described previously. Then we present the concept of applying particle swarm optimization to replace the learning algorithm and evolve both the weights and structure of a neural network. In the final section, we provide an illustration of the approach by means of a classification system for the Iris Data Set.

Review of Previous Work

Evolutionary computation methodologies have generally been applied to three main attributes of neural networks: network connection weights, network architecture, and network learning algorithms. A fourth area, the evolution of inputs (finding the optimal set of inputs), has received a relatively minor amount of attention.

With respect to the architecture of a neural network, evolutionary algorithms (EAs) have been applied to evolve the network weights, the network topology (structure), and the processing element (PE) transfer function. Occasionally, they have been used for more than one purpose, for instance, evolving the network weights and the structure simultaneously. Furthermore, EC methodologies are sometimes used in combinations and sometimes with other methodologies. For example, it is possible for an EA such as a GA to find a set of weights in the global minimum's basin of attraction. A greedy local search algorithm can then be used to find the globally optimal weight matrix (Yao, 1995). A number of approaches have been used to encode the weights into the chromosome of a GA. Included are direct encoding schemes, in which each weight is explicitly represented in the chromosome, and indirect schemes, in which a compression scheme is used that requires an expansion of the chromosome to derive the individual weights. We discuss specific examples of these approaches below. We intend that the examples chosen be only representative; an exhaustive survey is beyond the scope of this section.

As early as 1968, Bremmermann, a pioneer in the EC field, suggested that "we should be encouraged to try [evolutionary search] procedures on more complex problems, where no efficient algorithms are known (e.g., searching for strategies, optimizing 'weights' in a multilayer neural net, etc.)." Widespread efforts to evolve neural networks, however, did not occur until the popularization of the backpropagation algorithm.

One of the first published works that described use of a GA and included example applications was by Whitley (1989), in which a GA was used to evolve the weights in a feedforward neural network. He applied the technique to relatively small problems, such as the exclusive-or (XOR). Also in 1989, Montana and Davis described the use of a GA to train a neural network of approximately 500 weights. It wasn't a "traditional" GA in that, instead of replacing the entire population each generation, only one or two individuals were produced, which then had to compete to be included in the new population. Also, network weights were represented by real, rather than binary, numbers. This type of implementation is known as a "steady-state" GA. Furthermore, Montana and Davis' paradigm included an option for improving population members using backpropagation. This was thus a truly hybrid approach. (This hill-climbing capability, however, did not result in better results than when using the GA alone.)

Another promising early result was that of Schaffer, Carvana, and Eshelman (1990), which demonstrated that an evolved neural network had better generalization performance than one designed by a human and trained with backpropagation. A number of similar papers were also published. The network training times reported were sometimes faster and sometimes slower than backpropagation, but were generally not as fast as network training algorithms noted for their speed, such as quickprop.

Most of the work involving the evolution of network architecture has focused on the network topological structure. Relatively little has been done on the evolution of neural network processing element (PE) transfer functions and even less on evolving topological structure and PE transfer functions simultaneously.

Two of the general (nonevolutionary) approaches used to evolve network topology are *constructive* and *destructive* algorithms. A constructive algorithm starts with a minimal topology and evolves the appropriate topology by adding weights, PEs, and layers, as needed. The destructive approach starts with a large network and evolves the appropriate topology by removing weights, PEs, and/or layers. EAs have been shown to be superior to these approaches because of the large (often infinite) size, nondifferentiability, complexity, and multimodality of the search space (Yao, 1995).

Reduced (indirect) coding schemes have been developed in which parameters that specify the network topology are evolved. This approach often involves a discrete number (limited set) of architectures. Other times, the number of PEs and/or the number of hidden layers is encoded (Caudell, 1990). These approaches result in chromosome discontinuities between any two network configurations.

Another approach is to evolve developmental rules used to construct the network topology. Kitano (1990) evolved a graph generation grammar, or rules for generating weight connection matrices. His grammar included rules for obtaining 2×2 matrices from 1×1 matrices, 4×4 matrices from 2×2 matrices, and so on, until a matrix of the size necessary to specify the weight connectivity for the network was obtained. Although Kitano reported better results than some direct encoding methods, his method is not very good at fine-tuning connections among single nodes.

Perhaps the first publication reporting the evolution of both network topology and PE transfer functions using a GA was that of Stork et al. (1990). They were modeling a biological neuron in the tail-flip circuitry of a crayfish. Although the network had only seven PEs, the transfer function evolved was the very complex Hodgkin-Huxley equation for neuronal activity. Chromosomes included coded specifications for neuron type, cell surface molecules, neurotransmitter type, synapse receptor types, cell channel densities, and other functional properties of the network.

Koza and Rice (1991) used the genetic programming paradigm to find both the weights and the topology (number of layers, number of PEs per layer, and weight connectivity pattern) of a neural network. They encoded a tree structure of Lisp S-expressions in the chromosome. Special crossover and mutation operators were used that preserved the syntax. This may be the first published report of using genetic programming to evolve neural networks.

Some investigators have investigated the optimization of the EA operators used to evolve neural networks. Research reported by Whitley, Dominic, and Das (1991) indicated that hill-climbing capabilities of GAs using real-valued encoding for the network weights were increased significantly by a combination of increasing the mutation rate, decreasing the crossover rate, and decreasing the population size. Convergence was faster, too, but the probability of obtaining a usable solution decreased by about 10 percent. It should be noted that "steady-state" GAs similar to those of Montana and Davis (1989) were used, resulting in relatively monotonic searches. This type of GA is referred to as a "genetic hill climber" (Schaffer, Whitley, and Eshelman, 1992). GAs have thus been designed that emphasize either global *or* local search. The trick, of course, is knowing which to use for a particular problem, or, perhaps more importantly, how and when to switch from one to the other when solving a problem.

It seems that relatively few researchers, when reporting their work, provide quantitative comparisons with other approaches. For example, how well did their network with evolved weights perform compared to a network with weights trained using backpropagation? Included in the term *performance* are both a performance metric, such as percent correct, and the speed of computation, expressed in such a way as to be comparable with other approaches. How well was the network able to generalize? How long did it take to train the network? Information given should allow the reader not only to reproduce the work described in the paper but to allow meaningful comparisons with other techniques.

Furthermore, the results obtained depend on the specific algorithms being compared. It is unfair to compare speed of convergence between a standard EA and a fast backpropagation algorithm or between a fast EA and a standard backpropagation algorithm. It is also important to recognize that all of these algorithms are sensitive to the parameters and operators used in them, and that some, particularly backpropagation, are quite sensitive to initial conditions.

Comparisons are therefore valid only if the best available version of each algorithm is used, complete and quantitative results are reported, and sufficient information is given so that the work can be reproduced.

Advantages and Disadvantages of Previous Approaches

In this section, we briefly summarize some of the advantages and disadvantages that have been discussed in the literature and that have been experienced with respect to using EC techniques with artificial neural networks. The discussion is not meant to be thorough. Rather, our intent is to highlight the successes and to examine issues that should be addressed in order to make further progress. It is not our intent in this section to review the advantages and disadvantages of neural networks and evolutionary algorithms individually. Such reviews appear in a number of other places (Schaffer, Whitley, and Eshelman, 1992; Yao, 1995).

Advantages

The backpropagation neural network learning algorithm, as well as others, requires a differentiable PE transfer function. EAs can be used to train neural networks with nondifferentiable (even discontinuous) PE transfer functions. Step functions are an example. Additionally, not all of the transfer functions have to be identical in a network trained by an EA.

EAs can also be used in cases where gradient or error information is not available (Schaffer, Whitley, and Eshelman, 1992). (See, however, a statement from the same reference in the section below on disadvantages.) EAs can thus be applied to neural networks using many architectures and topologies. In addition to backpropagation, EAs have been applied to networks using a variety of learning algorithms, including reinforcement learning, recurrent learning, and higher-order learning. EAs have the capability to perform a global search in the problem space.

We can define the fitness of an architecture evolved by an EA in a way appropriate for the problem. For example, speed of learning, topological complexity, and performance on the test set can all be incorporated into the fitness function. Furthermore, the fitness function does not have to be continuous or differentiable.

Disadvantages

Schaffer, Whitley, and Eshelman (1992) state: "Using a genetic algorithm as a replacement for back propagation does not seem to be competitive with the best gradient methods (e.g., quickprop)." GAs are known to perform global search quite well, but to be relatively inefficient in fine-tuned local search (Yao, 1995).

Evolution of network topology is generally done in ways that result in discontinuities in the search space. Examples include removing and inserting connections (weights), discrete changes in connections (weights), from 1 to -1, for example, and removing and inserting PEs. These discontinuities usually require retraining of the network. Since the training of a backpropagation network is sensitive to the randomized initial weights, the fitness value used to measure the network's performance reflects noise as well as the network architecture. It is therefore usually necessary to train the network several times and compute an average fitness value, or partially train the network a number of times to get an indication of convergence rates. Either approach is computationally intensive.

Selection of a representation for the weights in a chromosome is often difficult. In addition to the basic decision whether to use binary or real representations, the ordering of the weights must be considered, especially if an EA that uses crossover or recombination is being used. For instance, should the heuristic (Yao, 1995) that weights connecting into the same hidden PE be adjacent in the chromosome be implemented? If binary encoding is selected, which encoding method should be selected (uniform, Gray, exponential, etc.)? Once the representation is selected, the genetic operators (crossover, mutation, etc.) and their parameter values must be selected or, in many cases, developed. Often, operators are designed specifically for a problem.

If a real-number representation for weights is used, a set of operators must be selected or developed. These must generally be tailored to the application. In addition, the criterion used for selection must be specified.

Finally, a problem that has consistently been reported in the literature is the *permutation problem* (Yao, 1995; Hancock, 1992), also referred to as the *competing conventions problem* (Schaffer, Whitley, and Eshelman, 1992) and the *isomorphism problem* (Hancock, 1992). This situation arises whenever there exist multiple chromosome configurations that represent equivalent optimum solutions. These configurations are called *permutations* or *competing conventions*, and the error surfaces are multimodal. For example, two networks that have a different order to their hidden PEs (and thus have a different representation on the chromosome) but are otherwise identical are equivalent. In fact, any permutation of the hidden PEs produces an equivalent network in this case.

Hancock's work was limited to the specification of the network connectivity, not the weights associated with the connections. Nonetheless, he reported that "the most unexpected result here was that permutations are apparently more of a help than a hindrance," and that "it appears that, in practice, the permutation or competing conventions problem is not as severe as had been supposed" (Hancock, 1992). We address this "problem" area in our conclusions.

The Particle Swarm Optimization Implementation Used Here

Following is a brief description of the particle swarm implementation used in this chapter to evolve artificial neural networks. For a more detailed description of particle swarm optimization, see Chapter 7.

Each particle is treated as a point in an n-dimensional space. The ith particle is represented as $x_i = (x_{i1}, x_{i2}, ..., x_{in})$. The best previous position *pbest* of the ith particle is recorded and represented as $p_i = (p_{i1}, p_{i2}, ..., p_{in})$. The index of the best particle among all the particles in the population (global model) is represented by the subscript g. The index of the best particle among all the particles in a defined topological neighborhood (local model) is represented by the subscript l. The rate of the position change (velocity) for particle i is represented by $v_i = (v_{i1}, v_{i2}, ..., v_{in})$. The particles are manipulated according to the following equations (global model):

$$v_{id} = w_i * v_{id} + c_1 * \text{rand}() * (p_{id} - x_{id}) + c_2 * \text{Rand}() * (p_{gd} - x_{id}) \qquad (8.1)$$

$$x_{id} = x_{id} + v_{id} \qquad (8.2)$$

where d is the dimension ($1 \leq d \leq n$), c_1 and c_2 are positive constants, rand() and Rand() are two random functions in the range $[0,1]$, and w is the inertia weight. (Note that we use the inertia weight version of the particle swarm algorithm here.) For the neighborhood (lbest) model, the only change is to substitute p_{ld} for p_{gd} in Equation 8.1. For the global model, Equation 8.1 is used to calculate a particle's new velocity according to its previous velocity and the distances of its current position from its own best experience (pbest) and the group's best experience (gbest).

The local model calculation is identical, except that the neighborhood's best experience is used instead of the group's best experience.

The performance of each particle is measured according to a predefined fitness function, which is related to the problem to be solved. The inertia weight w controls the impact of the previous histories of velocities on the current velocity, thus influencing the trade-off between global (wide-ranging) and local (nearby) exploration (exploitation) abilities of the "flying points." A larger inertia weight facilitates global exploration (searching new areas), while a smaller inertia weight tends to facilitate local exploration to fine-tune the current search area. Suitable choices of the inertia weight provide a balance between global and local exploration abilities and thus require fewer iterations on average to find the optimum (Shi and Eberhart, 1998a). Although experimentation with the inertia weight is still in progress, it appears that a good general approach is to decrease the inertia weight linearly from 0.9 to 0.4 over 1,000 generations (Shi and Eberhart, 1998b). The same inertia weight is used for all dimensions of all particles in a given generation.

Implementing Neural Network Evolution

The benefits of evolving attributes of neural networks are clear. Multilayer perceptrons (feedforward networks using the backpropagation algorithm as the learning algorithm) have been shown to be capable of being universal approximators (Hornick, Stinchcombe, and White, 1989). The most common transfer function used is the sigmoidal function:

$$output = 1 \,/\, (1 + e^{-input})$$

Radial basis function networks, which generally use a form of a Gaussian function as the transfer function, have also been shown to be capable of serving as universal approximators (Poggio and Girosi, 1990). The idea of being able to automatically evolve a universal approximator is quite attractive, especially if it can be done as (or more) quickly as training the network with back propagation.

One of the first uses of particle swarm optimization (PSO) was for evolving neural network weights. Eberhart, Simpson, and Dobbins (1996) reported using PSO to replace the backpropagation learning algorithm in a multilayer perceptron.

The development being reported in this section is the use of PSO to evolve the network weights and, indirectly, to evolve the structure. The methodology has the additional benefit of making the preprocessing (such as normalization or scaling) of input data unnecessary.

This is accomplished by evolving, in addition to the network weights, the slopes of the sigmoidal transfer functions of the hidden and output PEs of a feedforward network. In other words, if we now consider the transfer function to be

$$output = 1 / (1 + e^{-k*input})$$

then we are evolving k in addition to evolving the weights. (The method is quite general and can be applied to other network topologies such as recurrent networks and to other transfer functions such as radial basis functions.)

Slopes are allowed to be either positive or negative. The output of a transfer function with a negative slope is just one minus the output with a positive slope of the same absolute value. The effect of a transfer function with a negative slope is identical to that of a transfer function with a positive slope (with the same absolute value) if the signs of the input weights are reversed. There is thus no reason to constrain slopes to be positive, and by allowing them to take on negative values, the flexibility of the network evolution is increased, resulting in faster convergence.

This method can be used to evolve the network structure indirectly. If the evolved slope is sufficiently small (the exact amount depends on the application), then the output is essentially constant regardless of the input. (In the case of the sigmoidal transfer function, the output would be 0.5, or very nearly so.) If the PE is in a hidden layer, it can therefore be removed. Its effect can be replicated by increasing the weights from the bias PE in that hidden layer to each of the PEs in the next layer by one-half the value of each weight from the PE being removed to the next-layer PEs. The method therefore can be used to prune PEs from the network, reducing network complexity.

Additionally, if the slope is sufficiently large (the exact amount depends on the application) then the sigmoid transfer function can be replaced by a step transfer function. A sigmoid with a large positive slope is thus replaced by a step transfer function that has an output of 0 for inputs less than or equal to 0 and 1 for positive inputs. A sigmoid with a large negative slope is replaced by a step function with an output of 1 for inputs ≤ 0 and 0 for positive inputs. Sigmoidal function PEs can thus evolve to be step function PEs, reducing the computational complexity of the network significantly.

Since the slopes can evolve to large values (relative to 1, which is the slope used in traditional backpropagation networks), input normalization or scaling is not needed. Since data preprocessing requires a significant amount of effort in most applications, this methodology can simplify the applications process and shorten development time.

Another feature of this methodology is the continuous nature of the PSO algorithm. Transfer function slopes are evolved in a continuous way; that is, slopes can vary continuously from large negative to large positive values. This results in an evolution of network structures that is also essentially continuous in nature. For example, as a hidden PE's transfer function slope approaches zero, it is replaced with revised connection weights from the bias PE; as the slope becomes very large, the sigmoidal PE is replaced by a threshold PE. No sudden significant discontinuities exist in the evolutionary process such as those that plague other approaches to evolving network structures.

An Example Application

The methodology described above was first tested on the Iris Data Set introduced by Anderson (1935) and popularized by Fisher (1936). This data set is frequently used as a benchmark for classification algorithms (Eberhart, Simpson, and Dobbins, 1996). Measurements of four attributes of iris flowers are provided in each data set record: sepal length, sepal width, petal length, and petal width. Fifty sets of measurements are present for each of three varieties of iris flowers, for a total of 150 records, or patterns.

We used both normalized and unnormalized data versions of the data set. The unnormalized version was transcribed from Fisher's original paper (Fisher, 1936). In the results presented below, all 150 patterns were used to evolve a network. The issue of generalization was thus not addressed; it is being addressed in work currently ongoing. For all of the results reported in this chapter, values of $-k*input$ within the transfer function that exceeded 100 (in other words, values greater than e^{100} in the denominator of the transfer function) were clamped such that the PE transfer function output was zero. Stated another way, in these cases the denominator was arbitrarily set to infinity, so that the output would be zero. This was done to avoid computational overflow errors. This condition can arise due to a large negative slope with large positive inputs or with a large positive slope and large negative inputs.

Table 9.1 Performance variations with slope thresholds. For each threshold value, 40 runs of 1,000 generations each were made using the 150-pattern data set.

Slope threshold s (absolute value)	Total number correct in 40 runs	Average number correct per run	Variance
None	5914	147.85	1.57
80	5914	147.85	1.57
40	5911	147.78	1.77
20	5904	147.60	1.94
10	5894	147.35	2.08
5	5882	147.05	2.25
4	5814	145.35	62.75
3	5811	145.28	62.56
2	5782	144.55	69.43
1	5693	142.33	126.64

A normalized version of the data set was first used to test the concept of evolving both weights and slopes. A high degree of success in dozens of runs led to our next investigating the threshold value for slope at which the sigmoidal transfer function could be transitioned into a step function without significant loss of performance. The preliminary results of this effort are summarized in Table 9.1. The values for s in the table are the absolute slope values above which the slope was set to infinity; that is, a step function was substituted for the sigmoidal transfer function. For example, for the data corresponding to $s = 20$, each PE transfer function with an absolute value of evolved slope greater than 20 was changed into a step function.

Since there are 150 patterns, 150 minus the average number of correct classifications per run yields the average number of errors per run. The results in the table indicate that the average number of errors was 2.15 out of 150 patterns when both weights and slopes were evolved (no slope threshold was implemented). This is a very good result for this data set. This table also shows that the accuracy degrades gracefully as the slope threshold is decreased to 5. As the slope threshold decreases from 5 to 4, there is an increase in the average number of errors from 2.95 to 4.65 and an increase in the variance from 2.25 to 62.75.

Table 9.2 Performance for 40 runs of 1000 generations each with the unnormalized data set.

Number correct	149	148	147	146	145	144	100	99
Number of runs with this number correct	11	16	6	3	1	1	1	1

This indicates that there were a few of the 40 runs that did not arrive at good solutions. (It is recognized that variance is defined for data sets with normal distributions, and our data sets do not meet this criterion. It seems to provide a useful metric for our work despite our skewed data sets.)

This experiment thus provides a preliminary indication that the slopes can be evolved and that a slope threshold of around 10 to 20 would be a reasonable value (minimal impact on performance) for this problem. Other data sets are now being examined with the same idea in mind. Also, the transition of low values of slope to 0, thus enabling the elimination of the PE, are being examined. Although some low values for slope were obtained in this experiment, there were too few of them to provide a statistically meaningful sample from which conclusions can be drawn.

One set of runs was made with the unnormalized data set. No slope threshold was implemented. The results, summarized in Table 9.2, show that a reasonably good solution was obtained in 38 of the 40 runs, or 95 percent of the time. The average number correct was 145.45, and if the two worst solutions are ignored, the average number correct is about 148, a very good result. This means that 95 percent of the time, using an unnormalized data set, and evolving both slopes and weights, solutions with an average of only two errors were evolved. The results shown in Table 9.2 are also only an indication of the potential for the method. Statistically valid conclusions await further experiments.

Conclusions

A brief review of prior work in using EC techniques to evolve attributes of neural networks has been presented. Advantages and disadvantages of these approaches were summarized. A new methodology using particle swarm optimization for evolving neural network weights and simultaneously indirectly evolving network architecture was presented. The methodology seems to overcome the first four disadvantages to previous approaches listed earlier. With respect to the fifth "disadvantage," our work on this and other projects leads us to agree with Hancock's conclusion that permutations are more a help than a hindrance and, further, that multimodality is one reason why particle swarm optimization (and other evolutionary algorithms) works so well. Classification of the Iris Data Set (both normalized and unnormalized) was used to show that good results can be obtained by evolving weights and architecture of neural networks using particle swarm optimization.

Human Tremor Analysis

In this section we discuss methods for the analysis of human tremor using particle swarm optimization. Two forms of human tremor are addressed: essential tremor and Parkinson's disease. Particle swarm optimization is used to evolve a neural network that distinguishes between normal subjects and those with tremor. Inputs to the neural network are normalized movement amplitudes obtained from an actigraph system. The results from this preliminary investigation are quite promising, and work is continuing.

Tremor is defined as any involuntary, approximately rhythmic, and roughly sinusoidal movement (Elble and Koller, 1990). The analysis and diagnosis of human tremor is a very challenging area. Two of the most common types of tremor affecting the U.S. population are essential tremor and Parkinson's disease (Elble and Koller, 1990). Despite years of effort, relatively little seems to be known about these disorders.

Parkinson's disease (PD) is due primarily to the degeneration of dopaminergic neurons. The motor circuit is subsequently open and downstream circuit flow inhibited in the brain. Central nervous system commands for muscle control are blocked and as a result produce rigidity, bradykinesia (slowed movements), and tremor in the body. In addition, PD patients on long-term medical management often develop debilitating abnormal movements called dyskinesias (Worth, 2000).

Essential tremor is a familial tremor with onset at varying ages, usually at about 50 years of age, beginning with a fine rapid tremor of the hands, followed by tremor of the arms, tongue, head, legs, and trunk; it is aggravated by emotional factors and is accentuated by volitional movement.

Both of these disorders are initially managed by administration of a variety of pharmacologic agents, but long-term cases often become refractory to medical management. In these patients, surgical intervention either in the form of neuroablation (destruction) of hyperactive neurons or implantation of chronic deep brain stimulators (DBS) into various subcortical targets becomes a necessary option. In both situations, surgery is done under local anesthesia in awake patients, and intraoperative neurophysiologic recording of single neurons is essential in order to achieve the safest and most efficacious surgical intervention. Furthermore, test stimulation is often carried out in the operating room to assess the patient's response before permanent electrode placement (Worth, 2000).

Precise characterizations of these forms of pathologic tremor in terms of frequencies and amplitudes do not exist. Furthermore, differentiation between normal physiologic tremor and these pathologic tremors is often difficult, and precise characterizations of the ranges of normal physiologic tremors have not been defined.

This section presents the results of a preliminary study that used digital actigraphs to acquire data from normal and tremor subjects and particle swarm optimization to evolve a neural network to discriminate between tremor and normal subjects.

Data Acquisition Using Actigraphy

Actigraphy is the measurement of movement. Wrist-worn devices for measuring movement called actigraphs have been available since the 1970s. These actigraphs have been widely used in medicine for theraputic, drug, and diagnostic studies. Analysis of data from a wrist-worn actigraph provides an inexpensive and noninvasive method of movement assessment.

Most actigraphs use a piezoresistive accelerometer (a sensing element that changes its electrical resistance with changes in acceleration) as the sensor. Many actigraphs, however, do not provide the absolute value of acceleration as output. Rather, they provide the varying, or "AC," component, of acceleration as output. Additionally, although motion occurs in three dimensions, most actigraphs measure movement on only one axis. When worn on the wrist, this axis is generally oriented to be perpendicular to the inside or outside flat surface of the wrist.

Until recently, available actigraph systems recorded only limited, summarized data. For example, typical measurements have been limited to the number of zero crossings (above some threshold) that occur each time epoch. Time epochs may be as brief as 4 or 5 seconds or as long as a minute or more.

Recently, trimode actigraphs have become available from Precision Control Design, Inc. (PCD) in Ft. Walton Beach, Florida, that record zero crossings, time above threshold, and integrated amplitude for each time epoch. These units still do not, however, provide the sampling frequency and amplitude resolution necessary to quantitatively characterize human tremor.

Within the past year, however, digital signal processing (DSP) based actigraphs have been developed that provide the required sampling frequency and sensitivity. PCD's Tele-Actigraph system samples data at approximately 27 Hz with a resolution of about 12 bits. It can sense a change in acceleration as small as about 10 milligravities (mGs). Data are telemetered real-time on a 300-megahertz carrier from the wrist-worn unit to an ambulatory unit that can be worn on the belt. The belt unit can acquire data autonomously for up to 5 hours 20 minutes, after which it is downloaded into a PC. Alternatively, the PC can be connected directly to the belt unit to achieve continuous data acquisition. Using Labview on a PC, for example, data can be simultaneously acquired, viewed, and stored on the hard disk of the PC.

For this preliminary study, data were acquired with the Tele-Actigraph (TAG) worn on the outside of the subject's nondominant wrist. The data acquired were for what is known as postural tremor. The subject held his or her arm with the wrist and elbow unsupported. They were allowed to hold their arm in a comfortable position, with the elbow bent and the forearm approximately parallel to the floor. Data were acquired for approximately 60 seconds from each subject.

Figure 9.1 shows the three components of the Tele-Actigraph system. On the left is the TAG unit itself, which is usually worn on the wrist, but which may be attached to other parts of the body such as the leg. In the center of Figure 9.1 is the belt-worn unit that acquires the data from the TAG unit. On the right is the belt unit programmer that is used to load programs into the belt unit. In the current system configuration, the belt unit must be reinitialized by the belt unit programmer each time a new data session is started. In practice, the belt unit programmer is connected to a PC via the parallel port. The TAG unit can be programmed via a serial port on the PC.

Figure 9.2 shows the TAG unit being worn on the wrist and the belt unit being worn on the belt. This is the usual configuration for

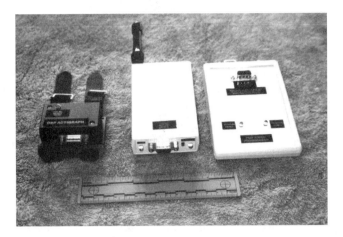

Figure 9.1 The Tele-Actigraph system.

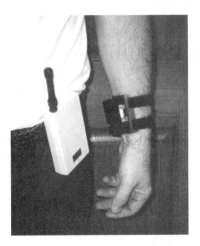

Figure 9.2 The TAG and belt units being worn.

ambulatory data acquisition sessions ranging up to 5 hours 20 minutes in length.

Data Preprocessing

The raw data acquired from the TAG is in a columnar ASCII format, with the high nybble (4 bits) followed by the low byte (8 bits) for each data sample. The first preprocessing entails adding 16 times the value of the high nybble to the value of the low byte to obtain each data sample value.

The resulting raw data file is then viewed using a Matlab script, and the file is shortened to remove data received during the warm-up period of the TAG, which can be up to 30 seconds (but which is usually less than 15 seconds). The shortened raw data file is then analyzed using the Matlab power spectral density routine. Various spectral resolutions were tried, from 512 points down to 64 points. It was decided to use 128-point transforms for this study, resulting in an amplitude value for each of 64 frequency bins.

The upper and lower two values are stripped from the files, resulting in 60-point data vectors. The square root is taken for each power value, and the resulting amplitude vectors are normalized such that the maximum value for each vector is one. These normalized 60-element amplitude vectors are then used as inputs to a neural network.

Analysis with Particle Swarm Optimization

Particle swarm optimization (PSO) was used for evolving the neural network weights and, indirectly, to evolve the network structure. This was accomplished by evolving, in addition to the network weights, the slopes of the sigmoidal transfer functions of the hidden and output processing elements (PEs) of a feedforward network. In other words, using the PE transfer function

$$output = 1/(1 + e^{-k*input})$$

the slope k was evolved in addition to evolving the weights.

The purpose of this preliminary study was to determine if pathologic tremor (essential tremor and Parkinson's disease) could be distinguished from normal physiologic tremor. No distinction was thus made between essential tremor and Parkinson's tremor subjects when evolving the neural network. A feedforward network with 60 inputs, 12 hidden PEs, and two outputs was evolved. Sigmoidal transfer functions were used in the hidden and output layers.

Data sets were available from 12 subjects with tremor and 10 normal subjects. The power spectral density plot for a subject with Parkinson's disease is shown in Figure 9.3, while Figure 9.4 depicts the spectrum for a normal patient. Neural networks were originally evolved using all 22 patterns; generalization was not the main object of this effort. However,

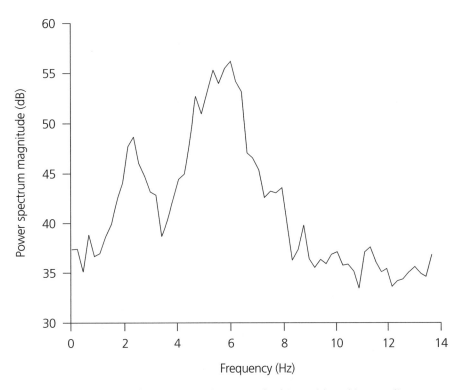

Figure 9.3 Power spectral density of wrist postural tremor of subject with Parkinson's disease.

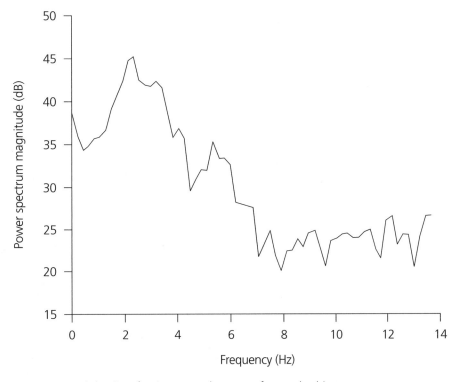

Figure 9.4 Power spectral density of wrist postural tremor of normal subject.

Table 9.3 Classification results with a 60-12-2 feedforward neural network.

Classification	Output 1	Output 2
Normal	0.053	0.948
Normal	0.027	0.973
Tremor	0.917	0.088
Tremor	0.981	0.019
Normal	0.181	0.813
Normal	0.025	0.975
Normal	0.038	0.962
Tremor	0.982	0.020
Tremor	0.932	0.067
Tremor	0.948	0.051
Tremor	0.968	0.036
Tremor	0.982	0.019
Normal	0.048	0.953
Tremor	0.986	0.015
Normal	0.066	0.935
Normal	0.070	0.930
Tremor	0.842	0.157
Tremor	0.944	0.058
Normal	0.028	0.972
Tremor	0.955	0.049
Tremor	0.990	0.011
Normal	0.038	0.961

subsequently, training on all but one pattern and testing on that remaining pattern has yielded an accuracy of 100 percent. Table 9.3 presents the outputs from a neural network evolved using all 22 patterns.

The outputs for the first processing element show outputs greater than 0.8 for all tremor subjects and under 0.2 for all normal subjects. Analogously, the second output has outputs greater than 0.8 for all normal subjects and under 0.2 for all tremor patients.

The particle swarm used to evolve the neural network had a population of 30 particles and a maximum velocity of 2.0. The initial damping weight was 0.9, and it was set to decrease to 0.4 over 2,000 iterations. However, only 38 iterations, or generations, were required to evolve the network. The process was thus extremely fast.

These results are very encouraging. Time has not permitted the evolution of other network topologies, but this is planned as part of the continuing work in this area. Also planned are attempts to distinguish between essential tremor and Parkinson's disease and between pathologic and physiologic tremor (at the early stages of pathologic tremor).

Summary

We successfully applied particle swarm optimization to evolve a neural network that classifies human tremor (Parkinson's disease or essential tremor) versus normal subjects. The method is extremely fast and highly accurate. The relatively small size of the data set indicates the need for further testing and development.

We gratefully acknowledge the assistance of Robert Worth, M.D., Neurological Surgery Department, and Joanne Wojcieszek, M.D., Neurology Department, at the Indiana University Medical Center. Without their help, this study could not have occurred.

Other Applications

Following are brief summaries of four recent applications of particle swarm optimization. They provide a snapshot of the rapidly increasing utilization of this technology.

Computer Numerically Controlled Milling Optimization

End milling is a fundamental and commonly encountered metal removal operation in manufacturing environments. While development of computer numerically controlled (CNC) machine tools has significantly improved productivity, the operation is far from optimized. Numerous predictive models are described in the literature, but none is sufficiently general to be applied in numerous situations with high

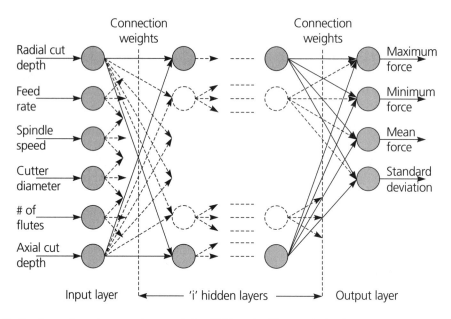

Figure 9.5 Configuration of a neural network for CNC end milling optimization.

accuracy. This is mainly due to the fact that the process fundamentals of multipoint metal cutting are not well understood. It is also due to the highly nonlinear nature of the process, which stems from numerous interdependent parameters such as spindle speed, feed rate, workpiece material characteristics, cutting depth, tool geometry, wear conditions, and tool rigidity.

A new and successful approach involves using artificial neural networks for process simulation and particle swarm optimization for multidimensional optimization. The implementation was accomplished using computer-aided design and computer-aided manufacturing (CAD/CAM) and other standard engineering development tools as the platform.

The configuration of the artificial neural network is illustrated in Figure 9.5. The evolution of the network weights using particle swarm optimization provides an approach that is accurate and reliable. A significant reduction in the time required to complete end milling operations is realized, which results in overall lower cost. Better milling quality is also achieved. The concept is being extended to other machining processes and to the prediction and optimization of a more comprehensive set of process parameters (Tandon, 2000).

Ingredient Mix Optimization

One of the most exciting applications of particle swarm optimization is that by a major American corporation to ingredient mix optimization. In this work "ingredient mix" refers to the mixture of ingredients—carbon sources, nitrogen sources, salts, trace minerals, and so on—that are used to grow production strains of microorganisms that naturally secrete or manufacture something of interest. Here, particle swarm optimization was used in parallel with traditional industrial optimization methods known commonly as statistical design or design of experiments (DOE). PSO provided an optimized ingredient mix that provided over twice the fitness as the mix found using DOE methods, at a very different location in ingredient space. PSO was shown to be robust: the occurrence of an ingredient becoming contaminated hampered the search for a few iterations but in the end did not result in poor final results. PSO, by its nature, searched a much larger portion of the design space as compared to DOE methods, which could be of value in other applications of ingredient mix optimization.

Reactive Power and Voltage Control

Another, completely different, application is the use of particle swarm optimization for reactive power and voltage control by a Japanese electric utility (Yoshida et al., 1999). Here, particle swarm optimization was used to determine a control strategy with continuous and discrete control variables, resulting in a sort of hybrid of binary and real-valued versions of the algorithm. Voltage stability in the system was achieved using a continuation power flow technique. One of the reasons the authors selected particle swarm optimization was its capability to be expanded to a nonlinear optimization problem using both continuous and discrete variables.

Battery Pack State-of-Charge Estimation

In the last application we summarize, particle swarm optimization was used in conjunction with the backpropagation algorithm to train a neural network as a state-of-charge estimator for a battery pack for electric vehicle use. Determination of the battery pack state of charge is an important issue in the development of electric and hybrid/electric vehicle technology. The state of charge is basically the fuel gauge of an electric vehicle. When it indicates empty, your batteries are dead; a "full tank" is a full charge on your battery pack. The main framework of the state-of-charge estimator is a three-layer neural network. Its inputs are the battery pack's terminal parameters: pack current, accumulated ampere hours (how much energy has been provided so far), average battery pack temperature, and the minimum voltage exhibited by any one of the individual batteries. A strategy was developed to train the neural network based on a combination of particle swarm optimization and the backpropagation algorithm. One innovation was to use this combination to optimize the training data set. We can't say much more about this, since the application is proprietary, but the results are more accurate than those provided by any other method.

Summary

These few examples of the widely expanding fields of applications for particle swarm optimization are intended to whet your appetite. More examples can be found in evolutionary computation–related conference proceedings, especially those with a special session on particle swarm optimization, and on a number of web sites.

chapter
ten

Implications and Speculations

This chapter reviews the arguments that have been made thus far and proposes some speculative conclusions. These include directions for further research and observations about the work itself and its place in the scientific endeavor generally. The particle swarm paradigm serves to provide a kind of framework for social-psychological theory, as well as a set of tools for engineering and computer science; in this chapter we begin to ponder the theoretical and practical implications of this new research. ∎

Introduction

A scientific approach to considering the mind is complicated by some factors that generally do not bother investigators in other fields. First of all is the well-known and widely discussed problem that minds are unobservable. This was the rationale behind behaviorism's rejection of things mental, but it has not stopped other sciences. Dinosaurs' eating habits are unobservable; black holes are unobservable; electricity is unobservable; gravity is unobservable—but all these things are considered appropriate subject matter for scientific research. Lots of scientific phenomena are statistically or mathematically inferred rather than observed; this is not really a good reason to deny them the attention of scientific investigation.

More importantly, it is difficult to study minds because we *are* mental beings. We have our own minds to maintain and protect, and may not wish to discover facts that force us to change, or make us question our own being in the world, or conflict with our sense of right and wrong. We have not discussed belief systems known as religions to any extent in this book. However, particularly threatening are facts that run counter to our religious beliefs, especially if those beliefs are strongly held. Further, scientists have hopes, standards, and ethical beliefs, and they—like anybody—are not eager to find that their beliefs are invalid. For instance, it is quite one thing to acknowledge that humans are animals; it is another thing to follow that belief to the logical conclusion that they could be euthanized by the millions when they are unwanted, as dogs and cats are, or eaten, as are cattle and chickens. Really, why shouldn't humans be raised for food? There's plenty of meat on some of them. It is not easy, even for scientists, to think dispassionately about people, and maybe this example suggests it is a good thing!

But that is exactly what must be done in order to "do science" about minds. The scientist needs to drop his or her romantic beliefs about minds, his or her sympathies and preferences, his or her preconceptions about what human nature is and—especially—what it should be, and instead needs to observe impartially. The "is" should not be sideswiped by the "ought to be." The intimate coloring of perception by interpretation makes pure objectivity impossible in any case, and this complication is magnified in consideration of something as personally relevant as the mind.

Assertions

We have presented the particle swarm paradigm as an interdisciplinary project, a kind of computer program and a kind of way to think about carbon- and silicon-based minds, life, and intelligence. In this chapter we focus in a bit on some arguments regarding the social-psychological and computational science assertions put forward here.

First, what are the assertions? Basically there are two major ones, with minor propositions dependent on them. The major assertions are

I. Minds are social.

II. Particle swarms are a useful computational intelligence (soft computing) methodology.

These two statements summarize the central themes of this book. From the perspective of the social and cognitive scientist trying to better understand human behavior, we emphasize the social interaction components of thought. Then, as AI researchers have done before us, we try to capitalize on intelligent human strategies by implementing mind as we know it—as a social phenomenon—in computer programs. This new method of problem solving has proved to be a very useful approach to optimization of complex functions.

Hierarchically dependent on these two major assertions are some corollaries that have provided the meat of the arguments filling this book.

I. *Mind is social.* We reject the cognitivistic perspective of mind as an internal, private thing or process and argue instead that both function and phenomenon derive from the interactions of individuals in a social world. Though it is a tenet of mainstream social science, the statement needs to be made explicit in this age where the cognitivistic view dominates popular as well as scientific thought.

A. *Human intelligence results from social interaction.* Evaluating, comparing, and imitating one another, learning from experience and emulating the successful behaviors of others, people are able to adapt to complex environments through the discovery of relatively optimal patterns of attitudes, beliefs, and behaviors. Our species' predilection for a certain kind of social interaction has resulted in the development of the species' inherent intelligence.

B. *Culture and cognition are inseparable consequences of human sociality.* Culture emerges as individuals become more similar through mutual social learning. The sweep of culture moves individuals toward more adaptive patterns of thought and behavior. The emergent and immergent phenomena occur simultaneously and inseparably.

II. *Particle swarms are a useful computational intelligence (soft computing) methodology.* There are a number of definitions of computational intelligence and soft computing. Computational intelligence and soft computing both include hybrids of evolutionary computation, fuzzy logic, neural networks, and artificial life. Central to the concept of computational intelligence is *system adaptation* that enables or facilitates intelligent behavior in complex and changing environments. Included in soft computing is the softening "parameterization" of operations such as AND, OR, and NOT.

A. *Swarm intelligence provides a useful paradigm for implementing adaptive systems.* In this sense, it is an extension of evolutionary computation. Included application areas are simulation, control, and diagnostic systems in engineering and computer science.

B. *Particle swarm optimization is an extension of, and potentially important new incarnation of, cellular automata.* We speak of course of topologically structured systems in which the members' topological positions do not vary. Each cell, or location, performs only very simple calculations.

These assertions are consistent with social-psychological and computational science findings and mainstream academic theorizing, with only a slight tilt of the head required to see some of the implications we have noted throughout this volume. In fact we admit that we bring almost no new facts to the discussion; it is mostly our interpretation and our computer programs that are new. Social scientists have always assumed that self and others comprise an integrated system, and no serious theorist would suggest that there was much to gain by attempting to study the individual out of context. As has been pointed out already, anthropologists since at least the time of Ruth Benedict have known that culture and personality are simply two views of the same phenomenon and have well understood the adaptiveness of cultural behaviors, even if the benefit is sometimes not obvious to a researcher from outside the community.

The recently fashionable view of cognitive science is a reductionist one that hopes to explain mind in terms of low-level neural events. New methods enable researchers to measure electrical and chemical changes in the brain as it performs various tasks, and it is hoped that the mind can be explained in terms of such things as synaptic dynamics and brain modularity. This is like trying to predict the weather based on the known behavior of gas molecules. It may be true that the weather is in fact a system of moving molecules, but forecasting must be based on molar patterns of air masses. Local weather is predicted by considering the state of the local weather in the context of the dynamics of weather patterns in other locations. Human conduct may one day be *explained* in terms of neural firings and the organization of the brain, but it will never be *understood* in those terms, just as the weather will never be understood by examining gas molecules. To understand people you have to know how they participate in their social context. You need to know who they care about and who they believe and who they want to be like, and you should know who taught them what they know if you want to understand their thoughts and actions.

The particle swarm view is a way to depict the movements of multiple individuals within a single mathematical frame. It is inherently a multivariate view of nonlinear systems including human behavior; it would be ludicrous to try to predict or describe the trajectories of cognitive variables without accounting for the trajectories of variables they depend on and interact with. For instance, a logical conclusion will change if its premises change; it is the *pattern* of propositions that must be optimized, not the premise or the conclusion, but both.

Some readers may feel that reducing warm-blooded, creative people to points in space is a little dry or demeaning. Let us point out that this is only a mathematical heuristic. Intelligence tests normally reduce people to points on a one-dimensional number line; personality tests like the Minnesota Multiphasic Personality Inventory reduce people to patterns on a printout that translate directly into points in a space of personality dimensions. By considering individuals as particles we are simply opening a view on the individual changing, searching for understanding in a complex space of thoughts, acts, feelings, and other people.

Up from Social Learning: Bandura

There seem to be two directions for approaching the particle swarm model as a psychological perspective: we can come "up" from the level of the individual or "down" from the level of the culture and the society.

The social learning theoretical perspective in social psychology arose out of the reinforcement theories of behaviorism and emerged as a challenge to them (Bandura, 1962). Behavioristic theorizing was mainly based on experiments with animals, especially fairly unintelligent and socially primitive animals such as rats and pigeons, where control of reinforcers produced predictable changes in behaviors. Social behavior was not very well understood in terms of operant or classical conditioning. Oddly, social learning is very rare among animals other than humans; even simple imitation is hard to demonstrate in species other than ours (Lorenz, 1973; Tomasello, 1999). You cannot teach a dog to sit, for instance, by letting it watch another dog sit—and don't even start with cats!

On the other hand, you very clearly *can* teach a human to do a trick by letting them watch you do it. Albert Bandura is the prime proponent of the social cognitive theoretical view of human observational learning:

> Observers can acquire cognitive skills and new patterns of behavior by observing the performance of others. The learning may take varied forms, including new behavior patterns, judgmental standards, cognitive competencies, and generative rules for creating behaviors (Bandura, 1986, p. 49).

In observational learning, the individual learns a behavior by watching a model perform it. In the case of cognitions, of course, direct observation is impossible, but models can express their thoughts verbally; plus, people are extremely good at inferring one another's thoughts and feelings from subtle cues. Research findings that cognitive processes can be learned observationally begin to lay down a bottom-up foundation for the development of a psychological particle swarm theory.

As social learning theory emerged from behaviorism, there has always been emphasis on the effect of reinforcement of models. The recurrent finding is that people are more likely to imitate models whose behavior is rewarded. In a general social context this might mean that individuals with status, prestige, friends, or other signs of success are more likely to be emulated—which everyday observation confirms. Specifically, in a cognitive domain it suggests that the belief patterns of individuals who appear to have made sense of some information will be emulated, assuming that it is rewarding to make sense.

Information and Motivation

Bandura (1986) has discussed two major kinds of advantages gained by emulating models; social learning has informative and motivational functions. The model's consequences give the observer information about what effects he or she would experience if they engaged in the same actions. In order to infer accurately, though, the observer needs to consider various aspects of the event. For instance, if the model is similar to the observer, then some kinds of similarities of outcomes might be expected. It is likely that some features of the context suggest whether imitation will result in similar outcomes. People are more likely to emulate a model when the requirements of a situation are ambiguous, that is, when they are not sure of what to do, or if the benefits of the modeled behavior far outweigh the observer's previous way of acting, the new one may be adopted. Interestingly, it appears that observers are much more likely to imitate others when the rules are very complex. Bandura points out that individuals can learn a modeled behavior that is punished—and in fact are more likely to learn punished behavior than behavior that does not have negative consequences—but they may be less likely to enact the behavior immediately; after time, though, the individuals tend to remember the act but forget the consequences, resulting in a kind of delayed perpetuation of undesirable behaviors.

Besides providing information, vicarious outcomes can motivate observers. Bandura reports that the frequency and magnitude of models' outcomes can affect the perseverance of behaviors that are learned through observation. Larger rewards are more motivating, especially when tenacious effort is required to obtain an infrequent reward. These effects are moderated by similarities between the observer and the model; if the two of them are very different from one another, then the observer is not as likely to expect similar outcomes.

Vicarious versus Direct Experience

Finally, Bandura has compared the effects of vicarious versus directly received outcomes. For one thing, others' outcomes provide a standard for judging if your own outcomes are just or fair. A number of studies have shown that observers learn more and faster than individuals whose own performances are reinforced, especially on tasks that are conceptual as compared to manual and complex as compared to simple. Vicarious or observational learning can take place in a large number of individuals simultaneously, unlike learning from direct experience; that is, one learner's example can provide a lesson for a large number of observers. Bandura theorizes that part of the relative advantage of observational learning derives from the fact that an actor must pay attention to the performance of his or her actions, but an observer may focus undivided attention on the behaviors and outcomes of a third-party actor.

The Spread of Influence

Bandura does note that observers who imitate a model's behavior may become models for other observers, resulting in the spread of an adaptive behavior through a community, but he does not dwell on the consequences of such an effect. The sociocognitive theory promulgated in the present volume suggests that the spread of behaviors from person to person through a population results in the dominance of adaptive attitudes, behaviors, and cognitions.

The social algorithms described in the previous chapters instantiate the assumptions of social learning theory within populations of individuals. Individuals' trajectories through the problem space carry them nearer to one another, or nearer to the successes of one another, which because of self-presentational concerns is more likely to be what models convey to observers—people are more likely to let others know about their successes than about their unproductive or failed explorations. (Selection of characteristics to promote publicly about oneself introduces a bias into social search; perfectly good problem solutions that are socially unacceptable may be hidden from others.)

The spread of adaptive behaviors through a population results in a clustering or convergence of individuals within a region of the problem space. This clustering, in the short term, comprises the formation of norms; on a longer time scale it is culture.

Machine Adaptation

Adaptation as used in particle swarm computer programs falls within the area of *computational intelligence,* which comprises practical *adaptation* concepts, paradigms, algorithms, and implementations that enable or facilitate appropriate actions (intelligent behavior) in complex and changing environments (Eberhart, Simpson, and Dobbins, 1996). For a discussion of complex adaptive systems that is applicable to intelligent systems, see Holland (1992).

Particle swarms can be used in many machine adaptation applications, such as evolving fuzzy expert systems, a process that relies on nonprogrammed emergent behavior to evolve fuzzy rule sets (Shi, Eberhart, and Chen, 1999). In turn, incidentally, the parameters of the particle swarm (such as constriction factors) can be adapted via fuzzy rules. The resulting system is like a self-referential Gordian knot that is impossible to classify as either evolutionary or fuzzy; hence the term *computational intelligence.* A fuzzy expert system is a powerful tool for control, diagnosis, classification, and optimization. Evolving such a system using particle swarm methodology can yield compact systems (low number of rules) that degrade gracefully. And these systems can be evolved in a small fraction of the time required to build traditional expert systems, which must use knowledge engineering to acquire all relevant rules from experts and which are inherently "brittle," failing catastrophically when presented with situations outside their rule domain.

An area that must currently be labeled as speculative (but this *is* a chapter titled "Implications and Speculations") is the use of particle swarms to evolve computer programs in a manner analogous to genetic programming. One approach could be to build a feedforward network using each member of the terminal set (the input variables and constants) as a node of the input layer and each member of the function set as a node in the hidden layer. More than one hidden layer could be used, and the terminal set could be incorporated in all but the last hidden layer. Then the binary particle swarm could be used to establish the optimal (near-optimal) connection matrix that defines a program to solve the problem. A potentially more powerful approach would use the particle swarm in real numbers to evolve a program that weighted each connection. The implementation would then be similar to a traditional feedforward neural network, but the solution could be represented in standard computer programming code.

Learning or Adaptation?

The words *learning* and *adaptation* tend to mean different things to different people. The field of psychology, for example, tends to use them somewhat differently from computer science and engineering.

According to *Webster's New Collegiate Dictionary* (1975), *adaptation* is

> *1: the act or process of adapting: the state of being adapted 2: adjustment to environmental conditions: as a: adjustment of a sense organ to the intensity or quality of stimulation b: modification of an organism or its parts that makes it more fit for existence under the conditions of its environment.*

The same source defines the word *adapt* as follows:

> *to make fit (as for a specific or new use or situation) often by modification.*

(To be fit is to be suitable, adapted so as to be capable of surviving, and acceptable from a particular viewpoint.)

These definitions essentially describe (and apply to) computational intelligence systems. Often, the process of altering structures such as neural networks, evolutionary computation tools, and fuzzy systems is described as *learning*. This usage is in accordance with that of a majority of researchers.

The first definition for learning, however, is "knowledge or skill acquired by instruction or study," and the synonym listed for learning is *knowledge*. Likewise, to learn is defined as "to gain knowledge or understanding of or skill in by study, instruction or experience" (*Webster's New Collegiate Dictionary*, 1975).

From the perspective of computer science, learning is what an entire *intelligent system* does. Learning thus applies to the entire intelligent system, while adaptation mainly applies to the portion of the system we are addressing in this book: the area where computational intelligence is relevant.

Adaptation must often overcome numerous barriers, including local optima and nonlinearities. The problem hyperspace landscape (topography, environment) may be constantly changing. The adaptive (computational intelligence) systems with which we are dealing are complex. The fitness or performance measure is often complicated and varying over time.

Adaptive systems answer this challenge by progressively modifying population structures, using a set of operators that themselves evolve (adapt) over time. These adaptive processes drastically shorten the time required to arrive at a solution when compared with enumerative methods that must explore significant portions of the problem space. In summary, from the perspective of computer science and engineering, adaptation is arguably the most appropriate term for what computational intelligence systems do. In fact, it is not too much of a stretch to say that, in computer science and engineering, *computational intelligence and system adaptation are synonymous.*

We are not trying to redefine either learning or adaptation in this chapter. We are simply calling your attention to the different ways these words tend to be used, particularly the differences in how they are used in psychology and the social sciences and in engineering and computer science.

Psychologists might prefer not to define learning in terms of knowledge, especially insofar as the term is used for machine learning. The word "knowledge" connotes a conscious experience as well as the correct processing or storage of information. Cognitive psychologists study "feeling of knowing," which is a reported subjective experience that you would be able to recall a fact, and which is not at all the same as the actual ability to recall the fact—and which is not the same as knowledge. Traditionally, psychologists have studied learning as an empirical phenomenon, as Webster's *second* definition puts it: "modification of a behavioral tendency by experience (as exposure to conditioning)." Scientific psychology recognizes the impossibility of the measurement of subjective events; the concept of "knowledge" can only be operationalized in terms of behavior and therefore cannot be studied directly. "Learning" therefore can only be studied as a measurable change in behavior; these interdisciplinary differences in definition of a commonly used term, philosophical on the one hand and empirical on the other, could possibly lead to confusion, which we hope has not been too great in this volume.

Cellular Automata

Cellular automata (CAs), referred to throughout the book, are topologically structured systems in which the members' topological positions do not vary. Each cell, or location, performs only very simple calculations.

CAs are a subset of automata, and like their big brothers, have undergone development (evolution) over time. The evolution of automata was recognized decades ago by Norbert Wiener (1961):

> At every stage of technique since Daedalus or Hero of Alexandria, the ability of the artificer to produce a working simulacrum of a living organism has always intrigued people. This desire to produce and to study automata has always been expressed in terms of the living technique of the age. In the days of magic, we have the bizarre and sinister concept of Golem, that figure of clay into which the Rabbi of Prague breathed life with the blasphemy of the Ineffable Name of God. In the time of Newton, the automaton becomes the clockwork music box, with the little effigies pirouetting stiffly on top. In the nineteenth century, the automaton is a glorified heat engine, burning some combustible fuel instead of the glycogen of the human muscles. Finally, the present automaton opens doors by means of photocells, or points guns to the place at which a radar beam picks up an airplane, or computes the solution of a differential equation.

The first use of the term "cellular automata" seems to have occurred in the writings of John von Neumann, as edited by Arthur Burks. Although von Neumann defined a system that was capable of reproducing itself, he didn't implement it. The concept was discussed and further developed by Arthur Burks (1970) and others. It was Chris Langton, academic son of Burks, who first developed a computer program simulation for a self-reproducing CA and popularized the concept of artificial life (Simon, 1996). In a traditional CA, each cell can sense the state of only immediately adjacent cells. Each cell then performs a very simple, usually rule-based, calculation using these immediately adjacent states and determines the new cell state. The CA is a variation of a finite state machine, with the next state dependent on the states of immediately adjacent cells and rules applied to these cells' states.

Particle swarms can thus be considered an extension and evolution of CAs that include four additional features. First is the dependence of a cell's (particle's) value on the state of a cell that may or may not be adjacent to it (the neighborhood or global best cell). (But it is dependent on only *one* such external value, rather than multiple adjacent cells in a CA.) Second is the dependence on the state (location) of the cell itself. Third is that generally real-number space is used for calculations, although binary versions of the particle swarm also exist—as do real-valued CAs. Fourth, and perhaps most significant, is that particle swarm calculations involve stochasticity. Otherwise, these particle swarm CAs mimic traditional CAs, including the dependence on unprogrammed emergent behavior (self-organization) to evolve a solution.

Down from Culture

In 1952 A. L. Kroeber and Clyde Kluckhohn published an invaluable and scholarly volume entitled *Culture: A Critical Review of Concepts and Definitions,* wherein they set out to collect and classify every social science definition and usage of the word "culture." (A new edition is expected in 2000.) Even though this list was compiled nearly fifty years ago, and a huge amount of social science research has been conducted since then, the compendium is definitely worth taking a look at; Table 10.1 summarizes the categories Kroeber and Kluckhohn used for the 161 definitions cited.

The sheer variety of definitions (each fitting comfortably within the scientific culture that produced it) by itself suggests that the concept is ambiguous, difficult—and important. All meaningful human behavior takes place within the nurturing context of a culture. The linguistic symbol systems that provide the primary medium of conscious thought are transmitted culturally; allies and enemies are defined in cultural terms, in-groups and out-groups. At the same time, a culture is nothing at all but the sum of the contributions of the individuals who participate in it. If X is a cultural norm, and all members of a community stop doing X, then X simply ceases to be a cultural norm. Culture comprises nothing but the attitudes, behaviors, and cognitions of people and the artifacts that are associated with them.

It may seem illogical to start with culture and try to work down to the individual, since individuals themselves make up the tissue of culture—minds are the chicken, culture is the egg. From the particle swarm perspective, though, the contribution of culture is unique and conceptually separate from the contributions of individuals or from their pairwise interactions. The effect of culture arises from the numerous interactions within and between neighborhoods; as various subgroups explore their regions of the space, interactions within the neighborhoods result in improvement, and when the communication is at least partially unblocked the interactions between subgroups contribute again to the extension and refinement of improved problem solutions.

Table 10.1 Categoriation of the definitions of culture.

Group		Description	Number	
A		Enumeratively descriptive		20
B		Historical		22
C		Normative		25
	C-I	Emphasis on rule or way	19	
	C-II	Emphasis on ideals or values plus behavior	6	
D		Psychological		38
	D-I	Emphasis on adjustment, on culture as a problem-solving device	17	
	D-II	Emphasis on learning	16	
	D-III	Emphasis on habit	3	
	D-IV	Purely psychological definitions	2	
E		Structural		9
F		Genetic		40
	F-I	Emphasis on culture as a product or artifact	20	
	F-II	Emphasis on ideas	10	
	F-III	Emphasis on symbols	5	
	F-IV	Residual category definitions	5	
G		Incomplete definitions		7
		Total		161

Kaiping Peng and Richard Nisbett (1999) argue that differences in the history of science in China and the West can be explained by differences in reasoning styles between the two cultures. When confronted with a contradiction between two propositions, individuals may choose to deny or ignore the logical conflict, discount its importance, choose one or the other, or they may engage in *dialectical thinking,* searching for the truth of both statements and the overarching proposition that reconciles them. Peng and Nisbett have shown differences in the ways that Chinese and American subjects deal with contradiction.

According to Peng and Nisbett, Chinese reasoning is based on three important principles:

- The *Principle of Change* holds that reality is a dynamical, constantly changing process. Thus the concepts that reflect reality must be subjective, active, flexible.

- The *Principle of Contradiction* holds that reality is full of contradictions and never clear-cut or precise. Opposites coexist on every level and in everything. The opposites exist in harmony with one another, opposed but connected.

- The *Principle of Relationship or Holism* states that in order to know something completely, it is necessary to know its relations, what it affects and what affects it. This, according to Peng and Nisbett, is the holistic principle that facilitates dialectical thinking.

Western reasoning, on the other hand, is based on these principles:

- The *Law of Identity* is the logical assertion that "everything is what it is." Thus it is a necessary fact that *A* equals *A*, no matter what *A* is.

- The *Law of Noncontradiction* holds that no statement can be both true and false; that is, *A* cannot equal not-*A*. According to Aristotle this was the most certain of the principles.

- The *Law of the Excluded Middle* says that every statement is either true or false. In Aristotle's words, "Between the two members of a contradiction, there is no middle term."

Peng and Nisbett found support for these two approaches to reasoning—which have long been noted by various writers—in their research with Chinese and white American research participants. Their (admittedly dialectical) argument is that each approach has its strength, depending on its application, and differences between the two approaches lend understanding to differences in the accomplishments of Eastern and Western science.

Peng and Nisbett's research provides a clear example of the relation between culture and the individual mind. These psychologists found that an individual's interpretation of a set of propositions, that is, their reasoning, depends on their culture. Where logicians have attempted to enumerate the processes that allow valid logical inference, it turns out that the information-processing methods people actually use in their daily lives are largely dependent on what the people around them are using, their culture.

Thus even though culture can be defined reductively as a kind of statistical summary of its participants, its top-down effect on those participants is both unitary and variable, is widespread and profound, is fundamental and intimate. Cultural truisms are reified into personal truths; the culture's way of interpreting facts is presumed to be the correct way, and in a sense it is, regardless of the culture and the method of the interpretation, because it enables humans to become human beings.

Soft Computing

As stated earlier, there are a number of definitions of "soft computing." Included are the softening "parameterization" of operations such as AND, OR, and NOT. Also included are hybrids of fuzzy logic, neural networks, and evolutionary computation, sometimes called *computational intelligence*. According to a web page devoted to the *Soft Computing Journal (owner-newjour@ccat.sas.upenn.edu),* soft computing technologies are a fusion of the following research areas:

- Evolutionary algorithms and genetic programming
- Neural science and neural network systems
- Fuzzy set theory and fuzzy systems
- Chaos theory and chaotic systems

Soft computing has a number of attributes that distinguish it from traditional (hard) computing:

1. Hard computing requires programs to be written; soft computing can evolve its own programs.
2. Hard computing uses two-valued logic; soft computing can use multivalued or fuzzy logic.
3. Hard computing is deterministic; soft computing incorporates stochasticity.
4. Hard computing requires exact input data; soft computing can deal with ambiguous and noisy data.
5. Hard computing is strictly sequential; soft computing allows parallel computations.

6. Hard computing produces precise answers; soft computing can yield approximate answers.

One net result is that soft computing systems are more robust and degrade more gracefully than hard computing systems. In addition, it is usually possible to develop an application using soft computing techniques in significantly less time than required for a similar application using hard computing.

We believe that the material in this book supports our conclusion that particle swarms and swarm intelligence are useful additions to soft computing methodologies. Certainly, particle swarms mesh nicely with the attributes of soft computing and are applicable to the listed research areas.

Interaction within Small Groups: Group Polarization

In 1961 a graduate student named James Stoner set out to test the widely held belief that groups make more conservative choices than individuals. He showed experimental participants some vignettes describing individuals faced with decisions and asked them to rate how much risk they would advise the protagonist to take. After making their individual ratings, five or six participants were gathered together and asked to discuss the situations until they reached agreement. Whereas it had been anticipated that the group ratings would regress toward the mean, resulting in more conservative ratings than the individuals had made on their own, it turned out that the groups' decisions were riskier than those selected by the members individually.

This phenomenon, called the *risky shift*, resulted in the publication of hundreds of papers replicating the effect and trying to understand how and why it occurred. A subsequent finding was that individuals tested after the discussions would also change their individual ratings; it was not just a group pressure effect, but an actual change in individual views.

As the research progressed and various hypotheses were tested and rejected, some investigators noted that the shift was not always in the risky direction; sometimes the group's decision was more cautious than the average of the individuals' positions. This was called, not surprisingly, the "cautious shift," and it occurred reliably on some of the items in the decision vignette set that was routinely administered in risky-shift experiments.

More careful experimentation revealed that groups tend to exaggerate the opinions of the individuals; that is, if the group members were initially cautious, then group discussion was even more cautious, and if they initially favored riskiness, group discussion would make them even more daring. This new phenomenon was labeled *group polarization,* to reflect the finding that group views moved toward the extremes. The effect especially occurs when group members initially agree; when two factions disagree, a tendency toward compromise known as "group depolarization" might be seen.

Social psychologists have provided several theoretical explanations for group polarization (e.g., Jones, 1998). For instance, the persuasive-arguments view suggests that individuals change their views because they are exposed to a greater number of arguments in favor of one position. When the group comes together, members mention more reasons supporting the position they like than the position they don't like, and they end up persuading one another. Further, active verbal participation in the discussion results in more change than does passive listening; having made the statement, a person might feel more inclined to find support for it.

A normative explanation suggests that individuals underestimate the group's position; when group members learn that others' opinions are as extreme as their own, they feel that they can win other members' approval, or even admiration, if they go out on a limb a little bit. The result is that other members perceive the group position as more extreme, and they too feel they can express more polarized opinions. This view is supported by the finding that members' opinions will shift after they have simply been exposed to the other members' views, with no discussion. It is also likely that, having learned what others' views are on the subject, individuals would want to go "one up" and present views that they expect to be perceived as unique or especially bold, thus further pushing the perceived norm out toward the limit.

The particle swarm theoretical view suggests that the effects of individual learning and social influence, modeled as movement through a space toward own and others' best positions, can maximize cognitive consistency. It has long been remarked that one feature of the particle swarm that helps it to optimize complex functions is that particles tend to "overfly" the previous bests, they go past them in their explorations. That is, in changing toward a position that has been found to be good, particles move past the known good points to more extreme positions. It often turns out that optima lie outside the average of best-known points; thus overflying results in improvement. Likewise, groups of humans "overfly" their previous belief positions as a result of interaction, discovering points of agreement beyond their initial locations in the belief space.

Does group polarization always result in improvement? Certainly not. Irving Janis' research into *groupthink,* where a group of individuals converges on a solution that is not optimal, demonstrates this very well (Janis, 1972). In groupthink, members talk themselves into agreeing to something that is often just plain stupid; the Kennedy cabinet's decision to attack the Bay of Pigs is given as a salient example. In particle swarm terminology, groupthink happens when solutions fail to spread from one neighborhood to another—it turns out that group insulation is a prerequisite for groupthink. As one part of the population might focus on a region of the parameter space, other parts are exploring elsewhere. Interaction among the neighborhoods results in the spread of innovations throughout the population. Thus where norms form in local neighborhoods, particle swarm cultures transcend these, and the moderating effect of culture tempers the tendencies of small groups to converge on suboptimal regions.

It would be very difficult to measure the oscillations of attitudes, beliefs, and opinions in the course of real time, and we would not venture to assert that human thought follows the exact trajectories we have proposed. There is some reason to believe, though, that as individuals consider a proposition, they entertain aspects of it that are slightly but not extremely different from the core idea; that is, they explore the region of the semantic space. An effect like group polarization occurs in particle swarms as well as in human groups.

Informational and Normative Social Influence

In 1955 Morton Deutsch and Harold Gerard (Deutsch was a student of Kurt Lewin) proposed that people are affected by normative social influence, defined as "an influence to conform with the positive expectations of others," and by informational social influence, "an influence to accept information obtained from another as evidence about reality" (p. 629). In later research, Chester Insko (e.g., Insko et al., 1983) referred to these as the "desire to be right" and the "desire to be liked," and demonstrated that they are in fact two distinct processes that can be disambiguated experimentally.

Social theorists have tended to talk about informational and normative influence as if they were two orthogonal, more or less equal, processes. The particle swarm is usually thought to be a model of informational influence, not directly addressed to the question of normative behavior. The algorithm, though, can be seen to consist of two aspects: a hard-coded set of behaviors and a flexible process of discovery. Every particle swarm program tells the particles to maintain a memory of their previous successes, to evaluate the relative achievements of their neighbors, and to adjust their trajectories toward previous successes; their social and cognitive propensities are specified in computer code. No particle swarm program tells the particles where the optimum is, or which particular neighbors will have found the best so far, and nothing tells them to converge on optima—or even to look for optimal regions of the space. Like the gliders in a cellular automaton, the optimization behavior of the system emerges (self-organizes) from interactions of parts whose behaviors are specified in the program's code.

In a similar way, human behavior and cognition consist of a hard-coded set of predispositions and an emergent process of discovery. And just as a naive researcher attempting to simulate the behaviors of gliders in a cellular automaton would be tempted to hard-code the patterns of the cells and their movement, investigators of human nature have had a difficult time disentangling the low-level causes from the emergent effects. It is rarely clear, with humans, what is learned and what is inherited. We offer the suggestion that people are predisposed to evaluate, compare, and imitate—normative social influence, *in general*. While this tendency itself has nothing to do with the development of knowledge, it can clearly be adapted to informational situations; the effect of this is Tomasello's ratchet effect (1999) and improved information-processing capabilities.

Self-Esteem

If there is anything that is universally motivating for people it must be the need for self-esteem; this seems as good a place as any to search for the biological base of human nature. Thousands of studies have supported the assumption that people tend to seek behaviors and situations that help them value themselves positively and to avoid those that make them feel bad about who they are. Several functions of self-esteem have been theorized. For instance, high self-esteem seems to help the individual deal with stress and other negative emotions, while low self-esteem is associated with depression. Further, high self-esteem appears to facilitate goal achievement by giving the individual confidence and persistence in the face of adversity. Some theorists have suggested that self-esteem is just a good thing; the motive to maintain it leads people to behave with integrity, to meet a high standard.

Why would evolution have produced a species with a need to maintain high self-esteem? What possible adaptive value could be provided by a motivation to feel good about yourself, to value yourself highly? Wake Forest social psychologist Mark Leary and his colleagues (Leary and Downs, 1995; Leary et al., 1995) have proposed that self-esteem is adaptive in that it facilitates the maintenance of social groups; self-esteem acts as a *sociometer,* a measure of how well the individual is succeeding at social inclusion, that is, how well they are accepted by their social group. According to Leary, people do not really have a need to maintain self-esteem itself; instead, they have a need for inclusion in the social group. The experience for the individual is that he or she feels good when maintaining a good relationship with the social group and feels bad when losing the approval of the group; the instrumental effect is that information flows effectively through the social network of individuals whose self-esteem, that is, social connectedness, is being constantly maintained, resulting in potential adaptation for all members of the group.

Self-esteem is a good example of a dissociation between the phenomenology of a behavior and its function—what it feels like versus what it does. Nature has provided us with an adaptive set of behavioral tendencies, but has not graced us with a good explanation for why we act as we do. We may interpret our biological requirements as something very different from what they are. Let us briefly put the shoe on the other foot and ask, what would it feel like to be a particle in a particle swarm program? You will be constantly attending to your performance—am I doing better than I was?—and constantly watching the particles around you to see how they are doing. You will see successful neighborhood members as being intensely attractive and will feel irresistibly drawn to act like them. There will be great pride in being the one that the others imitate and warm pleasure in emulating successful particles. To be stuck between optima, pulled this way and that by neighbors in different regions of the problem space—ugh! Being a particle, in other words, might be very much like being a person in some fundamental ways.

If we accept that people are hardwired as individuals to seek and maintain a high reading on the sociometer, then we see that the functional result is the emergence of groups and cultures and immergent cogitation. Evolution does not have the technology to program group formation directly, but can only motivate individuals in such a way that social groups result. Thus a low-level behavioral tendency, distributed across members of a population, can create a society or culture whose power and accomplishments far exceed the sum of the parts, just as low-level rules in the Game of Life result in complex higher-level dynamic patterns and patterns of patterns.

On a related note, human self-presentation is biased in order to display the self—to yourself as well as to others—in a positive light. We note the similarity between this human tendency and the particle swarm reliance on communication and memory of successes. Most function evaluations in a particle swarm trial are forgotten instantly, but the best ones are saved and dwelt upon, much as people dwell upon their accomplishments and the accomplishments of others, discounting their meandering fantasies and failed attempts.

Self-Attribution and Social Illusion

Daryl Bem (1967, 1972) has argued persuasively that we make attributions about ourselves on the basis of the same kinds of information we use to interpret the actions of others: we observe our own behavior and draw conclusions from the visible evidence to determine how we think and feel. Bem describes it as putting "the hypothetical communicator and the observer in the same skin" (Bem, 1967, p. 188). Bem's self-perception theory, which derives from the assumptions of radical behaviorism, offers an explanation for cognitive dissonance effects that requires no mention of cognition and that fits the research data very well. The individual may infer his or her own mental state from the same sorts of facts that would be used to make inferences about others.

For instance, in a common version of the induced compliance experiment, a subject may write a counterattitudinal essay, with the experimental result that his or her attitude shifts toward the view expressed in the essay. Dissonance theorists had argued that the conflict between the belief and the behavior, namely, "I believe A" versus "I said B," produces a tension—cognitive dissonance—that is aversive and can be relieved by shifting the attitude. In contrast, Bem asserted that the individual simply observes his or her own behavior; that is, they see what they have written and draw the same conclusion that anybody would: someone who writes an essay in favor of B probably believes B. Thus the individual infers his or her own beliefs from their behavior.

Bem is not unique in his interpretation of subjective experience—in fact his views are consistent with a long tradition in mainstream psychology. More than a century ago William James wrote:

Common sense says, we lose our fortune, are sorry and weep, we meet a bear, are frightened, and run; we are insulted by a rival, are angry and strike. The hypothesis here to be defended says that this order of sequence is incorrect . . . and that the more rational statement is that we feel sorry because we cry, angry because we strike, afraid because we tremble" (James, 1890/1948, pp. 375–376).

James notes, even while proposing the idea, that it goes against common sense; it contradicts the usual way we think of ourselves. Our experience tells us that we have direct, immediate knowledge of our thoughts and feelings—but perhaps we simply look at ourselves in the same way that other people look at us, make the same kinds of inferences about ourselves that we make about other people.

Consider Schachter and Singer's (1962) "misattribution" paradigm. In their experiments they injected subjects with epinephrine (the same as adrenaline), inducing, after several minutes, a state of arousal similar to an emotional state. Subjects who had been told they were given a "vitamin" waited for the drug to take effect in a room with another person who they believed had also been administered the substance, but who in fact was a confederate of the experimenter. In the "happy" condition, the confederate started behaving euphorically, laughing wildly and becoming quite silly, while in the "angry" condition the confederate began to become angry at having to wait and started slamming things around and shouting. As the epinephrine began to have its effect on the real experimental participants, they generally became exhilarated in the happy conditions and angry in the angry conditions; they interpreted their own arousal as elation when the other person was elated, or anger when the other person was angry. In other words, social cues enabled them to interpret their own sensations, gave meaning to their feelings, defined them to themselves.

In a classic paper, Nisbett and Wilson (1977) reviewed self-report research up to that time and conducted a series of experiments in which conditions were manipulated to influence subject's choices; subjects then were asked to explain *why* they had made that choice. All of them gave good rational reasons—but all of them were wrong. Nisbett and Wilson came to the conclusion that people are sometimes unable to report their own mental processes, not because they are being deceptive, but because they are not aware of how they think. They summarize by stating that "there may be little or no direct introspective access to higher order cognitive processes" (p. 231). According to them, people are sometimes able to report accurately about their cognitive processes when the causes of their behaviors are plausible and the stimuli are salient. People can infer their cognitive processes but are not directly aware of them.

People are usually quite confident in the explanations they give for what may have caused an event to occur. According to Nisbett and Wilson (1977, p. 255), people are more likely to believe that they have correctly identified the cause of an event

- when causal candidates are few in number

- when causal candidates are perceptually or memorially salient

- when causal candidates are highly plausible (especially when strong and explicit cultural rules apply)

- when causes have been associated with the same outcomes in the past

We have wandered a long, long way from the Enlightenment view of man as a rational being. And it is interesting to consider what this says about our attempts to "mimic" human intelligence and reason using approaches such as traditional rule-based AI.

Barry Schlenker's self-identity theory (1982) asserts that self-presentation is an attempt to control information about your identity before real or imagined audiences—including yourself. People try to provide explanations of their own conduct; they try to construct an identity that is satisfying to themselves and that explains their behavior in a favorable light. One of the criteria of a good explanation is believability; that is, explanations must fit with existing knowledge. Schlenker argues that people are not motivated to attain cognitive consistency as an end in itself; rather, they need to provide a believable and self-beneficial account of their conduct, and consistency is a by-product of that. The need to provide explanations for your conduct results in the construction of an internally consistent view of reality.

Thus self and mind are theorized to arise from the individual's participation in a social milieu. Constructing an identity requires believing our own explanations about ourselves and requires us to construct a model of the world that is believable and that justifies our behavior or casts it in a positive light when viewed by others.

These findings contradict our common experience of being human; we are usually unaware of our reliance on social information for constructing our own awareness of ourselves. We don't feel like we infer our emotions and cognitive processes; we feel like we have immediate awareness of them. The naive thinker tries to find explanations that are consistent with experience, while the social scientist tests hypotheses about how that experience might come to be. This is why a science of psychology is important.

Philosopher Wilfrid Sellars (1956) attacked what he called "the myth of the given." While some had assumed that the content of mental life is simply given, or immediately accessible, Sellars argued that our own mental states enjoy no special epistemic status. He theorized that early man may have developed, as an adaptation, a theory of inner events as causes for overt behaviors. The application of this theory to others allowed early humans to predict the behaviors of their peers. While the theory was originally used only to interpret the behaviors of other people, it was eventually altered to allow individuals to interpret their own behavior by attributing "mental states" to themselves, too.

Sellars' view is not inconsistent with that expressed by Theodore Sarbin and William Coe (1979). They point out that metaphors are used to indicate a resemblance between two things and not an identity: "A is like B." It is common, however, to stretch this kind of statement, known as a simile, into one with the form "It is as if A is B"—a standard expression of metaphor or analogy. It is even common to drop the "as if" part of the statement: "A is B"; for instance, you might say to a loved one, "You are a ray of sunshine." The context of the statement provides evidence for the interpretation—they will know they are not really a ray of sunshine. Occasionally, though, the context is not clear; this often provides material for humor and frequently results in confusion.

Sarbin and Coe argue that under some conditions metaphors become reified into myths. For instance, when the metaphor is opaque and hard to understand, it may be given the weight of a mythical fact; also when the metaphor maker is authoritative, people may accept metaphor as fact. Their argument is that the concept of mind is the product of just such a process: mind is "a reified region of mythical space" (p. 509).

Their discussion goes on to question the meaning of hypnotic states and some psychopathological categories. Individuals who believe in a metaphor, they suggest, are likely to play along with the script, as actors in a play. Thus, knowing what hypnosis "is" and what is expected of a hypnotized person, they proceed to play the role. This social-psychological explanation for hypnotic behavior is contrasted in the psychological literature with the cognitive view, which postulates that there are actual mechanisms at work in "trance" and suggestion, such as executive control processes and dissociations between various modular cognitive subprocesses.

Possibly the most articulate—and certainly the most prolific—proponent of the *sociocognitive* view of hypnotic responding was Nicholas Spanos (1986), who sadly died recently in an airplane crash. Spanos' sociocognitive theory of hypnosis and psychopathology held that these "abnormal" states are continuous with "normal" ones; individuals perceive some social cues about how they are supposed to behave, feel, or think in a situation and then respond accordingly. The interesting aspect of these two special cases—hypnosis and psychopathology, especially such forms as multiple personality disorder—is that they seem to violate our expectations about how people think. We don't expect people to have multiple personalities, and so when someone does appear to, it surprises us and we need an explanation. On the other hand, it does not surprise us when people have single personalities, and so we do not seek an explanation. The sociocognitive view theorizes that the same processes that give rise to hypnotic and psychopathological behaviors also underlie ordinary psychology. People detect the expectations or "demand characteristics" of their social context and attempt to do what they are supposed to do.

The paradigm of hypnosis research—a specialized subset of social influence research—offers an unusually poignant view of the subtle and profound interaction between cognition and social context. From all appearances, good hypnotic subjects who have been told not to remember really believe they cannot remember. The question of whether they "really can" remember details of the hypnosis session, which they had been instructed to forget, is moot—if they think they can't recall the events and they can't produce the memories, then effectively the answer is no, they can't remember (until they are told that they can recall, and then—presto!—the memory returns to them). Some good subjects are so intimately involved in fulfilling the role of the hypnotized person that they lose their sense of being deceptive; they are adapted to the demands of the context.

A particle swarm program gives simulated individuals a social predisposition to evaluate, compare, and imitate; individuals are only programmed to behave socially. Likewise, a social-psychological perspective would suggest that a great deal of humans' resources are dedicated to maintaining social relations. Intentionally contrasted with the cognitive view of the self-interested solipsistic individual, this perspective analyzes thinking and behavior in terms of interpersonal strategies.

Summary

This chapter has argued that mental events are rooted in social interactions. Why is this important? Because it shifts the explanation of mind away from the inner mechanisms of the individual—and especially from the brain, which is an entirely isolated piece of machinery—and out into the connections between people. The experience of thinking is contradicted by empirical evidence about what thinking really is. The prevailing myth is that of mind as an internal process, the myth of the given, the myth of consciousness, and we believe it as fact. As scientists we must penetrate the myth; it is our duty to look dispassionately at the evidence, if we are to understand human conduct.

The particle swarm paradigm works through dynamics that are overtly similar to processes found in human societies. As individuals interact with and are influenced by others, their cognitive structures are optimized. A great deal of human effort is expended in maintaining the social relations that allow this optimization to occur. Our experience of being human is naturally biased; there is no reason to think that nature would have provided us with a mechanism for transcending the perspective that comes with our biological package, no reason to think that introspection should be scientifically valid. While it may be difficult to imagine persons in a way that is contradicted by our everyday experience, it is necessary if we are to hold a correct scientific understanding of human behavior.

chapter
eleven

And in Conclusion . . .

This book is about a complex kind of information processing that can be performed by a simple computer program. The whole thing can be written in a few dozen lines of code, and most of the time it is used to solve tough math problems. Human minds and social behavior provide the metaphors for explaining how the program works. Alternatively, this book is about minds and intelligence and how they emerge from the dynamics of social interaction. This sociocognitive theoretical view is supported by simple computer simulations.

Several motifs have recurred throughout the telling of this story; some of them were intentional, and some just turned out to be important in a lot of ways. One important theme has been the mysterious relationship between simulation and metaphor. We run mental simulations to test the consequences of actions without danger, to understand things we can't control, to see where an argument would end up—simulation, or imagination, is a most important and useful kind of talent. Often the simulations—thoughts, imaginings, beliefs—take on greater importance than actual events in the physical environment; the simulacrum is more real than the referent. And just as easily as it happens in wetware minds, simulations of life and mind in computer programs can take on lives of their own and have thoughts of their own, perhaps as real in every sense as the physical events they model.

Sometimes the world provides explanation for informational processes that have no necessary referent. A mathematical system might be represented as a set of algebraic symbols and describe the behaviors of a wide range of systems; for instance, the formula for a sine wave can be used to describe the dynamics of many kinds of systems, from electromagnetic transmissions to astronomical orbits. The abstract system itself is just that, an abstraction, but its properties can be understood by comparison to events in the world, and people have an understanding of what a sine wave is and what it does because they know of sine-like phenomena in the world. The physical world provides material for understanding ideal phenomena.   ∎

Thus abstractions and observable things sustain one another, and it may be impossible to perceive anything that is entirely of one or the other; they are always interwoven. This theme of reciprocal imitators and explicators is central to cognitive science as well as the humanities in the current era; thus it threads through the present narrative almost uninvited.

We are balancing on a tightrope between psychology and computer science, a tightrope that is usually strung between poles under the tent of cognitive science, but this time we are performing under a different Big Top. Cognitivist reductionism has provided a great wealth of understanding about how brains work, and even about the relationships between some mental activities and the physical infrastructure that supports them, but it has not given us what we need in terms of explanation of mind in its ordinary sense. In this book we have not required that minds be human and have not even required that they do any profound information processing, as cognitive science and human vanity tend to presume. Our assumption is that individuals are components of a system that thrives on their participation and nurtures it, too. We assume that individual and culture can be embodied in abstract computer programs, where the individual's behavioral tendencies are hard-coded, adaptive culture emerges from those tendencies, and culture, the emergent effect, turns around to have a causative effect on the individuals who empower it.

Such an abstract social system can perform intelligent tasks for engineering purposes—and the simulation has a mind of its own. That is, we again are relying on the emergent attributes of the system to produce solutions; these solutions are not hard-coded into the program. And we cannot understand the solutions in terms of the coding. This tends to go against the grain of traditional engineering.

We have tried to remain vigilant about distinguishing between things that exist in the world and things that exist mentally only, which might paradoxically include what we call mind itself. What is a feature of a thing, and what features are attributed to the thing in order to expedite explanation and understanding? It is not always clear, and sometimes it matters. And as we have been walking through the fantastic world of minds and mental attributes such as intelligence, and of life and its attributes, it has been especially important and especially difficult to try to distinguish between attribution and qualities of things. We present a view that mind is socially created—a view that maps directly into the Eastern concept of *maya,* the veil of social illusion that separates humans from truth. In one sense we have been attempting to describe the warp and woof of maya, who is a powerful goddess capable of bestowing great powers. It is paradoxically through social illusion that we come to know our world; it is our mind, and yet it is seen to be built upon the shaky ground of social serendipity. We become what the people around us are. And in the long run, maya does not provide truth, but only facts and methods for seeking truth.

In the computational arena, we have embraced the basic tenets of soft computing. We recognize that we don't live in a world of ones and zeroes, of truth and falsehood. Real life, including real-life engineering and computer science applications, must deal successfully with vagueness and imprecision, with linguistic variables and noisy data. We must increasingly learn to live with near-optimum answers. We must realize that global optimization is usually a myth. We don't do it as humans; neither should we, in most cases, expect our machines to do it.

The particle swarm algorithm is a short sequence of algebraic steps that can solve hard mathematical problems easily. As a computer algorithm, the particle swarm is a new way to do business. It emerges easily from, and allies itself with, the evolutionary algorithms already in wide use. But it differs from them importantly in both metaphorical explanation and in how it works. Individuals that persist over time influence one another's search of the problem space. This is new, and we hope we have emphasized it enough. As the review of other paradigms should indicate (and we fear the review was too short; we would like to have gone on and on with it), the particle swarm is but an incremental advance beyond existing methods. There are socially intelligent agents, and there are population-based search methods, and there are swarms, and there are even simulations of populations of simple interacting agents that represent people. But none of these have shown the pure problem-solving power inherent in the simple act of imitating your betters. This one simple feature, when it occurs stochastically in a population of individuals who are all doing it, leads incessantly toward better and better ways of doing things.

Another theme that has come up repeatedly is that of the integration of individual experience with social learning. The particle swarm algorithm has two terms, and we have shown that a good number of social science models also contain some version of these same two terms. Recently, Eddie Harmon-Jones and Judson Mills (1999) edited a volume of papers on current cognitive dissonance research. The volume included an unpublished paper by Leon Festinger, the paper that eventually grew into *The Theory of Cognitive Dissonance* in 1957. Festinger had shared carbons of this draft with his graduate students as early as 1954. The paper was titled "Social communication and cognition: A very preliminary and highly tentative draft," suggesting that Festinger considered social communication to be fundamental to the theory (a feature that was not emphasized in the final publication). The paper is arranged as a series of hypotheses and their elaboration. The first one is

Hypothesis I: There are two major sources of cognition, namely, own experience and communication from others (p. 355).

The theme of dual fountainheads of cognition has not been widely explored by dissonance researchers, or at least has not been an important topic of inquiry, but as we have seen it is a view that comes up in a number of social theories. The particle swarm system shows that this two-term model is a very powerful way to process information. Even dumb dots on a computer screen are able to figure out extremely hard problems, just by remembering their own experiences and being influenced by the experiences of others.

A minor theme that has emerged in the writing of this book is that of distance; proximity is a most important aspect of causation and implication. We have come across at least three kinds of distances: Euclidean distance in real-number spaces, Hamming distance in binary spaces, and topological distance in networks of interconnected nodes. Euclidean and Hamming distances have been usually called upon to describe aspects of a problem, while topological distances have been used to describe the population of individuals working on the problem.

Societies of organisms, humans, and even abstract algebraic matrix elements can be depicted through use of a *sociogram,* a graph showing the connections between pairs of individuals. Distances between individuals can be measured in terms of the minimum number of links that must be traversed in order for information to pass from one to the other.

Cognitive variables such as beliefs and attitudes can be measured, too, and can be located in a binary or real-valued space, and members of a society can be plotted in a graph of that cognitive space. Again, distances between them can be measured in that parameter space; in this case, distance represents similarity and not necessarily physical proximity. In artificial societies these variables might be vectors that represent potential problem solutions; in any case they can be plotted within a single set of coordinates.

There is almost always a correspondence between location on the topological sociogram and location in the problem space. That is, there is a correlation between problem solution or cognitive vectors of individuals that are near each other in the social network. Social scientists know that the causes and effects go both ways: people are attracted to others who are similar to themselves—or perhaps more accurately, to their ideal selves—and (the greater effect) people who associate frequently and closely become more similar to one another. The result is that physical location, sociometric location, and semantic/problem space location tend to correlate rather highly. People who are near each other geographically will be more likely to have friends in common and more likely to share a worldview. Distance matters.

The topic of similarity-as-distance is not uncontroversial, and this ties into another motif that has woven throughout this book. Usually we have called it "linear separability," and it really refers to the difference between categories as used by persons, sets as used in mathematics, and classes as used in diagnostic systems. It seems that as humans we assign things to categories in a more or less messy, maybe incoherent, way, violating our own rules without justification or concern whenever it seems appropriate. Things that are near one another in some feature space may be assigned to different categories. People with many things in common may be worst enemies. Some of the most virulent hatred exists between factions of the same religion or members of the same family. Current research in categories suggests that people categorize objects not by imposing definitions, but rather by comparison of the object at hand with some category center, which might be an ultimate exemplar or some kind of prototypical average of examples of that category; a robin is "more of a bird" than a penguin is. This use of semantic proximity seems to require linear separability, or something like it. The tension is resolved, though, in connectionist models, where the qualities of stimuli are mapped to hidden nodes, which then link to the outputs. For instance, in some cases the function of an object determines its category (tomatoes are used more like a vegetable than a fruit), and other times appearance (dolphins look like fish) determines how it will be categorized.

Traditional computer systems categorize in a strict, rigid fashion. An object or a pattern is either totally a member of a class or totally not a member. You either have a myocardial infarction or nonspecific abdominal pain. Only in the past few years, with the acceptance of fuzzy logic, has the hard science world begun to recognize the validity of approximate reasoning and fuzzy category membership.

It is only a short leap from there to another theme that has insinuated itself into the current narrative, which is the bottom-up view of language. Language is a sticky topic, for many reasons. First, there are many theories of language and beliefs about what language is and how it works, and researchers are polarized into ideological camps; we have no desire to jump into that particular boiling vat of controversy. This situation exists because language is so important. It is how we communicate, how we think most of the time, what we notice about each other and present to each other. Our language tells the public private things about ourselves and lets us control the impression we make on others; it gives information and conceals it. Language encapsulates what being human is all about.

Human languages are not like computer languages. A variable in a computer program can have any name (within some limitations imposed by variable types and reserved names); it can be called "AAA" or "FutzGibble" or anything. This is technically true of human speech, and now and then there is an arbitrary word, but mostly words evolve from other words. The English language is very close to the Germanic and Romance languages that it came from, and they were close to the languages that preceded them. Human language is definitely not arbitrary; it is a part of tradition. It is a superindividual phenomenon that extends beyond lives and boundaries, beyond persons; individuals partake of it.

The bottom-up view of language has come up in two distinct ways in this book. First, it has been the explicit topic in several sections, for instance, the discussions of hyperspace analogues of language and Shannon's experiments with the transition probabilities of letters and words. But in another sense a view of language that *could* be called anarchic or antiauthoritarian has provided a backdrop for all that has been said here. Words mean whatever we agree they mean, though we will only agree with meanings that improve our feeling of understanding, that make sense—so meanings are not arbitrary, any more than sounds are, but fulfill some pragmatic ends.

Some words have been commandeered by ideologues who claim to have special understanding of meanings. Words such as "life," "intelligence," "mind," and others have been redefined by special interests who would prefer to discourage language users from imposing common sense on them. In this book we have tried to ignore the special interests, occasionally pointing out their influence, trying to see what these concepts would really mean if the "oughts" were pushed out of the way.

This book has been built, starting with the title, around the idea of swarms. The word "swarm," as has been noted, has a certain attractiveness to it beyond its scientific value, and we have already admitted being drawn to it for idiosyncratic reasons. But beyond that, there are some important connections between swarms, such as swarms of insects on the one hand and human thought and behavior on the other. Kerstin Dautenhahn (1997) distinguishes swarms from more complex societies on the basis of the propensity of individuals to recognize one another. Humans form lifelong alliances, while as far as we know bugs do not. So from her point of view we should not be calling this algorithm a "swarm," because individuals in the population interact repeatedly throughout the entire trial. But we note that while these enduring relationships exist between individuals in the topological lattice, the behavior of points in the parameter space is perfectly swarmlike. It seems to us that swarms provide the best metaphor for understanding the irregular oscillations of individuals in high-dimensional parameter spaces.

In computational science as well, swarms provide a very useful and insightful metaphor for the imprecise and robust field of soft computing. The idea of multiple potential solutions, even computer programs, interacting throughout a computer run is a relatively revolutionary and important one. Also new is the idea that potential problem solutions (system designs, etc.) can irregularly oscillate (with stochasticity, no less) in the system parameter hyperspace on their way to a near-optimal configuration.

It is almost unbelievable, even to us, that the computer program that started out as a social-psychology simulation is now used to optimize power grids in Asia; develop high-tech products in the United States; and to solve high-dimensional, nonlinear, noisy mathematical problems. To our knowledge, particle swarms are being researched and used for engineering purposes in the United States, Great Britain, Japan, China, Turkey, Greece, France, Germany, Canada, and Australia. If a user can define a function that can be evaluated, the particle swarm can most likely find the solution to it. This approach to problem solving is extremely versatile and robust. It can be applied to many kinds of problems with only minimal adjustments to the algorithm itself. Tell it how many variables you want, give it a way to evaluate its performance, and let it go.

We hope we have made it clear that this paradigm is in its youth. It may, in fact, still be at the toddler stage. We and our colleagues have accumulated some insights into ways to manage the population, but the topic is not nearly exhausted. We hope that readers will look at this work and say—as they say to us constantly at conferences and seminars—"I have an idea I want to try." This is how the paradigm unfolds, individuals collaborating to improve the state of knowledge for all.

Appendix A
Statistics for Swarmers

Stochastic algorithms, including evolutionary methods, some gradient descent algorithms, simulated annealing, and particle swarms—most of the approaches discussed in this book—can perform differently each time they are run. If a landscape has a number of hills, the algorithm may start in a better or a worse region and may proceed from there to a better or worse point in the problem space. A researcher is often interested in comparing the effects of various parameter levels, or different algorithms, or problem types, or some other causal factor, called an *independent variable,* on some effect or *dependent variable,* which might be the fitness after some number of iterations, number of survivors, variance or convergence of the population, or some other measure. In an experiment, an independent variable is one that is manipulated, and a dependent variable is one that is measured; an experiment then is a situation in which the effect of one variable or set of variables upon another can be isolated and identified.

The value of the dependent variable is typically different on every trial, at least where it is measured with floating-point precision. For instance, if fitness is measured after 10,000 iterations of an algorithm, it will be found that some trials result in fitter solutions to the problem than others. It will usually be found that these values are normally distributed around a mean, with some variance. If two or more levels of an independent variable are tested, then the distributions of the values for the two levels of the independent variable may be different from one another—the question is whether they are significantly different, that is, whether changing the independent variable really caused a change in the dependent variable. Unless the manipulation was unusually powerful, the researcher will discover that the distributions of the two conditions overlap; that is, the best trial of the worse condition might be better

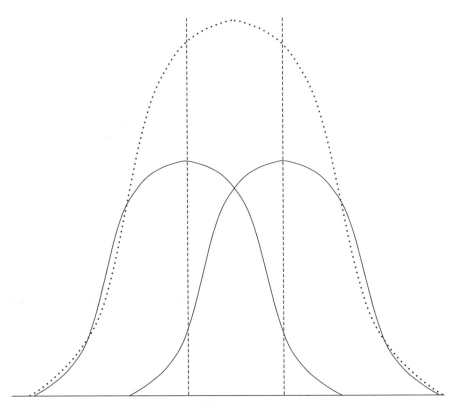

Figure A.1 Comparing two groups. Is the difference between two groups meaningful, or could the values have come from a single normal distribution?

than the worst trial of the other (see Figure A.1). In some cases there is considerable overlap of scores in the conditions that are being compared. The question is, then, whether the differences between the means are large relative to the variance of the scores.

Descriptive Statistics

There are two kinds of statistics: descriptive and inferential. *Descriptive statistics* give information about the particular sample of observations. For instance, if 20 measures are taken, say, 20 runs of an algorithm where fitness is recorded after 1,000 iterations, then a mean, usually symbolized as $\bar{x}$, can be calculated as the sum of the values divided by their number. This is a sample mean, descriptive of the particular 20 times the algorithm was run. Further, we can calculate the standard deviation:

$$s.d. = \sqrt{\frac{\sum_{i=1}^{n} (x_i - \bar{x})^2}{n}}$$

which summarizes the dispersion of values around the mean—for this particular sample.

A *sample* is a set drawn from some *population* of values. Scores in a sample can be *standardized* by scaling them to the mean and standard deviation. Having computed the mean and standard deviation of a sample, a researcher can subtract the mean from each score and divide the result by the standard deviation, producing a sample with a mean of zero and a standard deviation of one. Values in the sample are called *z-scores*.

An example is IQ, which is defined to have a mean of 100 and a standard deviation of 15. If you, our very intelligent reader, have an IQ of 130, then we can subtract the mean, $130 - 100 = 30$, and divide by 15 (the standard deviation) to get a z-score of 2. This information indicates that your IQ is two standard deviations above the average. If someone else had a different scale where IQ was scaled to a mean of 50 and a standard deviation of 10, we could easily transform those test scores to compare these two scales. Thus the z-score is very useful for describing a value in a sample.

The normal distribution is a probability distribution that is observed very widely in nature. In a perfect normal distribution there are more events near the average (think of the height of adult males in a population), somewhat fewer above and below the average, and fewer at more extreme distances above and below the average. An important aspect of normal distributions is that the z-score can be used to estimate the percentage of scores that lie below or above a sampled data point. For instance, it is known that 95 percent of the values in a normal distribution lie between plus and minus 1.96 standard deviations of the mean. Thus there is a 0.05 probability of sampling a point with a z-score less than -1.96 or greater than $+1.96$. As will be seen, this knowledge can be used to draw conclusions from data.

Descriptive statistics can be useful but contribute nothing to the development of scientific inferences. In fact no one is interested in reading a journal article about the 20 times that you ran your algorithm. Researchers need to know about the performance of the algorithm in general. It is necessary to understand how the sample represents the population about which inferences will be made.

Inferential Statistics

Inferential statistics, such as *t*-tests, analysis of variance, chi-square, and some nonparametric statistics, do more than just compare groups that are observed in the experiment; they allow inferences about the population. That is, they extrapolate from a small sample to estimate what would happen if all possible tests were run; this enables generalization from data with known certainty. For instance, to measure the performance of a new algorithm you might have the time and energy to run it using every possible random number seed. The values that describe such a population are called *parameters;* the numbers that estimate population parameters from a sample are called *statistics.* It is customary to represent parameters algebraically with Greek letters and statistics with Roman letters.

It is traditional to consider the population parameter to be a variable's "true" or expected value, and the difference between an observed value and its population value is called "error." Thus, if the population mean fitness value for a certain algorithm on a certain function, after a certain number of iterations, is 0.4567, and we run our algorithm and observe a fitness (after the same number of iterations) of 0.5678, then we would say that the error for that trial was 0.1111. Usually squared error, or its square root, is used in order to make the number positive.

Luckily it is possible to make an unbiased estimate of the population mean, that is, the average of all possible scores, which is usually represented as μ. ("Unbiased" just means that the statistic does not consistently overestimate or underestimate a parameter; errors are randomly distributed around the true value in such a way that the average value of the statistic over all possible random samples would equal the parameter being estimated.) This is simply done by taking the mean of a sample randomly drawn from that population. So if you started your random number generator from randomly determined seeds and ran it some number of times, the average of your dependent variable would be an unbiased estimate of the average you would obtain if you were able to run the algorithm on all possible seeds. The Law of Large Numbers says that the more times you measure your dependent variable, the more accurately your average estimate will approximate the population mean. Surprisingly, the size of the sample needed to make an accurate estimate is not related to the size of the population to which results will be generalized.

In order to compare experimental treatments it is necessary to estimate the population variance, or dispersion or scores, as we want to know if our effect is large relative to ordinary error variance. Unluckily this time, the sample variance is biased in estimating the population variance. But it turns out that the population variance *can* be estimated without bias, simply by changing the denominator of the formula for the mean. If σ^2 is the population variance, then by definition

$$\sigma^2 = \frac{\sum_{i=1}^{N}(\mu - x_i)^2}{N}$$

where N is the size of the population. For infinite or very large populations—which will be assumed in this discussion—this population parameter can be estimated by the sample statistic s^2:

$$\hat{\sigma}^2 = s^2 = \frac{\sum_{i=1}^{n}(\bar{x} - x_i)^2}{(n-1)}$$

where the hat over the sigma means that the population parameter is estimated, with n now representing the sample size and $\bar{x}$ being the mean of the sample. It is obvious that as the sample size increases, n and $(n-1)$ become closer together, and the true value of the population parameter and its estimate converge—which is as it should be.

One more concept is important for understanding inferential statistics. It would be possible to draw a sample of one single observation and try to use it to estimate the population mean, since the mean is the sum of observations divided by n, but of course we would not be very confident in its accuracy. We could improve our estimate considerably by averaging two observations—and it would be better, but you would not want to invest your money on the basis of such an estimate! The mean of a sample of three observations is even more likely to approximate the population mean, four is better, and so on. The mean of bigger random samples better approximates the population mean.

This fact suggests that if we take a number of random samples of some size, compute their means, and plot them, the distribution—called the *sampling distribution of the means*—will be narrower where *n* is larger. Obviously if the samples were as large as the population, they would all have the same mean, which would equal the population mean, and their variance would equal zero. Smaller samples will vary around the population mean. As you would think, the mean of the sampling distribution of the means equals the population mean. An interesting finding of this kind of exercise is that, if the samples contain about 30 or more observations, the sampling distribution will be normally distributed—whether the population values are normal or not.

Knowledge of the sampling distribution of the means of samples of size *n* can allow inferences about the population distribution. It turns out that the standard deviation of the sampling distribution of the means, called the *standard error,* can be estimated from the sample estimate of *σ:*

$$\hat{\sigma}_{\bar{x}} = \sqrt{\frac{s^2}{n}}$$

Now, with knowledge of the mean of the sampling distribution and an estimate of standard error, we are in a position to answer questions about the relationship of our sample's statistics and the population parameters they estimate.

Confidence Intervals

We have seen that the mean of a sample gives an unbiased estimate of a population mean. We have seen further that it is possible to estimate the standard error of the distribution of means of samples of size *n,* using the information acquired from one sample.

Because we know what percentage of a population lies within certain regions of the normal curve measured in standard deviations, we can use our sample to identify a range within which we are confident, with some known certainty, that the true population mean will be found. Thus we might mark, for instance, a 90 percent confidence interval around a sample mean; this means we are 90 percent certain that if we sampled every possible value in the population we would find that the true mean is between the bounds we have set.

If sample sizes are small ($n < 30$), then the estimate of the sampling distribution of the means is not normally distributed, but rather follows what is called a *t* distribution. The *t* distribution is generally flatter and wider than the normal, meaning that values are concentrated somewhat less in the center and more in the tails. As samples grow larger the distribution approaches the normal distribution: this is the *Central Limit theorem*.

Say we have a sample of 20 observations, that is, with 19 degrees of freedom, and wish to establish the bounds within which we are 90 percent certain that the true population mean will be found. We look up in a table and discover that, for $df = 19$, 5 percent of scores lie above a *t* value of 1.729. That suggests that another 5 percent lie below -1.729, accounting for 10 percent. The confidence interval is computed:

$$\bar{x} \pm t \cdot \hat{\sigma}_{\bar{x}}$$

It is a simple matter to add and subtract and put a bound around the sample mean. We can then say, with 90 percent certainty, that the true population mean, that is, the average of all possible observations, would lie within the interval we have just defined.

Student's *t*-Test

Now that we can estimate mean and variance population parameters from sample statistics, we are in a position to compare whether the effects of a manipulation result in a difference in population means of a dependent variable—that is, a true difference and not just one that turned up in our random samples. First we consider the case where the independent variable has two levels. Let us assume that a single measure of the dependent variable y is a function of the gth level of the causal factor and some unique deviation due to error ε. In other words:

$$y_{gi} = \beta_g + \varepsilon_{gi}$$

where g represents the group and i is the individual within the group.

We have noted that ε is the unique term that distinguishes each value from its group mean. An estimate of the population error in each group is made by summing the within-group squared error over all scores and dividing by $n - 1$; this estimates the average deviation of squared individual measures from their cell means. The square root of this is the sample standard deviation:

$$s = \sqrt{\frac{\sum_{i=1}^{n} \varepsilon^2}{(n - 1)}}$$

This statistic is used to estimate the population standard error for differences between the means of the two groups:

$$s_{\bar{x}_1 - \bar{x}_2} = \sqrt{s_{\bar{x}_1}^2 + s_{\bar{x}_2}^2}$$

Now we have a standard for deciding whether β is big or small. A t-statistic is defined as the difference between the group means divided by the standard error of the difference:

$$t_{df} = \frac{\bar{x}_1 - \bar{x}_2}{s_{\bar{x}_1 - \bar{x}_2}}$$

Note the "df" subscript to the t; for the two-group case, the *degrees of freedom* equal the total number of values minus two. Given the degrees of freedom, it is possible to look up the probability that the t-statistic would have occurred if there was in fact no difference between the two groups, that is, if the population mean of the difference equaled zero. This is sometimes expressed in null-hypothesis terminology: the p-value is the probability of rejecting the null hypothesis of no difference when the null hypothesis is in fact true. It is customary to say that if $p < 0.05$, the difference is *significant*, though more stringent or more relaxed levels may be used.

In sum, we can compare the causal effect of a manipulated variable upon a measured one—that is, the change in a phenomenon resulting from a hypothesized cause—by estimating the differences, in the population, between the means of the dependent variable in the groups receiving different treatments. The differences are compared to the standard error, which is an estimate of the amount of deviation to be expected in the population whether the groups differ or not; if the differences between the groups are big relative to the error within groups, then we can say with some known confidence that the independent variable causes a change in the dependent variable.

One-Way Analysis of Variance

The *t*-test can be adapted to a single-group design, where the question is whether the group mean equals some given value, usually zero; in this case, there are $n - 1$ degrees of freedom in the denominator. If we have more than two groups, it may seem tempting to perform pairwise *t*-tests between all the different conditions to see if any of them differ significantly from any of the others. This is clearly not okay, as it allows the analyst to capitalize on chance. That is, if you conduct 20 *t*-tests at a $p <$ 0.05 significance level, then by chance alone you would expect one of the tests to have a *t* greater than the critical value. There are some adjustments that can be made, for instance, those proposed by Bonferroni, Tukey, Sheffé, and Fisher, but these are all approximations and are as likely to penalize the researcher as to help. The correct approach when more than two groups are being compared is to conduct an analysis of variance, or ANOVA.

The transition from *t*-tests to ANOVA is very easy, as we can still think of it in terms of nearly the same model:

$$y_{gi} = \mu + \beta_g + \varepsilon_{gi}$$

where μ is the "grand mean" of all the observations.

As before, the sum of the betas equals zero, and the sum of the errors within each cell is zero. Again, as before, the question is the size of the betas relative to the error. In the analysis of variance, a slightly different statistical distribution is used for the hypothesis test. The *F*-distribution maps, as does the *t*, to a normal population distribution, and as before, allows calculation of a *p*-value that gives the probability that we will say there is no difference between groups when in fact there is. In fact, in the two-group case $F = t^2$.

The analysis of variance calculates the mean error variance, usually defined as the mean within-group variance, and the mean between-group variance, which estimates the amount of difference between groups receiving different manipulations. These are defined in terms of sums of squares. The mean error variance is defined as

$$MS_W = \frac{\sum_g \sum_i (\bar{x}_g - x_{gi})^2}{N - G}$$

where g is groups, i is individuals within groups, N is the number of measurements, and G is the number of groups. Thus this mean square represents the average deviation of values in each group around their group's mean.

The between-group mean square estimates the difference between groups:

$$MS_B = \frac{\sum_g \bar{x}_{tot} - \bar{x}_g)^2}{G - 1}$$

where $\bar{x}_{tot}$ is the grand mean and G is the number of groups.

The results of these two operations are used to compute an F-statistic, which is related to t:

$$F(df_B, df_W) = \frac{MS_B}{MS_W}$$

As with t, the F-statistic can be compared to a table to determine whether the observed difference is significant at a preselected probability level. If $F < 1$, there is certainly not a significant difference; the criterion value is a function of the degrees of freedom of the numerator and the denominator. The one-way analysis of variance tells you if there is any significant effect, that is, whether the average differences among the groups is greater than what would have been expected by chance alone. But it doesn't tell you *where* the difference lies.

Apologetic disclaimer: We hope you will understand that we are simplifying terribly here. Our hope is to describe in general terms how these analytical techniques can be used to determine the significance of effects; we seriously hope that the interested reader will pick up a thorough text on the subject, and perhaps take a course in statistics, before proceeding. For instance, we assume that group sizes are equal; we overlook problems with nonnormal data; we do not explain the critical importance of the philosophy of null-hypothesis testing; we ignore the measurement scale; we say nothing about effect size or statistical power; and there are many other gaps in this discussion. We are only trying to introduce the subject to an audience that may not otherwise have been exposed to it.

Factorial ANOVA

The one-way analysis of variance assumes that there is one independent variable with two or more levels. For instance, we may test an evolutionary algorithm's performance with 3 or 4—or 20—different population sizes or with several kinds of tweaks or features. And for these situations the one-way ANOVA is an extremely powerful way to determine if an independent variable causes a change in the dependent variable.

Often the more interesting experimental question has to do with the simultaneous effects of multiple independent variables. Some examples might be

- the effect of population size in GA with mutation, crossover, and both
- the effect of problem dimensionality when mutation rate is high, medium, or low
- the effect of using first-term, second-term, and both-term particle swarms when populations are initialized symmetrically around the optimum, or asymmetrically

and so on. If it were possible to list all potential experiments, they would not be worth conducting, would they?!

The factorial ANOVA can have two or more independent variables. *F*-statistics are computed for the effects of each of the independent variables and for interactions among them.

$$y_{ghi} = \mu + \alpha_h + \beta_g + \alpha_h \cdot \beta_g + \varepsilon_{igh}$$

Let us begin with an example where we have two independent variables—called *factors*—which we call A and B. Each of these factors is administered in two levels, which we will call here "high" and "low." Thus we can conceptualize the experiment in terms of a 2 × 2 design. We insert some fictional means into the cells, as in Table A.1.

In this design we can look for two kinds of effects, called *main effects* and *interactions*. These particular data have been cooked up to

Table A.1 A hypothetical pattern of means demonstrating a main effect for Factor B and none for Factor A.

| | | Factor A | |
		Low	High
Factor B	Low	0.5	0.5
	High	1.5	1.5

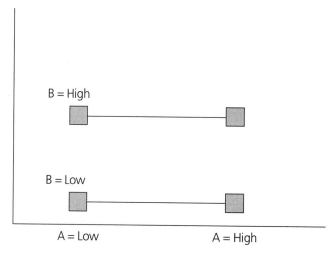

Figure A.2 Main effect for Factor B only. The value of the dependent variable depends on the level of B and not A.

demonstrate a main effect for Factor B. There is a large difference between the sums of the means for the Low level of B (0.5 + 0.5 = 1.0) and for the High level (1.5 + 1.5 = 3.0). Of course the significance of this difference can only be determined by reference to the error, which cannot be calculated from this table. In comparison, if we compare the sums for the High and Low levels of Factor A, which are 2.0 in both cases, there is obviously no difference. Thus this experiment demonstrates a main effect for Factor B and none for Factor A (see Figure A.2).

Besides main effects for Factor A and Factor B, there is a third important kind of test to be performed in this kind of simple experiment. To test interactions in a 2 × 2 table of results, we look at the diagonals.

Table A.2 demonstrates two main effects in a 2 × 2 experiment. Here the sums for Factor A are 0.75 and 2.25, a rather large difference, and the

Table A.2 A hypothetical pattern of means where both Factors, A and B, have main effects with no interaction.

		Factor A	
		Low	High
Factor B	Low	0.75	1.5
	High	0.0	0.75

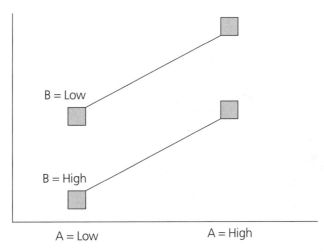

B = Low

B = High

A = Low   A = High

Figure A.3 The parallel lines indicate that both A and B have main effects on the dependent variable, and they do not interact.

sums for Factor B are also 0.75 and 2.25. Both diagonal sums are 1.5; thus there is no interaction between the two factors. This indicates that Factor A has a causal effect on the dependent variable, and that Factor B also has a causal effect, but the two effects are independent of one another (see Figure A.3).

Finally, Table A.3 demonstrates an interaction effect. In this case, the sums for the Factor A columns are both 2.25, as are the sums for B's rows. The diagonals, though, are not equivalent. Low-A plus High-B equals 1.5, while High-A plus Low-B equals 3.0. In this more interesting case we see that the independent variables do have an effect on the dependent variable, but the effect of each factor depends on the level of the other (see Figure A.4). Had we conducted our experiment with Low-A only, we would have concluded that the High level of B increases the dependent variable, relative to Low B. But if we had tested with High A we would

Table A.3 A hypothetical pattern of means showing an interaction without main effects. The effect of A depends on the level of B.

Factor B		Factor A	
		Low	High
	Low	0.75	1.5
	High	1.5	0.75

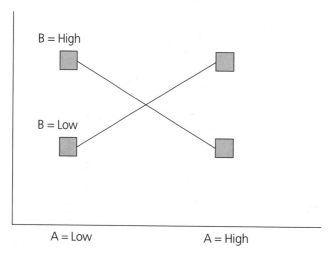

Figure A.4 The effect of A depends on the level of B: they interact.

have concluded the opposite. If we had tested either variable without controlling the level of the other, we might very well have concluded that it had no effect.

The test of interaction in analysis of variance is extremely powerful and allows the researcher to investigate subtle and interesting questions regarding a number of independent variables. Let us warn, though, that the 2 × 2 ANOVA is the simple case. If there are more than two levels of either factor and a significant test of interaction is obtained, it may not be obvious where the interaction lies. There are special tests, including contrast tests, for focusing the analysis on particular cells in the design. For instance, we may have wanted to test the question of whether the Low-A, Low-B condition was significantly different from all the other cells combined: analysis of variance includes a way to perform that test and many others.

One more warning: ANOVA can be conducted with any number of factors, but experience has shown that the human mind can rarely comprehend interactions of more than three independent variables.

Multivariate ANOVA

The analysis of variance can be extended to the analysis of the effects of one or more independent variables on more than one dependent variable. This is not a topic for the squeamish, and we will not dwell on it here. Most statistical software, such as SAS, SPSS, and others, will perform multivariate analysis of variance, or MANOVA, as it is called. Usually, a linear composite of the dependent variables is created, and a variety of ANOVA is conducted using that new combination as the dependent variable.

One example using MANOVA has been described already in this volume. Kennedy and Spears analyzed differences in the shapes of curves generated over time by various algorithms. For instance, if we measured two algorithms over three points in time (Kennedy and Spears looked at four algorithms over 20 points in time), we might find the means of the dependent variable for algorithm 1 after 1,000, 2,000, and 3,000 iterations were (0, 1, 2) and for algorithm 2 were (2, 1, 0). The averages for the two algorithms are the same, though that is really meaningless, since we want to compare how they changed over time. A multivariate analysis of variance will handle this by multiplying the raw data by a vector of weights that are designed to sum to zero and that maximally distinguish the groups, and taking the sum. It is easy to see that $(-1, 0, 1)$ will do the trick, as would $(+1, 0, -1)$—we will demonstrate with the first. The weighted means for the first group, that is, $(0, 1, 2) \times (-1, 0, 1) = (0, 0, 2)$, resulting in a sum of 2; for the second group we get $(-2, 0, 0)$, summing to -2. This makes it apparent that the two weighted groups will differ; weighting each value in the data set lets us perform a univariate ANOVA on the transformed data, to test the hypothesis that change over time is different for the two groups.

If you are considering using a sophisticated technique such as this, we hope it goes without saying—but we'll say it again—that a good text should be kept nearby, or preferably, a good graduate-level class should be taken. Calculation of multivariate composite weights can be done without effort in some of the powerful statistical programs that are available—kids, don't try this at home.

Regression Analysis

Regression analysis has a kind of bad name in the computational intelligence world because it is so, well, linear. In fact, all the analytic methods described so far are subsets of the so-called General Linear Model (GLM), so the accusation can be directed at ANOVA as well as regression. Perhaps snobbish computer scientists would feel better if we suggested that regression analysis and its kin can be considered as a linear univariate subset of a larger family of models, which we could call "neural networks." In fact, the argument can be made and has been made; neural nets and regression are very closely related methods. But beyond that, there is no constraint that regression models have to be linear; exponents can be placed on the terms of the equations, and any kind of data surface can be analyzed.

Regression analysis is typically used when both the independent and the dependent variables are numeric. For instance, we might want to vary population sizes in a GA, or neighborhood size in a particle swarm, and see its effect on fitness after 1,000 iterations. In this case we will use a familiar formula:

$$y_i = \mu + \beta x_i + \varepsilon_i$$

In ordinary quantitative regression the x's are the values of the independent variable and may fluctuate numerically as we assign our conditions; for instance, if we have a hypothesis about population size, then in some conditions population $= 100$, thus $x = 100$, and so on. In quantitative regression analysis, error is not calculated within cells as in ANOVA, but is calculated as a function of the deviation of observed values from values calculated on a regression line. Beta values are found that minimize squared error, and these can be plotted as a line through the data that minimizes the distance from data points. A line slanted upward toward the right, such as the one in Figure A.5, indicates that the independent variable has a positive effect on the dependent variable. If it had no effect, the line would be level, and a negative effect would slope upward toward the left.

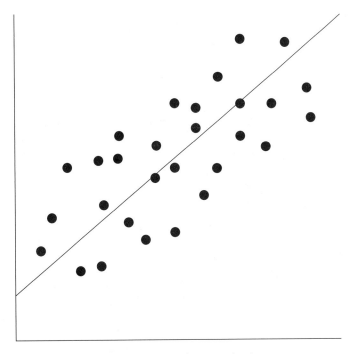

Figure A.5 A regression line is plotted through a set of points.

As the graph indicates, the system parameter μ functions as the intercept of a linear equation, that is, it gives the value of the dependent variable when the value of the independent variable equals zero; β represents the slope of the regression line. Where the independent variable has a large effect, the slope of the line will be steep and the β will be large; no effect results in a flat line, and $\beta \approx 0$. The size of β required for a significant effect depends, as with previous models, on the dispersion of the data; it is possible to have a very steep regression line but error variance very large, meaning that the effect is strong on average but highly variable. For regression, an F-statistic is computed, and significance tested, based on the ratio of the slope estimate divided by residual error.

Regression is sometimes misused, especially when independent variables are not assigned by the researcher but are simply collected observationally. In this case it cannot be said that an experiment has been conducted, and in fact all that exists is a correlation—for simple regression is very little more than a form of correlation equation. But if the researcher assigns values for the independent variables and measures effects on a dependent variable, then an experiment *has* been conducted, and the slope of the regression line—the β—estimates the strength of the effect of the independent variable. More will be said about the weakness of correlational research below.

Multiple regression allows testing of effects of multiple independent variables simultaneously. The variance in the dependent variable is "explained" by the values of the independent variable, with the effect of one variable removed before the next is tested. There are numerous ways to implement a multiple regression test, and some of the ways are not rigorous—regression is easily abused by untrained, untalented, or unscrupulous analysts.

The Chi-Square Test of Independence

In some experiments it is not possible to make a numeric measurement of the dependent variable, but only to count occurrences of certain kinds of results. For instance, a particle swarm experiment might look at how many individuals end up in a particular region of the parameter space, or solve the problem, or accomplish some objective; a researcher might measure the proportion of times that success, or failure, or some other event, is encountered. In these cases no quantitative measure is taken, other than simply counting events. (Note that proportions are only frequencies divided by a total; thus, the methods described here can be applied to both frequencies and proportions.)

Consider, for instance, a situation where High and Low levels of an independent variable are administered, and we count how many times the algorithm succeeds or fails to meet a criterion. In this case it may be hypothesized that the two variables are not independent; that is, we hypothesize that the number of observations associated with one level of one variable (say, the Fail-Succeed variable) will be affected by the level of the other (Low-High). The resulting data can be displayed in a *contingency table* such as in Table A.4.

This data set has 125 observations in it. From these simple frequencies we can calculate what the expected count in each condition should be. For instance, the first row contains 20 + 25 = 45 observations, and the first column contains 20 + 50 = 70 observations; thus we would expect to find, in the first (Low-Fail) cell,

$$E_{ij} = \frac{T_i \times T_j}{N} = \frac{45 \times 70}{125} = 25.2$$

Table A.4 A contingency table.

	Fail	Succeed
Low	20	25
High	50	30

Table A.5 Expected frequencies.

	Fail	Succeed
Low	25.2	19.8
High	44.8	35.2

where T_i is the total number of observations for the row, T_j is the total for the column, and N is the total number of observations. Thus we expect the cell to contain 25.2 observations. The same calculations can be done for all cells, resulting in the table of expected frequencies, shown in Table A.5.

We are now ready to compute the χ^2 (chi-square) test of independence for this experiment. The formula is very simple:

$$\chi^2 = \sum_i \sum_j \frac{(O_{ij} - E_{ij})^2}{E_{ij}}$$

that is, it is the sum of the squared differences between the expected and observed values of each cell, divided by the sum of the expected values. In our present example, this amounts to

$$\chi^2 = \frac{(20 - 25.2)^2}{25.2} + \frac{(25 - 19.8)^2}{19.8} + \frac{(50 - 44.8)^2}{44.8} + \frac{(30 - 35.2)^2}{35.2} \approx 3.81$$

We wanted to test the hypothesis that the two variables are not independent. We do this by looking up the probability that a chi-square this big or bigger, with this number of observations, would be found if—the big if—the two variables were in fact independent. The *p*-value gives us the probability of saying there is an effect when there actually isn't, or, more formally, it is the probability of rejecting the null hypothesis when the null hypothesis is in fact true. Degrees of freedom for χ^2 are defined as the number of rows minus 1 times the number of columns minus 1;

a 2×2 table such as this then has 1 degree of freedom. Looking up in a table of critical values of χ^2, we see that the value needs to be greater than 3.84 to be significant at the $p < 0.05$ level with 1 degree of freedom. Thus there is insufficient evidence to say that manipulating the level of the Low-High variable resulted in a real difference in the numbers of Successes and Failures.

The χ^2 distribution is an important one in inferential statistics and can be implemented in a number of kinds of tests. This example was presented here as a kind of experimental design that can be analyzed with a test that is simple and easy to understand.

Experimental Design

The analysis and design of experiments are just two sides of the same coin. Before an experiment is run the researcher should know how the analysis will be conducted. Are you running a $2 \times 3 \times 2$ factorial experiment? Then allocate your trials accordingly, and plan your analysis. Do you expect certain cells in your design to contrast with certain other cells? Then you should express this in the beginning, and plan to allocate your degrees of freedom to test your exact hypothesis. A rational, efficient research plan will allow you to find the answers to your questions with confidence.

An experiment forces you to state your hypothesis, at least to yourself, before you start collecting data. You don't just go fishing for something/anything; you should know what you're looking for and plan a rational way to test for it. Computer experiments are ridiculously easy to conduct, compared to experiments in the real world. The test of a stochastic algorithm should have equal cell sizes, carefully chosen levels of independent variables, and a well-planned research design in general. You can control every possible confounding variable (that is, one that correlates with an independent in such a way that it is not clear which one caused an observed effect); in fact a kind of perfection can be achieved in computer-generated data that is impossible in any other field.

One word: simplicity. The 2×2 factorial design is elegant, comprehensive, and easily understood. The hypothesis of an experiment should be plainly stated, and the statistical result should tell the reader whether the initial hypothesis has been supported or not. Where—as is usually the case—the result suggests further research, the further research should be conducted with as much care as the original study. There is nothing worse than attempting to answer all questions about a topic simultaneously, throwing every possible factor into the design and then trying to figure out what may have caused what. Two or three factors at a time can provide insights that can be built upon; where interactions among factors are expected, by all means test for those interactions, but in reality the interaction of more than a couple of variables is not only hard to understand but rare.

We have warned above that this appendix is to be considered an extremely superficial introduction. There are a great many controversies and challenges in the use of these research methods—another reason to keep studies simple. On the other hand, techniques such as analysis of variance provide very powerful tools for determining, with known certainty, whether a hypothesized cause really does produce a hypothesized effect.

Appendix B
Genetic Algorithm Implementation

In this appendix, a genetic algorithm implementation is presented. Software for it is on the web site for the book at *www.engr.iupui.edu/~eberhart/web/PSObook.html*. The implementation includes files necessary to train a neural network to classify the Iris Data Set introduced by Anderson (1935) and popularized by Fisher (1936). The genetic algorithm is used to evolve the weights of the network.

The data set contains 150 patterns, each with four inputs and one output. Each input is a feature of an iris flower: sepal length, sepal width, petal length, or petal width. The output is the species of flower. There are three species, each represented by 50 patterns in the data set. It is a notoriously hard data set to classify well.

The genetic algorithm (GA) implementation is basically a "plain vanilla" GA, but with a few interesting options available. It implements one-point, two-point, or uniform crossover, as well as mutation, with an interesting set of options available for mutation, one of which is reminiscent of evolution strategies. To learn more about the basics of genetic algorithms, refer to Chapter 4.

Source code is provided on the web site along with the executable code needed to evolve neural network weights. We provide the source code because of the somewhat unusual nature of genetic algorithm implementations.

In the case of genetic algorithm implementations the fitness function being optimized is the main user-supplied component. This fitness function is an integral part of the genetic algorithm, and so must be coded and compiled with it. For each major change in fitness function, a new executable GA must be compiled. Therefore, for you to apply the GA to your problem, you will need to substitute your fitness function into the source code and recompile.

The Run File

To run the genetic algorithm implementation on the web site requires the executable file ga.exe and an associated run file, for example, ga.run. To run the implementation from within the directory containing ga.exe and ga.run, at the system prompt type ga ga.run at the DOS prompt.

One way to present the genetic algorithm implementation is to examine and discuss the contents of a typical run file that can be invoked with the executable:

```
c:\ga\iris.dat
c:\ga\weight_1
c:\ga\corect_1
4
2000
16
4
3
150
70
0.60
0.01
0.02
1
0
0
2
```

The first entry, iris.dat, is the name of the data file. The number of targets appears on the first line of the file, and targets are included with each pattern. The next entry, weight_1, is the name of the file into which the "evolved" weights are placed at the end of the run. Next is the name of the file into which the results (outputs) are written, corect_1 in this case.

Following the file list is a list of numbers: the number of hidden neural network processing elements (PEs) (4), the maximum number of generations (2,000), the number of bits per variable (16), the number of input PEs (4), the number of output PEs (3), the number of patterns in the pattern file (150), the population size (70), the percent probability of crossover divided by 100 (0.60), the percent probability of mutation divided by 100 (0.01), the acceptable sum-squared error to which the network weights are to be evolved (0.02), the "mutate according to bit position" flag (1), the fitness shift flag (0), the "mutate according to variance" flag (0), and the crossover flag (2).

The numbers of input, hidden, and output PEs define the configuration of the neural network for which the weights are to be evolved. The activation function utilized in the software is $(\tanh() + 1)/2$, which has an output range from 0 to 1. The maximum number of generations is the maximum number of epochs, that is, the maximum number of times the entire pattern set will be evaluated for fitnesses of individuals.

The number of bits per variable allows the user to set the resolution for each vector element; in this case, each element represents one network weight. The trade-off here is that a relatively high number of bits provides the resolution needed to successfully adjust weights on a complex error surface, but also increases computational complexity significantly. A debate has raged for years with respect to weight resolution needed to guarantee successful network training. This GA implementation provides a tool to investigate this question with a variety of data sets representing various problems.

The size of the pattern file can of course be varied to investigate the effect on test sets. If this value n is less than the number of patterns actually in the designated pattern file, only the first n patterns will be utilized for evolving the network weights; all patterns are utilized, however, when the network is tested.

The number of population members (70 in this case) can be varied according to the problem. A higher number allows a more thorough exploration of the problem domain, but increases computing time. Typically, the value should be set between 20 and 200, but values outside the range may be appropriate for relatively simple problems that involve relatively short individuals (<20) or for highly complex problems involving very long chromosomes (>200).

The probability of crossover should be set between 60 and 80 percent for many problems. Straightforward two-point crossover as described in Chapter 4 can be implemented, as can one-point and uniform crossover.

The next value (0.01 in the list) is the probability of mutation. Options for mutation implemented in this GA are explained later. The value listed here is a sort of baseline mutation value, as it can be modified in up to two ways. If not modified, however, the value represents the chance that mutation will occur, determined bit by bit.

The next value, 0.02, is the sum-squared error target for the performance of the "evolved" neural network. The GA will terminate when this error level is achieved or when the maximum number of generations have been calculated, whichever occurs first. In either case, the result and weight files are written.

The next value (1) is the "mutation according to bit position" flag. When this flag is 0 (disabled), mutation is carried out in the normal way: mutation is done bit by bit with the probability of mutation listed earlier. When it is 1 (enabled), the probability of mutation m_b varies with the bit position in each variable. The variation in mutation across each variable approximates a Gaussian function; that is, it is much more probable that the least significant bit will be mutated than it is that the most significant bit will. It is implemented according to the following equation, where b is the bit position ($b = 0$ for the least significant bit, $b = 1$ for the next-to-least significant bit, etc.) and m_0 is the probability of mutation in the run file:

$$m_b = m_0 \frac{1}{\sqrt{2\pi}} e^{-b/2} \qquad \text{(B.1)}$$

Note that this is only an *approximation* of a Gaussian function. Note also that the calculation is done across each variable. So, in the above run file, for a variable represented by 16 bits, the resulting probability of mutation is $m_0(1/2\pi)^{1/2}$, or about $(.01)(.40)$ for the least significant bit and about $(.01)(.40)\exp(-7.5) = (.004)(.00055)$ for the most significant bit. The variance for the quasi-Gaussian function can thus be seen to depend upon the variables' dynamic range (-10 to 10 in this case) and how the variables are represented by the binary string.

The next value in the run file (0) provides an option to implement a shift in fitness values (for purposes of reproduction only). When this flag is 0, the option is disabled, and "raw" fitness values are used to calculate normalized fitness values for the reproduction step. When set to 1, all fitness values are shifted so that the minimum value is 0.1; the min_value to max_value fitness range is preserved. For example, if all raw fitness values are between 0.90 and 0.99, implementing fitness shift will scale them between 0.10 and 0.19. This provides an expanded fitness differential that is often useful when fitness values become bunched together near 1.0 late in a GA run.

The next-to-last item in the run file (0) allows a change in mutation rate if the variance of the fitness values drops below 0.05. This change can be in addition to (on top of, so to speak) the change caused by mutating according to bit position, if it is enabled. If this option is disabled (0), no change in mutation rate is triggered by a low variance value. If it is enabled, the mutation rate m_{lv} is increased for low variance values, according to the following equation, where m_{prev} is the previous (existing) mutation rate, modified by bit position or not, depending on the modify-by-bit-position flag:

$$m_{iv} = m_{prev}\{(\text{int})[(0.05 - \text{var}) * 100] + 1\} \text{ for } \text{var} < 0.05 \qquad (B.2)$$

This option allows the mutation rate to increase significantly toward the end of a run when there are often many identical population members and fitness value variation is quite low. In the limit as variance approaches 0, the mutation rate multiplier will approach 6. This significantly "bumps up" mutation, facilitating exploration of more of the problem space.

The last value in the list (2) is the crossover flag. The GA implementation allows the user to choose either of three kinds of crossover. If the flag is set to 0, one-point crossover is implemented; if it is set to 1, uniform crossover is implemented; and a value of 2 implements two-point crossover.

Recompiling

Because source code for the genetic algorithm is included on the web site, it is possible to use the GA to optimize functions other than neural network weights. To do so, the following alterations to the source code that are tailored to the problem must be made.

First, in the file ga.c the individual population length calculation must be changed to reflect the new problem. The length is currently calculated as

$$\text{length} = ((N + 1) \times H + (H + 1) \times P) \times K$$

which reflects the number of weights in a neural network.

Second, in the file fitness.c it is necessary to do the following:

1. Arrange and allocate memory to the parameters for the new problem. In the example provided on the web site, the connection weights are arranged into two weight matrices in accordance with the topology of the backpropagation network.

2. Change the getWeight() function source code to correspond with the new parameter arrangement.

3. Write a new evaluation function to replace the one (bp.c) that is provided.

4. Modify the final fitness calculation source code according to the meaning of the return value of the new evaluation function. In the existing bp() function, the return value is the mean sum-squared error, not the fitness. This can be incorporated into the new evaluation function source code.

Third, the run file must be modified as appropriate in light of the new variable structure.

Running the Program

When the program is run, values of the best, average, and worst individual fitnesses are plotted on the screen. (Be sure that the file egavga.bgi is in the same directory as the executable file ga.exe.) By observing the plot a number of interesting things can be learned. For example, with the variance flag set, which implements a higher mutation rate for low variance, it is possible to see an almost immediate effect on the best, average, and worst values when the variance drops below 0.05.

Glossary

ACE (agent-based computational economics) Leigh Tesfatsion's paradigm for the computational simulation of economic systems. Includes agents called "tradebots" that sometimes choose whether to interact with other agents.

Adaptation Any process whereby a structure is progressively modified to give better performance in its environment (Holland, 1992).

Allele Alternate forms or varieties of a gene.

Arity The number of arguments required by a function.

Artificial intelligence A name for a paradigm in which people attempt to elicit intelligence from machines.

Artificial life (sometimes "Alife") The study of alternative forms of life to biological.

Artificial selection The survival of the fittest, as defined by some humans.

Attitudes Usually defined in scientific research as units of integrated cognition and emotion, thoughts that are bundled with evaluations.

Attractor A set of points or states in state space to which trajectories within some volume of state space converge asymptotically over time (Kauffman, 1993).

Autocatalytic Causing positive feedback.

Autokinetic effect The apparent movement of a fixed point of light in a dark room; results from movements of the eyes as they try to adjust to an invariant stimulus.

Autonomous The ability of an agent to form decisions independently.

Backpropagation of error A learning rule for multilayer feedforward networks, in which weights are adjusted by backward propagation of the error signal from outputs to inputs.

457

Belief space In cultural algorithms, information about what kinds of proposed solutions have performed especially well, and sometimes ones that have performed poorly and should be avoided; interacts with the *population space.*

Biomorphs Evolvable graphical creatures created by Richard Dawkins.

Bitstring A string or series of ones and zeroes.

Bottom up System effects resulting from local interactions or emergent order.

Breadth-first search A search strategy that involves going into the tree one layer at a time, through all branches, evaluating the result of each partial search before proceeding to the next layer.

Building blocks In a genetic algorithm, this term generally refers to compact (short) schemata that are part of highly fit individuals and that, with high probability, appear in ever-increasing numbers in future generations.

Cellular automaton (CA) A very simple virtual machine that can result in complex, even lifelike, behavior.

Cellular robots A population of robots based on insect societies, where global intelligence emerges from the local interactions of individually unintelligent beings.

Central executive A hypothesized centralized control mechanism of a cognitive system.

Chatterbots Programs whose specialty is real-time verbal interaction with human users.

Chromosome A data structure that holds a "string" of task parameters, or genes (IEEE NNC, 1996).

Classical conditioning The behaviorist paradigm that describes learned autonomous nervous system conditioning of responses to stimuli that precede the response.

Closure property In genetic programming, for the closure property to be satisfied, each function must be able to successfully operate on any function in the function set and on any value of any data type assumable by a member of the terminal set.

Cognitive consistency The state that exists when cognitive elements fit together with one another, as well as with phenomena in the real world.

Cognitive dissonance A clash or inconsistency among cognitive elements, including beliefs, behaviors, and attitudes.

Combination Some number of elements are selected from a universe of possibilities, without regard to their order.

Combinatorial explosion The number of possible solutions, and hence the difficulty, of a problem increases exponentially with its dimensionality.

Communicative gene Buck and Ginsburg's hypothesis that social tendencies called displays and preattunements evolve together through the adaptive selection of communicative relationships.

Complexity The interaction of many parts of a system, giving rise to behaviors and/or properties that are not found in the individual elements of the system.

Compressible A series of symbols can be encoded in fewer bits of information than it contains.

Computational intelligence Comprises practical *adaptation* concepts, paradigms, algorithms, and implementations that enable or facilitate appropriate actions (intelligent behavior) in complex and changing environments.

Confirmation bias The propensity of people to irrationally seek confirmation for their beliefs, rather than falsification.

Conformist transmission A term used by Boyd and Richerson for cultural evolution with a heavy dose of frequency-dependent bias.

Constraints The state of a node in a network depends on, or is constrained by, the states of other nodes and the strengths of their connections to it.

Constriction coefficient One of a family of coefficients that enable control over the convergence properties of a particle swarm system.

Convergent evolution Occurs when similar or identical phenotypic structures evolve through different genetic pathways.

Correlation A form of statistical association between two or more variables that gives some evidence for predicting one from the other.

Crossover Term used by genetic algorithm researchers for sexual recombination of chromosomes.

Crossover rate In a genetic algorithm, the probability that, given two parent chromosomes, the crossover process will occur.

Cultural algorithms A search paradigm, promulgated by Robert Reynolds, that uses a *belief space* of stored information to enhance an evolving population's search.

Cultural transmission The process whereby phenotypic traits, such as learned behaviors, are passed from one individual to another.

Culture The full set of attitudes, beliefs, behaviors, and artifacts that are shared by a population over a considerable length of time. A pattern of persistent, profound, and widespread norms.

Cybernetics The study of control systems, based on the theory that intelligent living beings adapt to their environments and accomplish objectives primarily by reacting to feedback from their surroundings.

Deduction Inference that is concerned with beliefs that follow from one another by logical necessity.

Defecting In game theory, the decision to try to gain advantage over an opponent, rather than cooperate for mutual advantage.

Depth-first search A search strategy that comprises going from the start all the way to the end of a path, evaluating one entire proposed solution. Once the end is reached, depth-first search goes back to the first decision node and searches to the end of the next path, and so on.

Deterministic A situation in which effects follow certainly from causes.

Dialectical thinking Searching for the truth of both statements and the overarching proposition that reconciles them.

Direct bias Boyd and Richerson's term for influence that occurs when the individual selects a belief or behavior based on attributes of the variant itself; that is, it may appear to be the best choice for some reason. Direct bias implies that better solutions to problems will become prevalent in a population.

Discrete recombination In evolution strategies, an operation that comprises selection of a parameter value from either of two parents.

Dynamic social impact theory Revision of social impact theory that considers the simultaneous causal interaction of individuals as sources and targets of influence upon one another.

Element In evolutionary computation, a vector element generally corresponds to one parameter, or dimension, of the numeric vector, or chromosome.

Elitist strategy In a genetic algorithm, ensuring that the individual chromosome with the highest fitness is always copied into the next generation.

Embodiment The property of having a body; includes occupying an observable position in space, perceiving the environment from a central position, and exaggeration of proximal causes relative to distal ones.

Emergence Global behavior of a system is not evident from the local behavior of its elements.

Endosymbiont theory The theory that organelles were originally forms of bacteria that entered plant and animal cells at a very early period of evolution.

Entropy The tendency of systems to lose energy and order and to settle to more homogenous states.

Epistasis In evolutionary computation, any kind of strong interaction among genes, not just masking effects. It is the extent to which the contribution of fitness of one gene depends on the values of other genes. Highly epistatic problems are difficult to solve, even for genetic algorithms. High epistasis means that building blocks cannot form, and there will be deception (IEEE NNC, 1996).

Eugenics The belief, prevalent in the early-twentieth century, that it would be a good idea to improve the human race genetically, especially by encouraging the reproduction of "desirable" individuals and discouraging the reproduction of "undesirable" individuals.

Evolutionary computation Encompasses methods of simulating evolution on a computer. The field includes research in genetic algorithms, evolution strategies, evolutionary programming, genetic programming, particle swarm optimization, artificial life, and so on.

Evolutionary programming An evolutionary algorithm developed in the mid-1960s. It is a stochastic optimization strategy that is similar to genetic algorithms, but dispenses with both "genomic" representations and crossover as a reproduction operator (IEEE NNC, 1996).

Evolution strategies A type of evolutionary algorithm developed in the early 1960s in Germany. It employs real-coded parameters, and in its original form, it relied on mutation as the search operator and a population size of one. Since then it has evolved to share many features with genetic algorithms.

Evolution window In evolution strategies, that step-band size, or window, within which mutation operations result in fitness improvement.

Exploitation (exploit) A search strategy that focuses on knowledge that has already been gained.

Exploration (explore) A search strategy that searches widely by sampling new regions of the problem space.

Finite state machine A transducer that can be stimulated by a finite alphabet of input symbols, can respond in a finite alphabet of output signals, and possesses some finite number of different internal states (Fogel, 1991).

Fitness The measure of the probability that an organism will survive to reproduce. A measure of the degree to which an individual (candidate solution) successfully solves the problem being addresses by an EA.

Fitness landscape The topography of the fitness surface over the parameter space.

Fitness space A one-dimensional space containing the fitness values of patterns of parameters.

Fixed action pattern A behavior that an organism emits in response to a particular, often very specific, stimulus.

Frequency-dependent bias Boyd and Richerson's term for cultural transmission that exists when the individual adopts a variant because it is statistically prevalent in the population, or at least among the people that the individual interacts with.

Full approach In genetic programming, each limb of the program tree extends for the maximum allowed depth.

Function space Contains the ranges of results of functions.

Fuzziness Refers to nonstatistical imprecision and vagueness in information and data.

Gaia hypothesis "The Gaia hypothesis says that the temperature, oxidation, state, acidity, and certain aspects of the rocks and waters are kept constant, and that this homeostasis is maintained by active feedback processes operated automatically and unconsciously by the biota" (Lovelock, 1988).

Gbest In particle swarm jargon, a neighborhood comprising the entire population, so that every individual is influenced by the very best solution found so far.

Generation In an evolutionary algorithm, an iteration of the measurement of fitness and the creation of a new population by means of reproduction operators (IEEE NNC, 1996).

Generation gap In an evolutionary algorithm, the percentage of the population replaced each generation.

Genetic algorithm (GA) A computational method for adapting problem solutions that is based on genetic aspects of evolution. Implementations typically use fixed-length character strings to represent their genetic information, together with a population of individuals that undergo crossover and mutation in order to find promising regions of the search space (IEEE NNC, 1996).

Genetic drift Genetic changes that evolve under minimal selection pressure and thus do not affect adaptation.

Genotype The genetic composition of an organism or data structure; the information contained in the genome (IEEE NNC, 1996).

Global method In evolution strategies, uses the entire population of individuals as potential sources from which individual components for the new individual can be obtained.

Global optimum The most optimal point in the parameter space.

Graceful degradation The behavior of a complex adaptive system is not catastrophically affected by modification or removal of some of its components.

Gradient A kind of multidimensional slope, often considered to lead toward an optimum.

Gray coding A kind of binary encoding that removes the Hamming cliffs.

Greedy algorithm An algorithm that makes the best decision, as computed locally, at each point in the series.

Grounding problem In order to have meaning, symbols cannot refer only to one another, but must eventually refer, even if through chains of relations, to real things or events in the environment.

Group polarization Groups tend to exaggerate the opinions of the individuals. Thus, if the group members were initially cautious, then group decision would be even more cautious, and if they initially favored riskiness, then the group decision would be even more daring.

Group selection The hypothesis that species may engage in behaviors that are good for the group, even when they reduce the chances of survival of the individual.

Guided variation Boyd and Richerson's term for adaptation that occurs when evolution is informed by within-lifetime adaptation.

Halo effect A person or object with some positive characteristics is assumed to have other positive characteristics as well.

Hamming cliffs In order for a binary number to increment by one numeric unit, it is sometimes necessary to flip all its bits, for example, 01111 (15) to 10000 (16). A numeric algorithm may encounter difficulty moving over this "cliff."

Haplodiploidy A genetic strategy in which males are haploid, that is, their cells contain only a single set of chromosomes, and females are diploid, containing two homologous chromosomes and thus two copies of each gene. In many social insects, for instance, fertilized eggs (diploid) become females, while unfertilized eggs (haploid) develop into males.

Heuristics Shortcuts in a search strategy that reduce the number of points in the space that need to be examined.

Hidden nodes Nodes in a neural net that receive their inputs from other nodes and send their outputs to other nodes.

Hill The depiction of a local optimum on a fitness landscape.

Hill climbing A strategy in which search proceeds from a position that is updated only when a better position is found.

Hypercube The conceptual representation of points in binary space of more than three dimensions. A hypercube has the characteristics of a cube; that is, the corners of the hypercube can be conceived as bitstrings whose distance is the number of sites in which they differ.

Hyperspace A Cartesian coordinate space of high dimension, typically higher than three dimensions.

Immergence Local states of elements are affected by global system behavior.

Inclusive fitness The concept that organisms strive to ensure the survival of others whose genes most resemble their own.

Indirect bias Boyd and Richerson's term for cultural transmission that occurs when a variant is associated with other variants that are considered desirable.

Individual In evolutionary computation, a single member of a population. Each individual represents a possible solution to the problem being solved; it is a single point in the problem space.

Induced compliance An experimental paradigm in social psychology, used in cognitive dissonance research. In the induced compliance experiment an individual is compelled to behave in a way that contradicts his or her attitude; the person's opinion is often seen to change in the direction of an attitude they were persuaded to express.

Induction Inference that is supported by the accumulation of evidence.

Inertia weight A kind of coefficient used in particle swarms to control the trajectories of particles.

Inference engine A program or subroutine that draws conclusions from premises.

Information theory A branch of mathematics that analyzes the amount of information or entropy in a system.

Intelligence The quality of a good mind.

Intermediate recombination In evolution strategies, setting each parameter value for a child at a point between the values for the two parents. Typically, the value is set midway between those values.

Intractability The quality of a problem that is too difficult to solve.

Landscape correlation The distance between two points in a multidimensional space correlates with the difference in their fitness values.

Lbest In particle swarm jargon, a neighborhood comprising some number of adjacent neighbors in the population array.

Life space Lewin's conception of a topological field comprising the interaction of a person and his or her environment.

Linear separability Exists when some weighted additive combination of properties can be used to classify examples. When attributes of things are graphed, a straight line or surface can be drawn that perfectly separates the two kinds of things.

Local method of recombination In evolution strategies, forming one new individual using components from two parents that are randomly selected.

Local optima Relatively optimal solutions that are not the best.

Locomotion Kurt Lewin's term for the person's movement through the life space. Does not necessarily or usually mean movement through physical space; it might better be thought of as change of state than change of position.

Maximum The vector of input parameters that produces the greatest function output.

Means-end analysis A classical AI technique for comparing the current state to the goal state and searching for an operator that reduces the most important differences between the two.

Memes Richard Dawkins' proposition that ideas and other cultural symbols and patterns, called *memes,* are analogous to genes; they evolve through selection, with mutation and recombination, increasing their frequency in the population if they are adaptive.

Memetics A view originating with Richard Dawkins that suggests that cultural features such as ideas and artifacts evolve in a process similar to genetic evolution.

Meta-evolution Described by developers of evolution strategies as the evolution of evolution, or evolution of the second kind; said to be accomplished by evolution strategies.

Mind That which thinks; Sarbin and Coe (1979) call it "a reified region of mythical space."

Mindlessness The production of automatic behaviors, often comprising inappropriate responses to stimuli that superficially resemble familiar ones.

Minimum The vector of input parameters that produces the smallest output.

Modeling See *social learning.*

Moore neighborhood The topological region around an individual in a two-dimensional cellular automaton comprising adjacent cells above, below, to the left and right, and the four diagonals.

Multimodality Exists when a problem has multiple solutions, global optima, or good local optima.

Mutation rate The proportion of chromosome elements that will be altered on average by mutation.

Natural language processing Computation based on the understanding, analysis, and generation of natural (e.g., written or spoken) language.

Neural networks A family of statistical configurations that superficially resemble the arrangements of neurons in the brain.

NK landscapes The fitness surface over a parameter space of N nodes, with an average of K inputs to each.

Nodes A term for the vertices of a graph.

Norm A pattern of attitudes, behaviors, and/or cognition shared by members of a group.

Normalized fitness An individual's normalized fitness is its raw fitness value divided by the sum of the raw fitness values for all population members.

N-squared complexity A simulation program may have to loop through each of the elements one at a time, and for each one calculate the distance to all of the others, in order to identify nearest neighbors.

Objective function The statement of the goal of an optimization problem.

Observational learning See *social learning*.

One-point crossover In a genetic algorithm, one-point crossover involves selecting a single crossover point at random and exchanging the portions of the individual strings to the right of the crossover point.

Operant conditioning The behaviorist paradigm that describes learned skeletomuscular (operant) responses that enable the organism to act on its environment in order to obtain a reinforcement.

Optimization The adjustment of a system in order to minimize or maximize the result of some function.

Optimum A pattern of parameters that produces a minimum or maximum function result, generally.

Organelle Structures found in almost all eukaryotic cells that have their own DNA and are able to synthesize some of their own proteins; organelles may be descended from prokaryotic invaders.

Overselection In genetic programming, for very large populations of 1,000 or more, highly fit individuals are sometimes given an even greater probability of selection than their normalized fitness would indicate.

Parameter space Contains the legal values of all the elements that can be entered into a function to be optimized.

Particle An infinitesimal, moving, usually multidimensional representation of a vector.

Particle system A kind of computer graphics method that comprises large numbers of individual agents or objects, each having its own behavior.

Payoff matrix A depiction of the pattern of rewards and costs (payoffs) allocated to each participant in an interaction on the basis of their own and their opponents' choices.

Peak The depiction of an optimum on a fitness landscape.

Perceptron A simple neural network consisting of an input layer connected to a single processing element. The activation function of this unit is a linear threshold function, applied to the inner product of the input and weight vectors.

Periodic attractor A repeating, cyclic sequence of events.

Phase transition A movement between static, periodic, or chaotic states.

Phenotype The expressed traits of an individual (IEEE NNC, 1996). Also, the observed characteristics of a species.

Pheromones Chemicals emitted by individuals that can elicit a response from conspeciates.

Point attractor The dynamics of a system simply stop.

Polarization The tendency of attitudes in a population to cluster in groups of proximal individuals.

Population In evolutionary computation, a group of individuals that may interact together, for example, by mating and producing offspring. Typical population sizes in evolutionary computation range from one (for certain evolution strategies) to many thousands (for certain genetic programming problems) (IEEE NNC, 1996).

Population space In cultural algorithms, the space that is searched by an evolving population.

Prisoners' dilemma A situation where two interacting players choose whether to cooperate or compete with one another. If both cooperate, their payoffs are high, and if both compete, payoffs are low. If one competes while the other cooperates, the defector receives a very high reward while the cooperator's payoff is very low—the lowest in the game, called the "sucker's payoff."

Probability A form of uncertainty that exists when it is not known whether a discrete event will occur or not.

Probability threshold A number that sets the probability of an event. A random number can be chosen from the same range as the threshold; if the random number is smaller than the threshold, the event occurs, otherwise not.

Problem space According to Simon and Newell, the set of states, operators, goals, and constraints comprising a problem. In more general use, the space of potential parameter values that can be tried to solve a problem.

Quasirandom Processes that appear random, though the cause of the sequence is deterministic.

Ramped half-and-half method In genetic programming, a method of creating programs with evenly distributed depth parameters. Within each subpopulation of a given depth, one-half of the programs are created using the grow approach, one-half using the full approach.

Random search A strategy, also called "random generate and test," in which random problem solutions are generated, one after the other, each one is evaluated, and the best one is saved.

Random walk A bitstring is generated, a randomly chosen bit is flipped, the new bitstring is evaluated, another bit is flipped, and so on.

Reasoned Action Model A social-psychological theory of attitude, behavior, and cognition proposed by Ajzen and Fishbein.

Reciprocal causation Bandura theorized that behavioral, cognitive, and personal factors interact with environmental influences to operate as determinants of one another; that is seen, almost paradoxically, as the root of free will.

Recombination The type of crossover used in evolution strategies. It manipulates entire variable values.

Recurrent network A graph or matrix whose nodes can reciprocally affect one another. Also, a feedback network in which the current activation state is a function of the previous activation state as well as the current input.

Region A local area of a space.

Representations Hypothesized constructs that enable the manipulation of symbols.

Result sharing Social cooperation in which the individuals share the results of their individual tasks.

Risky shift The tendency for groups to make riskier decisions than would have been made by the members individually.

Robust In this field, robustness means that an algorithm can be used to solve many problems, and even many kinds of problems, with a minimum amount of special adjustment to account for special qualities of a particular problem. It also can mean that an algorithm works acceptably even with noisy and/or missing data.

Role theory The social-psychological theory of behaviors that are intended to enact social roles; the individual's objective is positive self-presentation, and the function of covert information processing is to support that presentation.

Satisficing Herbert Simon's term for the search for a solution that is good enough, as compared to a perfect solution.

Scaling To adjust a set of values in some orderly way (linear, logarithmic, etc.) so that they fall between two specified end-points. For example, 100 data points with a minimum value of 3.1 and a maximum value of 125.4 could be scaled to be between 1 and 2. Note that this is not the same thing as normalization.

Schemata (*singular:* schema) In a genetic algorithm, each schema defines a subset of strings with identical values at specified string locations. Schemata provide a means by which similarities among the individual population members can be described and exploited.

Schema theorem Holland's proof that selection plus genetic operators results in improvement of problem solutions in a population.

Selection A number of problem solutions (chromosomes or patterns of features) are proposed and tested; those that do well in the test produce offspring in the next generation, while those that perform poorly are eliminated. This is what is meant by "survival of the fittest."

Self-organization The ability of some systems to generate their form without external pressures, either wholly or in part. It can be viewed as a system's incessant attempts to organize itself into ever more complex structures, even in the face of the incessant forces of dissolution described by the second law of thermodynamics.

Semantic differential An attitude measurement scale that asked people to rate a word or concept on a number of scales.

Semantic space Some set of dimensions within which meaning can be represented.

Sexual selection The selection of sexual characteristics, whether or not they are adaptive for survival.

Sigmoid S-shaped. Refers to a kind of mathematical function that transforms a linear monotonic input series into an s-shaped curve.

Simple permutations All the possible orderings of an entire set of elements.

Simulated annealing Simulated annealing is based on the metaphor of molecules cooling into a crystalline pattern after being heated. A problem solution is modified, and if the modified solution performs better than the original, then it replaces it. If the modified solution performs worse, though, it can still be accepted if a probability test is passed. The probability threshold is a function of the system's "temperature," which decreases over time. Thus the probability of accepting a poorer problem solution decreases as the system cools.

Social impact theory Latané's theory that people are influenced by a group in proportion to the strength, immediacy, and number of the group members: $î = f(SIN)$. (See also *dynamic social impact theory.*)

Social learning An individual learns by observing and imitating the behavior of a model.

Sociocognition A social-psychological view that considers thinking to be a social phenomenon.

Sociogenetic learning An intelligent process in which the basic unit of mutability is the idea, with culture being the reservoir of learned behaviors and beliefs. (Atmar's concept, cited by D. Fogel.)

Sociometer A measure of how well the individual is succeeding at social inclusion, that is, how well they are accepted by their social group. Used by Leary to explain self-esteem.

Stigmergy Communication by altering the state of the environment in a way that will affect the behaviors of others for whom the environment is a stimulus.

Stochastic process Contains random components; the word usually refers to a phenomenon that is probabilistic in nature.

Strange attractor A complex nonrepeating (chaotic) pattern of events with no apparent patterns to be seen. Variables never repeat their values but always are found within a restricted area of state space.

Strategy parameters Usually standard deviations or variances, used to scale the mutation of feature vectors in evolution strategies and evolutionary programming.

Subjective norm In the Reasoned Action Model, a sum of the products of an individual's beliefs that certain others think they should or should not perform the behavior, multiplied by the motivation to comply with each of those others; predicts intent, which in turn predicts behavior.

Subsumption architecture A robotic intelligence approach built from the bottom up; simple robot behaviors are developed, and then these are fitted loosely together.

Sufficiency property In genetic programming, for the sufficiency property to be satisfied, the set of functions and set of terminals must be sufficiently extensive to allow a solution to be evolved.

Surprisal value (surprise) The informational value of a particular event or kind of event, as a function of its probability.

Tabu search A kind of algorithm that maintains a list of points in the problem space that have been evaluated most poorly. The search algorithm then steers away from those points.

Task sharing Social cooperation that occurs when individuals share the computational load for performing the subtasks required to solve the problem.

Teleological Implying purpose.

Test function A mathematical function that is intended to be hard to optimize, used to test and compare algorithms.

Top down Local effects that result from systemwide causes or imposed order.

Tradebots Interacting autonomous agents in Tesfatsion's ACE. They have behavioral functions that enable them to trade with other tradebots and memories that allow them to identify the others that they have traded with.

Trade network game (TNG) One kind of choice-and-refusal model in Tesfatsion's ACE paradigm.

Trial and error A family of remarkably robust general approaches to problem solving.

Turing test If the subject can't tell if the computer's responses were generated by a human or a machine, then the computer is considered intelligent.

Tweaking Technical term for the adjustment of system parameters by a researcher in order to optimize the performance of a system.

Unbiased Boyd and Richerson's term for transmission of cultural information when individuals simply adopt the variants they are exposed to, for instance, by their parents.

Uniform crossover In a genetic algorithm, a random decision is made at each bit position in the string as to whether or not to exchange (crossover) bits between the parent strings.

Vicarious reinforcement According to Bandura, if a model is successful at performing a task, and in particular if they are reinforced for the behavior, then the observer is more likely to learn from them.

von Neumann neighborhood The topological region around an individual in a cellular automaton comprising adjacent cells above, below, and to the left and right.

References

Abelson, R. P. (1968). Simulation of social behavior. In G. Lindzey and E. Aronson (Eds.), *The Handbook of Social Psychology, Volume Two,* 274–356. Reading, MA: Addison-Wesley.

Abelson, R. P., Aronson, E., McGuire, W. J., Newcomb, T. M., Rosenberg, M. J., and Tannenbaum, P. H. (Eds.). (1968). *Theories of Cognitive Consistency: A Source Book.* Chicago: Rand McNally and Co.

Ajzen, I., and Fishbein, M. (1980). *Understanding Attitudes and Predicting Social Behavior.* Englewood Cliffs, NJ: Prentice Hall.

Anderson, E. (1935). The IRISes of the Gaspé Peninsula. *Bulletin of the American IRIS Society, 59,* 2–5.

Angeline, P. J. (1998a). Using selection to improve particle swarm optimization. *Proceedings of the 1998 International Conference on Evolutionary Computation,* 84–89. Piscataway, NJ: IEEE Press.

Angeline, P. J. (1998b). Evolutionary optimization versus particle swarm optimization: Differences in philosophy and performance differences. In V. W. Porto, N. Saravanan, D. Waagen, and A. E. Eiben (Eds.), *Evolutionary Programming VII: Proceedings of the 7th Annual Conference on Evolutionary Programming.* Berlin: Springer-Verlag.

Arthur, B. (1994). Inductive reasoning and bounded rationality. *American Economic Association Papers, 84,* 406–411.

Asch, S. E. (1956). Studies of independence and conformity. I: A minority of one against a unanimous majority. *Psychological Monographs, 70(9),* (Whole Number 416).

Atmar, W. (1976). Speculation on the evolution of intelligence and its possible realization in machine form. Sc.D. dissertation, New Mexico State University, Las Cruces. Cited in D. B. Fogel (1995), *Evolutionary Computation: Toward a New Philosophy of Machine Intelligence.* Piscataway, NJ: IEEE Press.

Attneave, F. (1959). *Applications of Information Theory to Psychology.* New York: Holt, Rinehart, and Winston.

Axelrod, R. (1980). Effective choice in the Prisoner's Dilemma. *Journal of Conflict Resolution, 24,* 3–25.

Axelrod, R. (1984). *The Evolution of Cooperation.* New York: Basic Books.

Axelrod, R. (1997). The dissemination of culture: A model with local convergence and global polarization. *Journal of Conflict Resolution, 41,* 203–226. Reprinted in R. Axelrod (1997), *The Complexity of Cooperation.* Princeton, NJ: Princeton University Press.

Baeck, T., and Schwefel, H.-P. (1993). An overview of evolutionary algorithms for parameter optimization. *Evolutionary Computation, 1*(1), 1–23.

Bagley, J. D. (1967). The behavior of adaptive systems which employ genetic and correlation algorithms. Ph.D. dissertation, University of Michigan, Ann Arbor.

Baker, J. A. (1987). Reducing bias and inefficiency in the selection algorithm. *Proceedings of the Second International Conference on Genetic Algorithms: Genetic Algorithms and Their Applications.* Hillsdale, NJ: Lawrence Erlbaum Associates.

Bandura, A. (1962). Social learning through imitation. In M. R. Jones (Ed.), *Nebraska Symposium on Motivation, 10,* 211–274, Lincoln, NE: University of Nebraska Press.

Bandura, A. (1965). Vicarious processes: A case of no-trial learning. In L. Berkowitz (Ed.), *Advances in Experimental Social Psychology, 2,* 1–55.

Bandura, A. (1974). Behavior theory and the models of man. *American Psychologist, 29,* 859–869.

Bandura, A. (1986). *Social Foundations of Thought and Action: A Social Cognitive Theory.* Englewood Cliffs, NJ: Prentice Hall.

Bateson, G. (1979). *Mind and Nature.* New York: E. P. Dutton.

Baumeister, R. F. (1982). A self-presentational view of social phenomena. *Psychological Bulletin, 91,* 3–26.

Bem, D. (1972). Self-perception theory. In L. Berkowitz (Ed.), *Advances in Experimental Social Psychology, Vol. 6.* New York: Academic Press.

Bem, D. J. (1967). Self-perception: An alternative interpretation of cognitive dissonance phenomena. *Psychological Review, 74,* 183–200.

Benedict, R. (1934). *Patterns of Culture.* New York: Penguin Books.

Ben-Jacob, E. (1998). Bacterial wisdom, Gödel's theorem and creative genomic webs. *Physica A, 248,* 57–76.

Ben-Jacob, E., and Levine, H. (1998). The artistry of microorganisms. *Scientific American, 279,* 82–87.

Ben-Jacob, E., Shochet, O., Tenenbaum, A., Cohen, I., Czirók, A., and Vicsek, T. (1994). Generic modeling of cooperative growth in bacterial colonies. *Nature, 368,* 46–49.

Bentley, P. J. (Ed.). (1999). *Evolutionary Design by Computers*. San Francisco: Morgan Kaufmann.

Bezdek, J. C., Boggavarapu, S., Hall, L. O., and Bensaid, A. (1994). Genetic algorithm guided clustering. *Proceedings of the International Conference on Evolutionary Computation*. Piscataway, NJ: IEEE Service Center, 34–39.

Bloom, H. (2000). *Global Brain: The Evolution of Mass Mind from the Big Bang to the 21st Century*. New York: John Wiley and Sons.

Bonabeau, E., Dorigo, M.,and Theraulaz, G. (1999). *Swarm Intelligence: From Natural to Artificial Systems*. New York: Oxford University Press.

Boring, E. G. (1923). Tests test it. *New Republic* (June 6), 35.

Boyd, R., and Richerson, P. J. (1985). *Culture and the Evolutionary Process*. Chicago: University of Chicago Press.

Breder, C. M. (1954). Equations descriptive of fish schools and other animal aggregations. *Ecology, 35,* 361–370.

Bremmermann, H. J. (1968). Numerical optimization procedures derived from biological evolution processes. In H. L. Oestreicher and D. R. Moore (Eds.), *Cybernetic Problems in Bionics,* 597–616. New York: Gordon and Breach.

Brewer, W. (1974). There is no convincing evidence for operant or classical conditioning in adult humans. In W. Weimer and D. Palermo (Eds.), *Cognition and the Symbolic Processes,* New York: Halstead Press.

Brogan, D. C., and Hodgins, J. K. (1997). Group behaviors for systems with significant dynamics. *Autonomous Robots, 4,* 137–153. Earlier version appeared as Brogan, D., and Hodgins, J. (1995). Group behaviors for systems with significant dynamics. *Proceedings of the 1995 IEEE/RSJ International Conference on Intelligent Robots and Systems, 3,* 528–534.

Brooks, R. A. (1990). Elephants don't play chess. In P. Maes (Ed.), *Designing Autonomous Agents*. Cambridge, MA: The MIT Press.

Brothers, L. (1997). *Friday's Footprint: How Society Shapes the Human Mind*. New York: Oxford University Press.

Buck, R., and Ginsburg, B. (1997). Communicative genes and the evolution of empathy. In W. Ickes (Ed.), *Empathic Accuracy*. New York: The Guilford Press.

Burgess, C. (1998). From simple associations to the building blocks of language: Modeling meaning in memory with the HAL model. *Behavior Research Methods, Instruments, and Computers, 30,* 188–198.

Burgess, C., Livesay, K., and Lund, K. (1998). Explorations in context space: Words, sentences, discourse. *Discourse Processes, 25,* 211–257.

Burke, E. K., and Smith, A. J. (1997). A memetic algorithm for the maintenance scheduling problem. *Proceedings of the International Conference on Computational Intelligence for Modeling Control and Automation,* 469–474.

Burke, E. K., and Smith, A. J. (1999). A memetic algorithm to schedule planned grid maintenance. *Proceedings of the International Conference on Computational Intelligence for Modeling Control and Automation.*

Burks, A. W. (Ed.). (1970). *Essays on Cellular Automata.* Urbana-Champaign, IL: University of Illinois Press.

Byrne, D. (1971). *The Attraction Paradigm.* New York: Academic Press.

Campbell, D. T. (1960). Blind variation and selective retention in creative thought as in other knowledge processes. *Psychological Review, 67,* 380–400.

Campbell, D. T. (1965). Ethnocentrism and other altruistic motives. In D. Levine (Ed.), *Nebraska Symposium on Motivation,* 282–311. Lincoln, NE: University of Nebraska Press.

Campbell, D. T. (1965). Variation and selective retention in sociocultural evolution. In H. R. Barringer, G. I. Blanksten, and R. W. Mack (Eds.), *Social Change in Developing Areas: A Reinterpretation of Evolutionary Theory,* 19–49. Cambridge, MA: Schenkman.

Campbell, D. T. (1974). Evolutionary epistemology. In P. A. Schilpp (Ed.), *The Philosophy of Karl Popper,* 412–463. LaSalle, IL: Open Court.

Campbell, D. T. (1990). Levels of organization, downward causation, and the selection-theory approach to evolutionary epistemology. In E. Tobach and G. Greenberg (Eds.), *Scientific Methodology in the Study of the Mind: Evolutionary Epistemology.* Hillsdale, NJ: Erlbaum.

Candland, D. K. (1993). *Feral Children and Clever Animals: Reflections on Human Nature.* New York: Oxford University Press.

Cartwright, D. (1979). Contemporary social psychology in historical perspective. *Social Psychological Quarterly, 42,* 82–93.

Casti, J. (1997). Keynote address to the IEEE International Conference on Evolutionary Computation, Indianapolis, IN.

Caudell, T. P. (1990). Parametric connectivity: Feasibility of learning in constrained weight space. *Proceedings of the IEEE International Joint Conference on Neural Networks, I,* 667–675. Hillsdale, NJ: Lawrence Erlbaum.

Chialvo, D. R., and Millonas, M. M. (1995). How swarms build cognitive maps. In *The NATO ASI Series, Series F, Computer and System Sciences, 144,* 439.

Chomsky, N. (1959). Review of Skinner's *Verbal Behavior. Language, 35,* 26–58.

Clerc, M., and Kennedy, J. (2000, under review). The particle swarm: Explosion, stability, and convergence in a multi-dimensional complex space.

Coello, C. A. C. (1999). An updated survey of evolutionary multiobjective optimization techniques: State of the art and future trends. *Proceedings of the 1999 Congress on Evolutionary Computation.* Piscataway, NJ: IEEE Press.

Cooley, C. H. (1902). *Human Nature and the Social Order.* New York: Scribner's.

Cowan, G. S., and Reynolds, R. G. (1999). The metrics apprentice: Using cultural algorithms to formulate quality metrics for software systems. *Proceedings of the 1999 Congress on Evolutionary Computation,* 1664–1671. Piscataway, NJ: IEEE Service Center.

Crutchfield, R. S. (1955). Conformity and character. *American Psychologist, 10,* 191–198.

Dautenhahn, K. (1997). Ants don't have friends—thoughts on social intelligent agents. *AAAI Technical Report FS 97–02, Working Notes Socially Intelligent Agents, AAAI Fall Symposium,* MIT.

Dautenhahn, K. (1998). The art of designing socially intelligent agents—science, fiction, and the human in the loop. *Applied Artificial Intelligence Journal, 12,* 573–617.

Dautenhahn, K. (1999). Embodiment and interaction in socially intelligent life-like agents. In C. L. Nehaniv (Ed.), *Computation for Metaphors, Analogy and Agent, Springer Lecture Notes in Artificial Intelligence, 1562,* 102–142.

Davis, L. (Ed.). (1991). *Handbook of Genetic Algorithms.* New York: Van Nostrand Reinhold.

Dawkins, R. (1976). *The Selfish Gene.* New York: Oxford University Press.

Dawkins, R. (1987). *The Blind Watchmaker: Why the Evidence of Evolution Reveals a Universe without Design.* New York: W. W. Norton and Company, Inc.

De Jong, K., Potter, M., and Spears, W. (1997). Using problem generators to explore the effects of epistasis. In T. Bäck, (Ed.), *Proceedings of the Seventh International Conference on Genetic Algorithms,* 338–345. San Francisco: Morgan Kaufmann.

De Jong, K. A. (1975). An analysis of the behavior of a class of genetic adaptive systems. Doctoral dissertation, University of Michigan, Ann Arbor.

Deneubourg, J.-L., Goss, S., Franks, N., Sendova-Franks, A., Detrain, C., and Chretien, L. (1991). The dynamics of collective sorting: Robot-like ant and ant-like robot. In J. A. Meyer and S. W. Wilson (Eds.), *Pro-

ceedings of the First Conference on Simulation of Adaptive Behavior: From Animals to Animats. Cambridge, MA: The MIT Press.

Dennett, D. C. (1995). *Darwin's Dangerous Idea: Evolution and the Meanings of Life.* New York: Simon and Schuster.

Deutsch, M., and Gerard, H. B. (1955). A study of normative and informational social influences upon individual judgment. *Journal of Abnormal and Social Psychology, 51,* 629–636.

Dorigo, M., and Gambardella, L. M. (1997). Ant colony system: A cooperative learning approach to the traveling salesman problem. *IEEE Transactions on Evolutionary Computation, 1* (1), 53–66.

Dorigo, M., Maniezzo, V., and Colorni, A. (1996). The ant system: Optimization by a colony of cooperating ants. *IEEE Transactions on Systems, Man, and Cybernetics, Part B, 26,* 1–13.

Dürr, S., Nonn, R., and Rempe, G. (1998). Origin of quantum-mechanical complementarity probed by a "which-way" experiment in an atom interferometer. *Nature, 394,* 33–37.

Eberhart, R. C., and Kennedy, J. (1995). A new optimizer using particle swarm theory. *Proceedings of the Sixth International Symposium on Micro Machine and Human Science,* Nagoya, Japan, 39–43. Piscataway, NJ: IEEE Service Center.

Eberhart, R. C., and Shi, Y. (2000). Comparing inertia weights and constriction factors in particle swarm optimization. *Proceedings of the 2000 Congress on Evolutionary Computation,* 84–88. Piscataway, NJ: IEEE Service Center.

Eberhart, R. C., Simpson, P. K., and Dobbins, R. W. (1996). *Computational Intelligence PC Tools.* Boston: Academic Press.

Edmonds, B. (1999). Gossip, sexual recombination and the El Farol Bar: Modelling the emergence of heterogeneity. *Journal of Artificial Societies and Social Simulation, 3.* www.soc.surrey.ac.uk/JASSS/2/3/2.html.

Elble, R. J., and Koller, W. C. (1990). *Tremor.* Baltimore, MD: The Johns Hopkins University Press.

Epstein, J. M., and Axtell, R. (1996). *Growing Artificial Societies: Social Science from the Bottom Up.* Cambridge, MA: The MIT Press.

Farmer, J. D. (1991). A Rosetta Stone for connectionism. In S. Forrest (Ed.), *Emergent Computation.* Cambridge, MA: The MIT Press.

Festinger, L. (1954). A theory of social comparison processes. *Human Relations, 7,* 117–140.

Festinger, L. (1957). *A Theory of Cognitive Dissonance.* Evanston, IL: Row, Peterson.

Fisher, R. A. (1936). The use of multiple measurements in taxonomic problems. *Annals of Eugenics, 7,* 179–188.

Fogel, D. (1998). *Evolutionary Computation: The Fossil Record.* Piscataway, NJ: IEEE Press.

Fogel, D., Chellapilla, K., and Angeline, P. J. (1999). Inductive reasoning and bounded rationality reconsidered. *IEEE Transactions on Evolutionary Computation, 3,* 142.

Fogel, D. B. (1990). A brief history of simulated evolution. Technical report, ORINCON Corporation, San Diego, CA.

Fogel, D. B. (1991). *System Identification through Simulated Evolution: A Machine Learning Approach to Modeling.* Needham Heights, MA: Ginn Press.

Fogel, D. B. (1995). *Evolutionary Computation: Toward a New Philosophy of Machine Intelligence.* Piscataway, NJ: IEEE Press.

Fogel, D. B. (2000). What is evolutionary computation? *IEEE Spectrum,* February 2000, *37*(2), 26–32.

Fogel, L. J. (1994). Evolutionary programming in perspective: The top-down view. In J. M. Zurada, R. J. Marks, II, and C. J. Robinson (Eds.), *Computational Intelligence: Imitating Life,* 135–146. Piscataway, NJ: IEEE Press.

Fogel, L. J., Owens, A. J., and Walsh, M. J. (1966). *Artificial Intelligence through Simulated Evolution.* New York: John Wiley and Sons.

Franklin, S., and Graesser, A. (1996). Is it an agent, or just a program?: A taxonomy for autonomous agents. *Proceedings of the Third International Workshop on Agent Theories, Architectures, and Languages.* New York: Springer-Verlag.

Fraser, A. S. (1957). Simulation of genetic systems by automatic digital computers. *Australian Journal of Biological Science, 10,* 484–499.

Fraser, A. S. (1960). Simulation of genetic systems by automatic digital computers: 5-linkage, dominance and epistasis. In O. Kempthorne (Ed.), *Biometrical Genetics,* 70–83. New York: Macmillan.

Fraser, A. S. (1962). Simulation of genetic systems. *Journal of Theoretical Biology, 2,* 329–346.

Friedberg, R. M. (1958). A learning machine: Part I. *IBM Journal of Research and Development, 2,* 2–13.

Friedberg, R. M., Dunham, B., and North, J. H. (1959). A learning machine: Part II. *IBM Journal of Research and Development, 3,* 282–287.

Fukuda, T. (1998). Plenary address to the World Congress on Computational Intelligence, Anchorage, Alaska.

Gabora, L. (1995). Meme and variations: A computational model of cultural evolution. In L. Nadel and D. L. Stein (Eds.), *1993 Lectures in Complex Systems,* Reading, MA: Addison-Wesley.

Gabora, L. (1996). Culture, evolution, and computation. In T. Furuhashi (Ed.), *Proceedings of the Second Online Workshop on Evolutionary Computation. Society of Fuzzy Theory and Systems.*

Gabora, L. (1998). Memetics: A bridge between science and spirituality. *Quest, 12,* 25–27.

Gardner, H. (1985). *The Mind's New Science: A History of the Cognitive Revolution.* New York: Basic Books.

Gardner, M. (1970). Mathematical games: The fantastic combinations of John Conway's new solitaire game "life." *Scientific American,* 120–123.

Gardner, M. (1986). The binary Gray code. In *Knotted Doughnuts and Other Mathematical Entertainments.* New York: W. H. Freeman.

Goffman, E. (1959). *The Presentation of Self in Everyday Life.* Garden City, NJ: Doubleday and Company.

Goldberg, D. E. (1983). Computer-aided gas pipeline operation using genetic algorithms and rule learning. Doctoral dissertation, University of Michigan. *Dissertation Abstracts International, 44*(10), 3174B.

Goldberg, D. E. (1989). *Genetic Algorithms in Search, Optimization, and Machine Learning.* Reading, MA: Addison-Wesley.

Goss, S., Aron, S., Deneubourg, J.-L., and Pasteels, J. M. (1989). Self-organized shortcuts in the Argentine ant. *Naturwissenschaften, 76,* 579–581.

Gould, S. J. (1991). *Bully for Brontosaurus: Reflections in Natural History.* New York: W. W. Norton and Company, Inc.

Gould, S. J. (1999). Message from a mouse. *Time,* September 13, 1999.

Grassé, P.-P. (1959). La reconstruction du nid et les coordinations inter-individuelles chez bellicositermes et cubitermes sp. La théorie de la stigmergie: Essai d'interprétation du comportement des termites constructeurs. *Insect Societies, 6,* 41–80.

Gray, F. (1953). Pulse code communication. United States Patent Number 2,632,058. March 17.

Grefenstette, J. J. (1984a). GENESIS: A system for using genetic search procedures. *Proceedings of the 1984 Conference on Intelligent Systems and Machines,* 161–165.

Grefenstette, J. J. (1984b). A user's guide to GENESIS. Technical Report CS-84-11. Computer Science Dept., Vanderbilt University, Nashville, TN.

Grefenstette, J. J. (Ed.). (1985). *Proceedings of an International Conference on Genetic Algorithms and Their Applications.* Hillsdale, NJ: Lawrence Erlbaum Associates.

Halder, G., Callaerts, P., and Gehring, W. J. (1995). Induction of ectopic eyes by targeted expression of the eyeless gene in drosophila. *Science, 267,* 1788–1792.

Hall, J. S. (1994). Utility fog part 1. *Extropy, 6*(2).

Hall, J. S. (1995). Utility fog part 2. *Extropy, 7*(1).

Hamilton, W. D. (1964). The genetical theory of social behavior, I and II. *Journal of Theoretical Biology, 7,* 1–16, 17–32.

Hamilton, W. D. (1971). Geometry for the selfish herd. *Journal of Theoretical Biology, 31,* 295–311.

Hancock, P. J. B. (1992). Genetic algorithms and permutation problems: A comparison of recombination operators for neural net structure specification. In L. D. Whitley and J. D. Schaffer (Eds.), *COGANN-92: International Workshop on Combinations of Genetic Algorithms and Neural Networks,* 108–122. Los Alamitos, CA: IEEE Computer Society Press.

Harmon-Jones, E., and Mills, J. (Eds.). (1999). *Cognitive Dissonance: Progress on a Pivotal Theory in Social Psychology.* Washington, DC: American Psychological Association.

Haupt, R., and Haupt, S. (1998). *Practical Genetic Algorithms.* New York: John Wiley and Sons.

Henrich, J., and Boyd, R. (1998). The evolution of conformist transmission and the emergence of between-group differences. *Evolution and Human Behavior, 19,* 215–242.

Heppner, F., and Grenander, U. (1990). A stochastic nonlinear model for coordinated bird flocks. In S. Krasner (Ed.), *The Ubiquity of Chaos.* Washington, DC: AAAS Publications.

Herrnstein, R. J., and Murray, C. (1994). *The Bell Curve: Intelligence and Class Structure in American Life.* New York: The Free Press.

Hilgard, E. R. (1977). *Divided Consciousness: Multiple Controls in Human Thought and Action.* New York: John Wiley and Sons.

Hilgard, E. R. (1979). Divided consciousness in hypnosis: The implications of the hidden observer. In E. Fromm and R. E. Shor (Eds.), *Hypnosis: Developments in Research and New Perspectives, second edition.* New York: Aldine.

Hodgins, J., and Brogan, D. (1994). Robot herds: Group behaviors for systems with significant dynamics. *Proceedings of Artificial Life IV,* 319–324.

Hofstadter, D. R. (1979). *Gödel, Escher, Bach: An Eternal Golden Braid.* New York: Basic Books.

Holland, J. (1998). *Emergence: From Chaos to Order.* Reading, MA: Perseus Books.

Holland, J. H. (1962). Outline for a logical theory of adaptive systems. *Journal of the Association for Computing Machinery, 3,* 297–314.

Holland, J. H. (1975/1992). *Adaptation in Natural and Artificial Systems.* Cambridge, MA: The MIT Press.

Hölldobler, B., and Wilson, E. O. (1990). *The Ants.* Cambridge, MA: Bellknap Press of Harvard University Press.

Hopfield, J. J. (1982). Neural networks and physical systems with emergent collective abilities. *Proceedings of the National Academy of Sciences (Biophysics), 79* (8), 2554–2558.

Hopfield, J. J. (1984). Neurons with graded response have collective computational properties like those of two-state neurons. *Proceedings of the National Academy of Sciences, 81,* 3088–3092.

Horgan, J. (1996). *The End of Science.* Reading, MA: Addison-Wesley.

Hornick, K., Stinchcombe, M., and White, H. (1989). Multilayer feedforward neural networks are universal approximators. *Neural Networks, II,* 359–366.

Hoskins, D. A. (1995). An iterated function systems approach to emergence. In *Evolutionary Programming IV: Proceedings of the Fourth Annual Conference on Evolutionary Programming,* 673–692. Cambridge, MA: The MIT Press.

Hutchins, E. (1995). *Cognition in the Wild.* Cambridge, MA: The MIT Press.

Hutchinson, C. A., Peterson, S. N., Gill, S. R., Cline, R. T., White, O., Fraser, C. M., Smith, H. O., and Venter, S. C. (1999). Global transposon mutagenesis and a minimal mycoplasma genome. *Science, 286,* 2165–2169.

IEEE Neural Networks Council. (1996). Glossary of evolutionary computation terms (working draft). Standing Committee on Standards, IEEE, Piscataway, NJ.

Insko, C. A., Smith, R. H., Alicke, M. D., Wade, J., and Taylor, S. (1983). Conformity and group size: The concern with being right and the concern with being liked. *Personality and Social Psychology Bulletin, 11,* 41–50.

Jacoby, L. L., and Dallas, M. (1981). On the relationship between autobiographical memory and perceptual learning. *Journal of Experimental Psychology: General, 110,* 306–340.

James, W. (1890/1948). *Psychology.* Cleveland, OH: World Publishing Co.

Janis, I. L. (1972). *Victims of Groupthink.* Boston: Houghton Mifflin.

Jones, E. E. (1998). Major developments in five decades of social psychology. In D. T. Gilbert, S. T. Fiske, and G. Lindzey (Eds.), *The Handbook of Social Psychology, Fourth Edition, Vol. 1,* 3–57. Boston: McGraw-Hill.

Kauffman, S. (1995). *At Home in the Universe: The Search for Laws of Self-Organization and Complexity.* New York: Oxford University Press.

Kauffman, S. A. (1993). *The Origins of Order : Self-Organization and Selection in Evolution.* New York: Oxford University Press.

Kauffman, S. A. (1995). *At Home in the Universe: The Search for the Laws of Self-Organization and Complexity.* New York: Oxford University Press.

Kennedy, J. (1997). The particle swarm: Social adaptation of knowledge. *Proceedings of the 1997 International Conference on Evolutionary Computation (Indianapolis, Indiana),* 303–308. Piscataway, NJ: IEEE Service Center.

Kennedy, J. (1998). Methods of agreement: Inference among the ele-Mentals. *Proceedings of the 1998 IEEE International Symposium on Intelligent Control (ISIC).* Piscataway, NJ: IEEE Service Center.

Kennedy, J. (1998). The behavior of particles. In V. W. Porto, N. Saravanan, D. Waagen, and A. E. Eiben (Eds.), *Evolutionary Programming VII: Proceedings of the 7th Annual Conference on Evolutionary Programming.* Berlin: Springer-Verlag.

Kennedy, J. (1998). Thinking is social: Experiments with the adaptive culture model. *Journal of Conflict Resolution, 42,* 56–76.

Kennedy, J. (1999). Small worlds and mega-minds: Effects of neighborhood topology on particle swarm performance. *Proceedings of the 1999 Congress on Evolutionary Computation,* 1931–1938. Piscataway, NJ: IEEE Service Center.

Kennedy, J. (2000). Stereotyping: Improving particle swarm performance with cluster analysis. *Proceedings of the 2000 Congress on Evolutionary Computation,* 1507–1512. Piscataway, NJ: IEEE Service Center.

Kennedy, J., and Eberhart, R. C. (1995). Particle swarm optimization. *Proceedings of the IEEE International Conference on Neural Networks, IV,* 1942–1948. Piscataway, NJ: IEEE Service Center.

Kennedy, J., and Eberhart, R. C. (1997). A discrete binary version of the particle swarm algorithm. *Proceedings of the 1997 Conference on Systems, Man, and Cybernetics,* 4104–4109. Piscataway, NJ: IEEE Service Center.

Kennedy, J., and Spears, W. M. (1998). Matching algorithms to problems: An experimental test of the particle swarm and some genetic algorithms on the multimodal problem generator. *Proceedings of the 1998 International Conference on Evolutionary Computation,* 78–83. Piscataway, NJ: IEEE Service Center.

Kitano, H. (1990). Designing neural networks using genetic algorithm with graph generation system. *Complex Systems, 4,*461–476.

Klayman, J., and Ha, Y.-W. (1987). Confirmation, disconfirmation, and information in hypothesis-testing. *Psychological Review, 94,* 211–228.

Koza, J. R. (1992). *Genetic Programming: On the Programming of Computers by Means of Natural Selection.* Cambridge, MA: The MIT Press.

Koza, J. R., and Rice, J. P. (1991). Genetic generation of both the weights and architecture for a neural network. *IEEE International Joint Conference on Neural Networks, II,* 397–404. Piscataway, NJ: IEEE Press.

Koza, J. R., Bennett, F. H., III, Andre, D., and Keane, M. A. (1999). *Genetic Programming III: Darwinian Invention and Problem Solving.* San Francisco: Morgan Kaufmann.

Kroeber, A. L., and Kluckhohn, C. (1952). *Culture: A Critical Review and Definitions.* New York: Vintage Books.

Kuhn, T. S. (1970). *The Structure of Scientific Revolutions,* second edition. Chicago: University of Chicago Press.

Kunda, Z., and Thagard, P. (1996). Forming impressions from stereotypes, traits, and behaviors: A parallel-constraint-satisfaction theory. *Psychological Review, 103,* 284–308.

Kurzweil, R. (1999). *The Age of Spiritual Machines,* 145–146. New York: Viking Penguin.

Landauer, T. K., and Dumais, S. T. (1997). A solution to Plato's problem: The latent semantic analysis theory of acquisition, induction and representation of knowledge. *Psychological Bulletin, 104,* 211–240.

Lane, H. (1976). *The Wild Boy of Aveyron.* Cambridge, MA: Harvard University Press.

Langer, E. J. (1989). *Mindfulness.* Reading, MA: Addison-Wesley.

Langton, C. G. (1988). Artificial life. In C. G. Langton (Ed.), *Artificial Life,* 1–48. Reading, MA: Addison-Wesley.

Langton, C. G. (1991). Computation at the edge of chaos: Phase transitions and emergent computation. In S. Forrest (Ed.), *Emergent Computation: Self-Organizing, Collective, and Cooperative Phenomena in Natural and Artificial Computing Networks.* Cambridge, MA: The MIT Press.

Latané, B. (1981). The psychology of social impact. *American Psychologist, 36,* 343–356.

Latané, B., and Darley, J. M. (1970). *The Unresponsive Bystander: Why Doesn't He Help?* New York: Appleton-Century-Crofts.

Latané, B., and L'Herrou, T. (1996). Spatial clustering in the conformity game: Dynamic social impact in electronic groups. *Journal of Personality and Social Psychology, 70,* 1218–1230.

Latané, B., and L'Herrou, T. (1996). Spatial clustering in the conformity game: Dynamic social impact in electronic groups. *Journal of Personality and Social Psychology, 70,* 1218–1230.

Leary, M. R., and Downs, D. L. (1995). Interpersonal functions of the self-esteem motive: The self-esteem system as a sociometer. In M. Kernis (Ed.), *Efficacy, Agency, and Self-Esteem,* 123–144. New York: Plenum.

Leary, M. R., Tambor, E. S., Terdal, S. K., and Downs, D. L. (1995). Self-esteem as an interpersonal monitor: The sociometer hypothesis. *Journal of Personality and Social Psychology, 68,* 518–530.

Levine, H. (1998). The dynamics of Dictyostelium development. *Physica A, 249,* 53–63.

Levine, J. M., Resnick, L. B., and Higgins, E. T. (1993). Social foundations of cognition. *Annual Review of Psychology, 44,* 585–612.

Levy, S. (1992). *Artificial Life.* New York: Random House.

Lewin, K. (1935). *A Dynamic Theory of Personality.* New York: McGraw-Hill.

Lewin, K. (1936). *Principles of Topological Psychology.* New York: McGraw-Hill.

Lewin, K. (1938). The conceptual representation and measurement of psychological forces. *Contributions to Psychological Theory, 1.* Durham, NC: Duke University Press.

Liepins, G. E., and Potter, W. D. (1991). A genetic algorithm approach to multiple-fault diagnosis. In L. Davis (Ed.), *Handbook of Genetic Algorithms.* New York: Van Nostrand Reinhold.

Lorenz, K. (1973). *Behind the Mirror: A Search for a Natural History of Human Knowledge.* New York: Harcourt Brace Jovanovich.

Lovelock, J. E. (1972). Gaia as seen through the atmosphere. *Atmospheric Environment, 6,* 579–580.

Lovelock, J. E. (1979). *Gaia.* Oxford, England: Oxford University Press.

Lovelock, J. E. (1988). *The Ages of Gaia: A Biography of Our Living Earth.* New York: W. W. Norton and Company, Inc.

Lund, K., Burgess, C., and Atchley, R. A. (1995). Semantic and associative priming in high-dimensional semantic space. *Proceedings of the Cognitive Science Society,* 660–665. Hillsdale, NJ: Erlbaum.

Mandelbrot, B. B. (1977). *The Fractal Geometry of Nature.* New York: W. H. Freeman and Company.

May, R. M. (1976). Simple mathematical models with very complicated dynamics. *Nature, 261,* 459–467.

McClelland, J. L., Rumelhart, D., and the PDP Group (1986). *Parallel Distributed Processing: Explorations in the Microstructure of Cognition. Vol. 2: Psychological and Biological Models.* Cambridge, MA: The MIT Press.

McKee, J. K. (2000). *The Riddled Chain: Change, Coincidence, and Chaos in Human Evolution.* Piscataway, NJ: Rutgers University Press.

Mead, G. H. (1934). *Mind, Self, and Society.* Chicago: University of Chicago Press.

Merz, P., and Freisleben, B. (1999a). A comparison of memetic algorithms, tabu search, and ant colonies for the quadratic assignment problem. In *Proceedings of the 1999 Congress on Evolutionary Computation,* 2063–2070. Piscataway, NJ: IEEE.

Merz, P., and Freisleben, B. (1999b). Fitness landscapes and memetic design. In D. Corne, M. Dorigo, and F. Glover, *New Ideas in Optimization.* London: McGraw-Hill.

Michalewicz, Z., and Michalewicz, M. (1995). Pro-life versus pro-choice strategies in evolutionary computation techniques. In M. Palaniswami, Y. Attikiouzel, R. Marks, D. Fogel, and T. Fukuda (Eds.), *Computational Intelligence: A Dynamic System Perspective,* 137–151. Piscataway, NJ: IEEE Press.

Michalewicz, Z., Schaffer, J. D., Schwefel, H.-P., Fogel, D. B., and Kitano, H. (Eds.). (1994). *Proceedings of the First IEEE Conference on Evolutionary Computation.* Piscataway, NJ: IEEE Service Center.

Millonas, M. M. (1993). Swarms, phase transitions, and collective intelligence. In C. Langton (Ed.), *Artificial Life III,* 417–445. Reading MA: Addison-Wesley.

Millonas, M. M. (1994). Swarms, phase transitions, and collective intelligence. In C. G. Langton (Ed.), *Artificial Life III,* 417–445. Reading, MA: Addison-Wesley.

Minsky, M. (1985). *The Society of Mind.* New York: Simon and Schuster.

Minsky, M., and Papert, S. (1968). *Perceptrons.* Cambridge, MA: The MIT Press.

Mitchell, M. (1996). *An Introduction to Genetic Algorithms.* Cambridge, MA: The MIT Press.

Montana, D. J. (1991). Automated parameter tuning for interpretation of synthetic images. In L. Davis (Ed.), *Handbook of Genetic Algorithms.* New York: Van Nostrand Reinhold.

Montana, D. J., and Davis, L. (1989). Training feedforward neural networks using genetic algorithms. *Proceedings of the 11th Annual Joint Conference on Artificial Intelligence,* 762–767. San Francisco: Morgan Kaufmann.

Moscato, P. (1989). On evolution, search, optimization, genetic algorithms and martial arts. Caltech Concurrent Computation Program, Report 790.

Moscato, P., and Norman, M. G. (1992). A "memetic" approach for the traveling salesman problem: Implementation of a computational ecology for combinatorial optimization on message-passing systems.

In M. Valero, E. Onate, M. Jane, J. L. Larriba, and B. Suarez (Eds.), *Parallel Computing and Transputer Applications,* Amsterdam: IOS Press.

Newell, A., and Simon, H. A. (1956). The logic theory machine: A complete information processing system. *Transactions on Information Theory (Institute of Radio Engineers), IT-2,* 61–79.

Newell, A., and Simon, H. A. (1977). *Human Problem Solving.* Englewood Cliffs, NJ: Prentice Hall.

Nisbett, R. E., and Wilson, D. W. (1977). Telling more than we can know: Verbal reports on mental processes. *Psychological Review, 84,* 231–259.

Nowak, A., Szamrej, J., and Latané, B. (1990). From private attitude to public opinion: A dynamic theory of social impact. *Psychological Review, 97,* 362–376.

Osgood, C. E., Suci, G. J., and Tannenbaum, P. H. (1957). *The Measurement of Meaning.* Urbana-Champaign, IL: University of Illinois Press.

Ostrom, T. M. (1984). The sovereignty of social cognition. In R. S. Wyer and T. K. Srull (Eds.), *Handbook of Social Cognition* (Vol. 1, 1–38). Hillsdale, NJ: Erlbaum.

Ozcan, E., and Mohan, C. K. (1999). Particle swarm optimization: Particles surfing the waves. *Proceedings of the 1999 Congress on Evolutionary Computation,* 1939–1944. Piscataway, NJ: IEEE Service Center.

Partridge, B. (1982). The structure and function of fish schools. *Scientific American,* June, 114–123.

Pedrycz, W. (1998). *Computational Intelligence: An Introduction.* Boca Raton, FL: CRC Press.

Peirce, C. S. (1931–1935). *Collected Papers.* C. Hartshorne and P. Weiss (Eds.). Cambridge, MA: The Belknap Press of the Harvard University Press.

Peng, K., and Nisbett, R. E. (1999). Culture, dialectics, and reasoning about contradiction. *American Psychologist, 54,* 741–755.

Picker, R. C. (1997). Simple games in a complex world: A generative approach to the adoption of norms. *The University of Chicago Law Review, 64,* 1225–1288.

Poggio, T., and Girosi, F. (1990). Networks for approximation and learning. *Proceedings of the IEEE, 78*(9), 1481–1497.

Popper, K. R. (1959). *The Logic of Scientific Discovery.* New York: Basic Books.

Popper, K. R. (1972). *Objective Knowledge: An Evolutionary Approach.* Oxford, England: Clarendon Press.

Press, W. H., Teukolsky, S. A., Vetterling, W. T., and Flannery, B. P. (1993). *Numerical Recipes in C : The Art of Scientific Computing,* second edition. Cambridge, MA: Cambridge University Press.

Read, S. J., Vanman, E. J., and Miller, L. C. (1997). Connectionism and Gestalt principles: (Re)Introducing Cognitive Dynamics to Social Psychology. *Personality and Social Psychology Review, 1,* 26–53.

Rechenberg, I. (1965). Cybernetic solution path of an experimental problem. Royal Aircraft Establishment, library translation 1122, Farnborough, Hants, U.K.

Rechenberg, I. (1973). *Evolutionsstrategie: Optimierung technischer Systeme nach Prinzipien der biologischen Evolution.* Stuttgart, Germany: Frommann-Holzboog Verlag.

Rechenberg, I. (1994). Evolution strategy. In J. Zurada, R. Marks II, and C. Robinson (Eds.), *Computational Intelligence—Imitating Life,* 147–159. Piscataway, NJ: IEEE Press.

Reed, R. D., and Marks, R. J., II. (1999). *Neural Smithing.* Cambridge, MA: The MIT Press.

Reeves, W. T. (1983). Particle systems—a technique for modeling a class of fuzzy objects. *ACM Transactions on Graphics, 2*(2), 91–108.

Resnick, L. B. (1987). Learning in school and out. *Educational Researcher, 16*(9), 13–20.

Resnick, M. (1998). *Turtles, Termites, and Traffic Jams.* Cambridge, MA: The MIT Press.

Reynolds, C. W. (1987). Flocks, herds, and schools: A distributed behavioral model. *Computer Graphics, 21,* 25–34.

Reynolds, R. G. (1994). An introduction to cultural algorithms. *Proceedings of the Third Annual Conference on Evolutionary Programming,* 131–139.

Rich, E. (1983). *Artificial Intelligence.* New York: McGraw-Hill.

Ringelmann, M. (1913). Research on animate sources of power: The world of man. *Annales de l'institute national agronomique,* 2e serie—tome XII, 1–40.

Rosenblatt, F. (1958). The perceptron: A probabilistic model for information storage and organization in the brain. *Psychological Review, 65,* 386–408.

Rucker, R. (1999). *Seek!* New York: Four Walls Eight Windows.

Rumelhart, D. E., McClelland, J. L., and the PDP Group (1986). *Parallel Distributed Processing: Explorations in the Microstructure of Cognition. Vol. 1: Foundations.* Cambridge, MA: The MIT Press.

Rumelhart, D. E., Smolensky, P., McClelland, J. L., and Hinton, G. E. (1986). Schemata and sequential thought processes in PDP models. In J. L. McClelland and D. E. Rumelhart (Eds.), *Parallel Distributed Processing: Explorations in the Microstructure of Cognition, Vol. 2,* 7–57. Cambridge, MA: The MIT Press.

Sarbin, T. R., and Coe, W. C. (1979). Hypnosis and psychopathology: Replacing old myths with fresh metaphors. *Journal of Abnormal Psychology, 88,* 506–526.

Saxe, J. G. (1869). The blind men and the elephant. In H. Felleman (Ed.) (1936), *The Best Loved Poems of the American People.* New York: Doubleday and Company.

Schachter, S., and Singer, J. (1962). Cognitive, social, and physiological determinants of emotional state. *Psychological Review, 69,* 379–399.

Schaffer, J. D. (1984). Some experiments in machine learning using vector evaluated genetic algorithms. Unpublished doctoral dissertation, Vanderbilt University, Nashville, TN.

Schaffer, J. D., Caruana, R. A., and Eshelman, L. J. (1990). Using genetic search to exploit the emergent behavior of neural networks. In S. Forrest (Ed.), *Emergent Computation,* 244–248. Amsterdam: North Holland.

Schaffer, J. D., Whitley, L. D., and Eshelman, L. J. (1992). Combinations of genetic algorithms and neural networks: A survey of the state of the art. In L. D. Whitley and J. D. Schaffer (Eds.), *COGANN-92: International Workshop on Combinations of Genetic Algorithms and Neural Networks,* 1–37. Los Alamitos,CA: IEEE Computer Society Press.

Schlenker, B. R. (1982). Translating actions with attitudes: An identity analytic approach to the explanation of social conduct. In L. Berkowitz (Ed.), *Advances in Experimental Social Psychology, Vol. 15.* New York: Academic Press.

Schwefel, H.-P. (1965). *Kybernetische Evolution als Strategie der experimentellen Forschung in der Stromungstechnik.* Diploma thesis, Technical University of Berlin, Germany.

Schwefel, H.-P. (1994) On the evolution of evolutionary computation. In J. M. Zurada, R. J. Marks II, and C. J. Robinson (Eds.), *Computational Intelligence: Imitating Life.* Piscataway, NJ: IEEE Press.

Segerstråle, U. (2000). *Defenders of the Truth: The Battle for Science in the Sociobiology Debate and Beyond.* Oxford, England: Oxford University Press.

Sellars, W. S. (1956). Empiricism and the philosophy of mind. In H. Feigl and M. Scriven (Eds.), *The Foundations of Science and the Concepts of Psychoanalysis, Minnesota Studies in the Philosophy of Science, Vol. I.* Minneapolis, MN: University of Minnesota Press.

Shannon, C. E. (1948). A mathematical theory of communication. *Bell System Technical Journal, 27,* 379–423 and 623–656.

Shepard, R. N., and Metzler, J. (1971). Mental rotation of three-dimensional objects. *Science, 171,* 701–703.

Sherif, M. (1936). *The Psychology of Social Norms.* New York: Harper Brothers.

Shi, Y., and Eberhart, R. C. (1998a). Parameter selection in particle swarm optimization. *Proceedings of the 1998 Annual Conference on Evolutionary Programming,* San Diego, CA.

Shi, Y. and Eberhart, R. C. (1998b). A modified particle swarm optimizer. *Proceedings of the IEEE International Conference on Evolutionary Computation,* 69–73. Piscataway, NJ: IEEE Press.

Shi, Y., and Eberhart, R. C. (1999). Empirical study of particle swarm optimization. *Proceedings of the 1999 Congress on Evolutionary Computation,* 1945–1950. Piscataway, NJ: IEEE Service Center.

Shi, Y., Eberhart, R. C., and Chen, Y. (1999). Implementation of evolutionary fuzzy systems. *IEEE Transactions on Fuzzy Systems, 7*(2), 109–119.

Simon, H. A. (1969). *The Sciences of the Artificial.* Cambridge, MA: The MIT Press.

Simon, H. A. (1979). Rational decision making in business organizations. *American Economic Review, 69,* 493–513.

Simon, H. A. (1996). *The Sciences of the Artificial, Second Edition.* Cambridge, MA: The MIT Press.

Sims, K. (1994). Evolving 3D morphology and behavior by competition. *Artificial Life, 1,* 353–372.

Skinner, B. F. (1957). *Verbal Behavior.* Englewood Cliffs, NJ: Prentice Hall.

Skinner, B. F. (1971). *Beyond Freedom and Dignity.* New York: Bantam/Vintage.

Smith, E. R. (1996). What do connectionism and social psychology offer each other? *Journal of Personality and Social Psychology, 70,* 893–912.

Smith, S. F. (1980). A learning system based on genetic adaptive algorithms. Unpublished doctoral dissertation, University of Pittsburgh, Pittsburgh, PA.

Smolensky, P. (1986). Information processing in dynamical systems: Harmony theory. In D. E. Rumelhart and J. L. McClelland (Eds.), *Parallel Distributed Processing: Explorations in the Microstructure of Cognition, Vol. 1,* 194–281. Cambridge, MA: The MIT Press.

Smolensky, P. (1986). Information processing in dynamical systems: Harmony theory. In J. L. McClelland, D. Rumelhart, and the PDP Group (Eds.), *Parallel Distributed Processing: Explorations in the Microstructure of Cognition. Vol. 2: Psychological and Biological Models.* Cambridge, MA: The MIT Press.

Spanos, N. P. (1982). A social psychological approach to hypnotic behavior. In G. Weary and H. L. Mirels (Eds.), *Integration of Social and Clinical Psychology,* 231–271. New York: Oxford University Press.

Spanos, N. P. (1986). Hypnotic behavior: A social-psychological interpretation of amnesia, analgesia, and "trance logic." *Behavioral and Brain Sciences, 9,* 449–502.

Stoner, J. A. F. (1961). A comparison of individual and group decisions involving risk. Unpublished M.A. thesis, MIT.

Stork, D. G., Walker, S., Burns, M., and Jackson, B. (1990). Preadaptation in neural circuits. *Proceedings of the International Joint Conference on Neural Networks, I,* 202–205. Hillsdale, NJ: Lawrence Erlbaum Associates.

Syswerda, G. (1989). Uniform crossover in genetic algorithms. In J. D. Schaffer (Ed.), *Proceedings of the Third International Conference on Genetic Algorithms.* San Francisco: Morgan Kaufmann.

Tajfel, H. (1978). Social categorization, social identity and social comparison. In H. Tajfel (Ed.), *Differentiation between Social Groups.* London: Academic Press.

Tandon, V. (2000). Closing the gap between CAD/CAM and optimized CNC end milling. Master's thesis, Purdue School of Engineering and Technology, Indiana University Purdue University, Indianapolis, IN.

Tesfatsion, L. (1995). A trade network game with endogenous partner selection. In H. Amman, B. Rustem, and A. B. Whinston (Eds.), *Computational Approaches to Economic Problems.* Boston: Kluwer Academic, 249–269.

Tesfatsion, L. (1997). How economists can get Alife. In B. Arthur, S. Durlauf, and D. Lane (Eds.), *The Economy as an Evolving Complex System II,* 533–564.

Thagard, P. (1989). Explanatory coherence. *Behavioral and Brain Sciences, 12,* 435–502.

Thagard, P., and Verbeurgt, K. (1995). *Coherence.* Unpublished manuscript, University of Waterloo.

Thomas, L. (1974). *Lives of a Cell: Notes of a Biology Watcher.* Toronto: Bantam Books.

Tomasello, M. (1999). *The Cultural Origins of Human Cognition.* Cambridge, MA: Harvard University Press.

Toner, J., and Tu, Y. (1999). Flocks, herds, and schools: A quantitative theory of flocking. *Physical Review E., 58,* 4828–4858.

Tuddenham, R. D., and Macbride, P. D. (1959). The yielding experiment from the subject's point of view. *Journal of Personality, 27,* 259–271. Quoted in D. Krech, R. S. Crutchfield, and E. L. Ballachey (1962), *Individual in Society,* New York: McGraw-Hill.

Turing, A. M. (1937). On computable numbers, with an application to the Entscheidungs-problem. *Proceedings of the London Mathematical Society (serv. 2), 42,* 230–265; correction *43,* 544–546.

Turing, A. M. (1950). Computing machinery and intelligence. *Mind: A Quarterly Review of Psychology and Philosophy*, Vol. 59, 434–460.

von Neumann, J. (1951). The general and logical theory of automata. In L. A. Jeffress (Ed.), *Cerebral Mechanisms in Behavior*. New York: John Wiley and Sons.

von Neumann, J. (1958). *The Computer and the Brain*. New Haven, CT: Yale University Press.

Waldrop, M. M. (1992). *Complexity: The Emerging Science at the End of Order and Chaos*. New York: Touchstone Books.

Webster's New Collegiate Dictionary. (1975). Springfield, MA: Merriam-Webster.

Weizenbaum, J. (1967). Contextual understanding by computers. *Communications of the ACM, 10*, 474–480.

Weizenbaum, J. (1976). *Computer Power and Human Reason: From Judgment to Calculation*. San Francisco: W. H. Freeman and Company.

Werbos, P. (1974). Beyond regression. Ph.D. dissertation, Harvard University, Cambridge, MA.

Wetzel, C. G., and Insko, C. A. (1982). The similarity-attraction relationship: Is it an ideal one? *Journal of Experimental Social Psychology, 18*, 253–276.

Whitley, D. (1989). Applying genetic algorithms to neural network learning. *Proceedings of the Seventh Conference of the Society of Artificial Intelligence and Simulation of Behavior*, 137–144. Sussex, England: Pitman Publishing.

Whitley, D., Dominic, S., and Das, R. (1991). Genetic reinforcement learning with multilayer neural networks. In R. K. Belew and L. B. Booker (Eds.), *Proceedings of the Fourth International Conference on Genetic Algorithms*, 562–569. San Francisco: Morgan Kaufmann.

Wiener, N. (1950/1954). *The Human Use of Human Beings: Cybernetics and Society*. Garden City, NJ: Doubleday Anchor Books.

Wiener, N. (1961). *Cybernetics*. Cambridge, MA: The MIT Press.

Wilson, E. O. (1978). *On Human Nature*. Cambridge, MA: Harvard University Press.

Winston, P. H. (1992). *Artificial Intelligence,* third edition. Reading, MA: Addison-Wesley.

Wolfram, S. (1984). Universality and complexity in cellular automata. *Physica D, 10*, 1–35.

Wolfram, S. (1994). *Cellular Automata and Complexity: Collected Papers*. Reading, MA: Addison-Wesley.

Wolpert, D. H., and Macready, W. G. (1996). No free lunch theorems for search. Technical Report SFI-TR-05-010, February 1996. Santa Fe Institute, Santa Fe, NM.

Wolpert, D. H., and Macready, W. G. (1997). No free lunch theorems for optimization. *IEEE Transactions on Evolutionary Computation, 1* (1), 67–82.

Wooldridge, D. E. (1968). *Mechanical Man: The Physical Basis of Intelligent Life*. New York: McGraw-Hill.

Worth, R. (2000). Personal communication. March 15.

Wynne-Edwards, V. C. (1962). *Animal Dispersion in Relation to Social Behaviour*. Edinburgh: Oliver and Boyd.

Yager, R. R., Ovchinnikov, S., Tong, R. M., and Nguyen, H. T. (Eds.). (1987). *Fuzzy Sets and Applications: Selected Papers by L. A. Zadeh*. New York: John Wiley and Sons.

Yao, X. (1995).Evolutionary artificial neural networks. In A. Kent and J. G. Williams (Eds.), *Encyclopedia of Computer Science and Technology*. New York: Marcel Dekker.

Yao, X. (1999). Evolving artificial neural networks. *Proceedings of the IEEE, 87*(9), 1423–1447.

Yoshida, H., Kawata, K., Fukuyama, Y., and Nakanishi, Y. (1999). A particle swarm optimization for reactive power and voltage control considering voltage stability. In G. L. Torres and A. P. Alves da Silva (Eds.), *Proceedings of the International Conference on Intelligent System Applications to Power Systems*, 117–121.

Index